MONOGRAPHS ON
STATISTICS AND APPLIED PROBABILITY

General Editors

D.R. Cox, D.V. Hinkley, D.Rubin, B.W. Silverman

1 Stochastic Population Models in Ecology and Epidemiology *M.S. Bartlett* (1960)

2 Queues *D.R. Cox and W.L. Smith* (1961)

3 Monte Carlo Methods *J.M. Hammersley and D.C. Handscomb* (1964)

4 The Statistical Analysis of Series of Events *D.R. Cox and P.A.W. Lewis* (1966)

5 Population Genetics *W.J. Ewens* (1969)

6 Probability, Statistics and Time *M.S. Bartlett* (1975)

7 Statistical Inference *S.D. Silvey* (1975)

8 The Analysis of Contingency Tables *B.S. Everitt* (1977)

9 Multivariate Analysis in Behavioural Research *A.E. Maxwell* (1977)

10 Stochastic Abundance Models *S. Engen* (1978)

11 Some Basic Theory for Statistical Inference *E.J.G. Pitman* (1979)

12 Point Processes *D.R. Cox and V. Isham* (1980)

13 Identification of Outliers *D.M. Hawkins* (1980)

14 Optimal Design *S.D. Silvey* (1980)

15 Finite Mixture Distributions *B.S. Everitt and D.J. Hand* (1981)

16 Classification *A.D. Gordon* (1981)

17 Distribution-free Statistical Methods *J.S. Maritz* (1981)

18 Residuals and Influence in Regression *R.D. Cook and S. Weisberg* (1982)

19 Applications of Queueing Theory *G.F. Newell* (1982)

20 Risk Theory, 3rd edition *R.E. Beard, T. Pentikainen and E. Pesonen* (1984)

21 Analysis of Survival Data *D.R. Cox and D. Oakes* (1984)

22 An Introduction to Latent Variable Models *B.S. Everitt* (1984)

23 Bandit Problems *D.A. Berry and B. Fristedt* (1985)

24 Stochastic Modelling and Control *M.H.A. Davis and R. Vinter* (1985)

25 The Statistical Analysis of Compositional Data *J. Aitchison* (1986)

26 Density Estimation for Statistical and Data Analysis *B.W. Silverman* (1986)

27 Regression Analysis with Applications *G.B. Wetherill* (1986)

28 Sequential Methods in Statistics, 3rd edition *G.B. Wetherill* (1986)

29 Tensor Methods in Statistics *P. McCullagh* (1987)

30 Transformation and Weighting in Regression *R.J. Carroll and D. Ruppert* (1988)

31 Asymptotic Techniques for use in Statistics *O.E. Barndoff-Nielson and D.R. Cox* (1989)

32 Analysis of Binary Data, 2nd edition *D.R. Cox and E.J. Snell* (1989)

33 Analysis of Infectious Disease Data *N.G. Becker* (1989)

34 Design and Analysis of Cross-Over Trials *B. Jones and M.G. Kenward* (1989)

35 Empirical Bayes Method, 2nd edition *J.S. Maritz and T. Lwin* (1989)

36 Symmetric Multivariate and Related Distributions *K-T. Fang, S. Kotz and K. Ng* (1989)

37 Generalized Linear Models, 2nd edition *P. McCullagh and J.A. Nelder* (1989)

38 Cyclic Designs *J.A. John* (1987)

39 Analog Estimation Methods in Econometrics *C.F. Manski* (1988)

40 Subset Selection in Regression *A.J. Miller* (1990)

41 Analysis of Repeated Measures *M. Crowder and D.J. Hand* (1990)

42 Statistical Reasoning with Imprecise Probabilities *P. Walley* (1990)

43 Generalized Additive Models *T. J. Hastie and R. J. Tibshirani* (1990)

(Full details concerning this series are available from the Publishers.)

Generalized Additive Models

T.J. HASTIE
AT&T Bell Laboratories
New Jersey

and

R.J. TIBSHIRANI
Department of Preventive Medicine and Biostatistics
and Department of Statistics
University of Toronto

CHAPMAN & HALL/CRC

Boca Raton London New York Washington, D.C.

Library of Congress Cataloging-in-Publication Data

Catalog record is available from the Library of Congress

Visit the CRC Press Web site at www.crcpress.com

Originally Published by Chapman & Hall
First edition 1990
First CRC reprint 1999

International Standard Book Number 0-412-34390-8
Printed in the United States of America 2 3 4 5 6 7 8 9 0
Printed on acid-free paper

TO

LYNDA, SAMANTHA AND TIMOTHY

CHERYL, CHARLIE AND RYAN

Contents

Preface xiii

1 Introduction 1
- 1.1 What's in this book 1
- 1.2 A word of caution 6
- 1.3 How to read this book 6
- 1.4 Data sets used in this book 6
- 1.5 Notation 7

2 Smoothing 9
- 2.1 What is a smoother? 9
- 2.2 Scatterplot smoothing: definition 13
- 2.3 Parametric regression 14
- 2.4 Bin smoothers 14
- 2.5 Running-mean and running-line smoothers 15
- 2.6 Kernel smoothers 18
- 2.7 Running medians and enhancements 20
- 2.8 Equivalent kernels 20
- 2.9 Regression splines 22
- 2.10 Cubic smoothing splines 27
- 2.11 Locally-weighted running-line smoothers 29
- 2.12 Smoothers for multiple predictors 32
- 2.13 Bibliographic notes 34
- 2.14 Further results and exercises 2 35

3 Smoothing in detail 39
- 3.1 Introduction 39
- 3.2 A formal model for scatterplot smoothing 39
- 3.3 The bias-variance trade-off 40

3.4 Automatic selection of smoothing parameters 42
3.4.1 Cross-validation 42
3.4.2 The bias-variance trade-off for linear smoothers 44
3.4.3 Cross-validation for linear smoothers 46
3.4.4 The C_p statistic 48
3.4.5 Variations on cross-validation and a comparison with C_p 49
3.4.6 Discussion 52
3.5 Degrees of freedom of a smoother 52
3.6 A Bayesian model for smoothing 55
3.7 Eigenanalysis of a smoother and spectral smoothing 57
3.8 Variance of a smooth and confidence bands 60
3.8.1 Pointwise standard-error bands 60
3.8.2 Global confidence sets 61
3.9 Approximate F-tests 65
3.10 Asymptotic behaviour of smoothers 68
3.11 Special topics 70
3.11.1 Nonlinear smoothers 70
3.11.2 Kriging 71
3.11.3 Smoothing and penalized least-squares 72
3.11.4 Weighted smoothing 72
3.11.5 Tied predictor values 74
3.11.6 Resistant smoothing 74
3.12 Bibliographical notes 76
3.13 Further results and exercises 3 79

4 Additive models 82
4.1 Introduction 82
4.2 Multiple regression and linear models 82
4.3 Additive models 86
4.4 Fitting additive models 89
4.5 Generalized additive models: logistic regression 95
4.6 Bibliographic notes 102
4.7 Further results and exercises 4 103

5 Some theory for additive models 105
5.1 Introduction 105
5.2 Estimating equations for additive models 106
5.2.1 L_2 function spaces 107
5.2.2 Penalized least-squares 110

5.2.3 Reproducing-kernel Hilbert-spaces 112
5.3 Solutions to the estimating equations 114
5.3.1 Introduction 114
5.3.2 Projection smoothers 115
5.3.3 Semi-parametric models 118
5.3.4 Backfitting with two smoothers 118
5.3.5 Existence and uniqueness: p-smoothers 121
5.3.6 Convergence of backfitting: p-smoothers 122
5.3.7 Summary of the main results of the section 123
5.4 Special topics 124
5.4.1 Weighted additive models 124
5.4.2 A modified backfitting algorithm 124
5.4.3 Explicit solutions to the estimating equations 126
5.4.4 Standard errors 127
5.4.5 Degrees of freedom 128
5.4.6 A Bayesian version of additive models 129
5.5 Bibliographic notes 130
5.6 Further results and exercises 5 132

6 Generalized additive models 136
6.1 Introduction 136
6.2 Fisher scoring for generalized linear models 137
6.3 Local scoring for generalized additive models 140
6.4 Illustrations 141
6.4.1 Clotting times of blood 141
6.4.2 Warm cardioplegia 143
6.5 Derivation of the local-scoring procedure 148
6.5.1 L_2 function spaces 148
6.5.2 Penalized likelihood 149
6.6 Convergence of the local-scoring algorithm 151
6.7 Semi-parametric generalized linear models 152
6.8 Inference 155
6.8.1 Analysis of deviance 155
6.8.2 Standard error bands 156
6.8.3 Degrees of freedom 157
6.9 Smoothing parameter selection 159
6.10 Overinterpreting additive fits 161
6.11 Missing predictor values 166
6.12 Estimation of the link function 166

6.13 Local-likelihood estimation 167
6.14 Bibliographic notes 169
6.15 Further results and exercises 6 171

7 Response transformation models 174
7.1 Introduction 174
7.2 The ACE algorithm 175
7.2.1 Introduction 175
7.2.2 ACE in L_2 function spaces 179
7.2.3 ACE and penalized least-squares 181
7.2.4 Convergence of ACE with linear smoothers 182
7.2.5 A close ancestor to ACE: canonical correlation 183
7.2.6 Some anomalies of ACE 184
7.3 Response transformations for regression 187
7.3.1 Introduction 187
7.3.2 Generalizations of the Box-Cox procedure 187
7.3.3 Comparison of generalized Box-Cox and ACE 189
7.4 Additivity and variance stabilization 190
7.4.1 Some properties of AVAS 193
7.5 Further topics 194
7.5.1 Prediction from a transformation model 194
7.5.2 Methods for inference 195
7.6 Bibliographical notes 196
7.7 Further results and exercises 7 197

8 Extensions to other settings 201
8.1 Introduction 201
8.2 Matched case-control data 202
8.2.1 Background 202
8.2.2 Estimation 203
8.2.3 Maximizing the conditional likelihood for the linear model 204
8.2.4 Spline estimation for a single function 205
8.2.5 Algorithm for the additive model 207
8.2.6 A simulated example 209
8.3 The proportional-hazards model 211
8.3.1 Background 211
8.3.2 Estimation 211
8.3.3 Further details of the computations 213
8.3.4 An example 214

8.4 The proportional-odds model 219
8.4.1 Background 219
8.4.2 Fitting the additive model 220
8.4.3 Illustration 223
8.5 Seasonal decomposition of time series 224
8.5.1 The STL procedure 224
8.5.2 Eigen-analysis and bandwidth selection 230
8.6 Bibliographic notes 231
8.7 Further results and exercises 8 232

9 Further topics 235
9.1 Introduction 235
9.2 Resistant fitting 236
9.2.1 Resistant fitting of additive models 236
9.2.2 Resistant fitting of generalized additive models 240
9.2.3 Illustration 242
9.2.4 Influence and resistance 244
9.3 Parametric additive models 245
9.3.1 Regression splines 246
9.3.2 Simple knot-selection schemes for regression splines 247
9.3.3 A simulated example 247
9.3.4 Adaptive knot-selection strategies 249
9.3.5 Discussion of regression splines 251
9.3.6 Generalized ridge regression — pseudo additive models 254
9.3.7 Illustration: diagnostics for additive models 256
9.4 Model selection techniques 259
9.4.1 Backward and forward stepwise selection techniques 260
9.4.2 Adaptive regression splines: TURBO 261
9.4.3 Adaptive backfitting: BRUTO 262
9.5 Modelling interactions 264
9.5.1 Simple interactions 264
9.5.2 Interactions resulting in separate curves 265
9.5.3 Hierarchical models 266
9.5.4 Examining residuals for interactions 268
9.5.5 Illustration 268
9.5.6 Regression trees 271

9.5.7 Multivariate adaptive regression splines: MARS 275
9.6 Bibliographic notes 277
9.7 Further results and exercises 9 278

10 Case studies 281
10.1 Introduction 281
10.2 Kyphosis in laminectomy patients 282
10.3 Atmospheric ozone concentration 294

Appendices 301
A Data 301
B Degrees of freedom 305
C Software 307

References 311

Author index 325

Subject index 329

Preface

Man can learn nothing unless he proceeds from the known to the unknown.

Claude Bernard

God is subtle but he is not malicious.

Albert Einstein

The linear model holds a central place (the "plane") in the toolbox of the applied statistician. Simple in structure, elegant in its least-squares theory and interpretable to its user, it is an invaluable tool. However, it need not work alone. With the recent explosion in the speed and size of computers, we can now augment the linear model with new methods that assume less and therefore potentially discover more. In this book, we describe one of these new methods, the *additive model*, a generalization of the linear regression model. The central idea is to replace the usual linear function of a covariate with an unspecified smooth function. The additive model consists of a sum of such functions. This model is nonparametric in that we don't impose a parametric form on the functions but instead estimate them in an iterative manner through the use of *scatterplot smoothers.*

The estimated model consists of a function for each of the covariates. This is useful as a predictive model and can also help the data analyst to discover the appropriate shape of each of the covariate effects.

In the hope that God is not malicious, the additive model retains some of the interpretability of the linear model by assuming additivity of effects. It also requires a fast computer for its estimation, and is an example of a procedure that would have been computationally unthinkable twenty years ago but today is feasible on a personal computer. It is also a prime example of how the power of the computer can be used to free the data analyst from making unnecessarily rigid assumptions about the data.

Additive models can also be applied to data other than the usual regression data. Examples of this are binomial or binary response data, survival data, and data from case-control studies. We discuss all of these in this monograph and also look at techniques that provide a nonparametric transformation of the response, in addition to those for the covariates.

We hope that this book will provide an up-to-date survey of current research in additive modelling. Our emphasis is on the practical rather than the theoretical aspects and much of this book will hopefully be comprehensible to a person with a Masters-level knowledge of statistics. Other scientists with less training in statistics should be able to peruse this book and pick up ideas that could be useful in their research.

We introduced the name "generalized additive models" in 1986 to denote additive extensions of the class of generalized linear models. In recent years the scope of topics that falls under the umbrella of generalized linear models has expanded, as the recent revision of McCullagh and Nelder's monograph reveals. Similarly, this book covers a wider scope than our 1986 paper, with discussion of topics such as multi-predictor smoothing, ordinal, case-control and survival data, transformation models, and time series.

There are many researchers whose cumulative work is reflected in this monograph. Indeed, this is such an active area that it seemed impossible to mention only a few names when introducing topics such as kernel smoothing or cubic smoothing splines. Thus to enhance readability of the text, we have deferred most references to the bibliographic notes at the end of each chapter. Exceptions are made in cases where a specific detail or section was taken from a single outside source. We have tried our best to be thorough and fair in the bibliographic notes; omissions and inaccuracies are inevitable, however, and we apologize for any that occur.

We were introduced to this area by Andreas Buja, Jerome Friedman and Werner Stuetzle, while at Stanford University. Tony Almudevar, Leo Breiman, David Brillinger, Sir David Cox, Michael Greenacre, Nancy Heckman, Keith Knight, Colin Mallows, Roy Mendelsohn, Duncan Murdoch, John Nelder, John Rice, Sami Tibshirani and Larry Wasserman read all or part of the manuscript and gave us valuable feedback. Many other colleagues have contributed to the writing of this monograph, either directly or indirectly, they include David Andrews, Richard Becker, Rollin Brant, Penelope Brasher, John Chambers, William Cleveland, Geoff Collyer, Paul Corey, Ruth Croxford, Bradley Efron, Peter Green, Allen Herman, June Juritz, Terry O'Neill, Art Owen, Daryl Pregibon, Bernard Silverman, Grace Wahba, Allan Wilks, and Brian Yandell. We thank Virginia Penn and Rosie Luisi for their secretarial help, Lorinda Cherry for assisting us with the computation of the subject index, and Howard Trickey for maintaining a healthy TEX environment. Thanks also to our editor Elizabeth Johnston for her constant encouragement during this project. We thank colleagues and staff at both AT&T Bell laboratories and the University of Toronto for support, both technical and moral. Finally, a special thank you to Peter McCullagh for his typesetting macros and his very thorough reading of the entire manuscript. If statistics becomes too easy for Peter, he has a real future in copy-editing!

Murray Hill and Toronto
March 1990

Trevor Hastie
Robert Tibshirani

CHAPTER 1

Introduction

1.1 What's in this book

One of the most popular and useful tools in data analysis is the linear regression model. In the simplest case, we have n measurements of a response (or dependent) variable Y and a single predictor (or independent) variable X. Our goal is to describe the dependence of the mean of Y as a function of X. For this purpose we assume that the mean of Y is a linear function of X,

$$E(Y \mid X) = \alpha + X\beta. \tag{1.1}$$

The parameters α and β are usually estimated by least-squares, that is, by finding the values $\hat{\alpha}$ and $\hat{\beta}$ that minimize the residual sum of squares. If the dependence of the mean of Y on X is linear or almost linear, the linear regression model is very useful. It is simple to compute and provides a concise description of the data.

It is easy to envisage data for which the linear regression model is inappropriate. If the dependence of $E(Y)$ on X is far from linear, we wouldn't want to summarize it with a straight line. We can extend straight line regression very simply by adding terms like X^2 to the model but often it is difficult to guess the most appropriate functional form just from looking at the data. The point of view that we take here is: *let the data show us the appropriate functional form.* That is the idea behind a *scatterplot smoother.* It tries to expose the functional dependence without imposing a rigid parametric assumption about that dependence.

To make the ideas more concrete, let's look at some data. Figure 1.1 shows a plot of a response variable `log(C-peptide)` versus a predictor `age`, two variables from some diabetes data

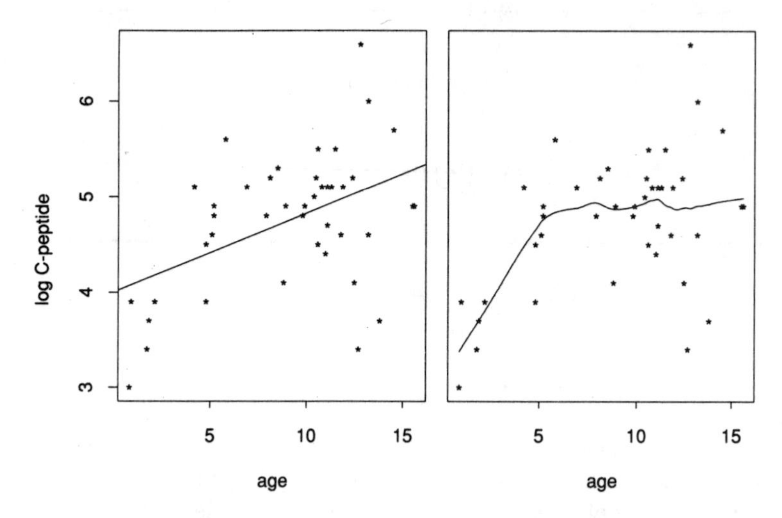

Fig. 1.1. *Left figure shows a scatterplot of* `log(C-peptide)` *versus* `age` *for the diabetes data, together with the least-squares regression line. The right figure shows the locally-weighted running mean smooth of* `log(C-peptide)`*. It summarizes the relationship with* `age`*, and can be regarded as a nonparametric estimate of the regression of* `log(C-peptide)` *on* `age`.

that are studied in this book (these data, and some other data sets, are described at the end of this chapter). It is quite clear from the graph that we wouldn't want to fit a straight line to these data. Instead let's apply a simple scatterplot smoother, the *locally-weighted running-mean.* Consider some fixed data point (X_0, Y_0). We find the 11 data points closest in X-value to (X_0, Y_0), and assign weights to them according to their distance in X from X_0. We compute the weighted average of the Y-values of these 11 data points to produce $\hat{Y}_0$, our estimate of the mean of Y at X_0. Doing this for all the data points produces the curve in Fig. 1.1. The curve is smooth and follows the trend of the data fairly well. The underlying assumption that we have made here is that the dependence of the mean of Y on X should not change much if X doesn't change much. This assumption is very often reasonable. We call the output of a scatterplot smoother a *scatterplot smooth* or simply a *smooth.* The smoother used here is discussed in more detail in section 2.11.

Chapter 2 describes some basic scatterplot smoothers, like the running mean, locally-weighted running-line, kernel and cubic-spline smoothers, and also looks briefly at smoothers for multiple predictors.

In **Chapter 3** we discuss some important issues such as how to choose the smoothing parameters (like the "10" above) for a given smoother, and how to make inferences about the fitted smooth.

One can think of a smooth as simply a description of the dependence of Y on X but if one is interested in building models, as we are here, a slightly more formal definition is in order. If we generalize the linear regression model (1.1) to

$$E(Y \mid X) = f(X), \tag{1.2}$$

where $f(X)$ is an arbitrary unspecified function, then a smooth may be defined as an estimate of $f(X)$.

More often than not, we have more than one predictor variable at our disposal. For example in the diabetes data there are five predictors and 43 observations. Let's consider just two of them; `age` (X_1) and `base deficit` (X_2). The usual way to model the dependence of Y on both X_1 and X_2 is through the multiple linear regression model

$$E(Y \mid X_1, X_2) = \alpha + X_1\beta_1 + X_2\beta_2. \tag{1.3}$$

The three parameters α, β_1 and β_2 are usually estimated by least-squares.

The multiple linear regression model will be inappropriate if the regression surface $E(Y \mid X_1, X_2)$ is not well approximated by a plane. As in the single-predictor case we can add polynomial terms to the model but it will be difficult to guess which terms will be appropriate, and even more so if there are many predictors in the model.

There is a heuristic way to apply a scatterplot smoother to the multiple predictor case. The model that we envision is

$$E(Y \mid X_1, X_2) = f_1(X_1) + f_2(X_2). \tag{1.4}$$

Given an estimate $\hat{f}_1(X_1)$, an intuitive way to estimate $f_2(X_2)$ is to smooth the residual $Y - \hat{f}_1(X_1)$ on X_2. With this estimate $\hat{f}_2(X_2)$ we can get an improved estimate of $\hat{f}_1(X_1)$ by smoothing

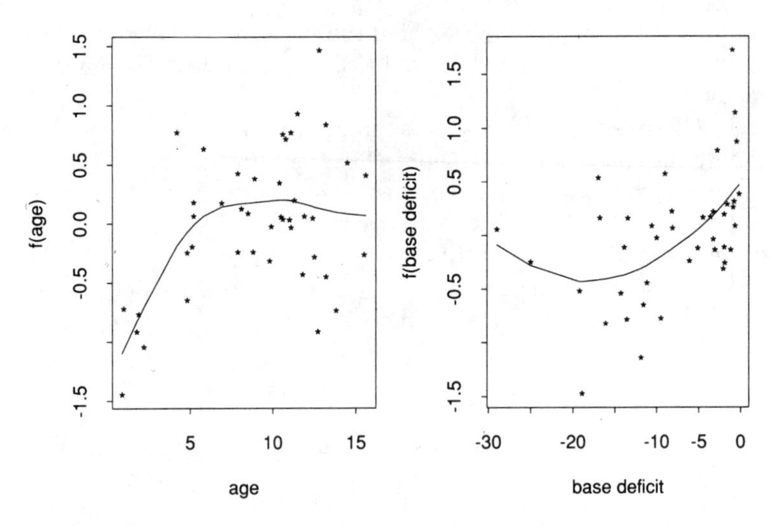

Fig. 1.2. *The fitted functions for* `age` *and* `base deficit` *in the additive model with response* `log(C-peptide)`. *The functions are fitted jointly by iteratively smoothing partial residuals, also known as backfitting.*

$Y - \hat{f}_2(X_2)$ on X_1. We continue this process until our estimates $f_1(X_1)$ and $f_2(X_2)$ are such that the smooth of $Y - \hat{f}_1(X_1)$ on X_2 is $\hat{f}_2(X_2)$ and the smooth of $Y - \hat{f}_2(X_2)$ on X_1 is $\hat{f}_1(X_1)$. Applying this algorithm to the diabetes data produces the curves in Fig. 1.2. The curves are far from linear (as a linear regression model would force on the data) and the shapes can help to discover important facts about the underlying phenomena. In this instance, the shape of the age estimate confirmed a prior belief of the investigators.

The model (1.4) is an example of an *additive model*, the central topic of this book. The iterative smoothing process described above is an example of *backfitting*, the main tool used for estimating additive models. **Chapter 4** describes the additive model and backfitting procedures. The term *additive* means that the model is a sum of terms, but such models can in fact be quite general. For example, some terms may be nonparametric, others linear, and others even a function of more than one predictor variable. Thus a product interaction like X_1X_2 may be incorporated into a model simply by defining a new predictor $X_3 = X_1X_2$ and including the term $X_3\beta_3$. More generally, one may allow terms like $f(X_1, X_2)$ where $f(\cdot, \cdot)$ is an unspecified function, estimated by a bivariate smoother. We also discuss in **Chapter 4** the extension of additive models to binary data, for which the response takes on only two values such as survived or died. Such a model is an example of a *generalized additive model.*

Chapter 5 is somewhat more technical in nature, and places additive models and backfitting on more solid theoretical ground. We make more precise the sense in which backfitting estimates the functions $f_1(X_1)$ and $f_2(X_2)$, and discuss a generalization of collinearity, something we call *concurvity.*

Chapter 6 describes the generalization of additive models to the exponential family, a class of models that we call *generalized additive models.* The idea is the same as in the usual regression setting: in this case, the *linear predictor* is replaced by an *additive predictor.* The estimation procedure for these models uses the backfitting idea in a procedure similar to the standard Fisher scoring method.

In **Chapter 7** we discuss extensions of the additive model that provide a transformation of the response Y in addition to the transformations of the predictors. In the standard linear model, response transformations can sometimes be helpful in finding a more natural scale on which to express the model. The same is true in the additive model setting. In particular we describe the "ACE" (Alternating Conditional Expectation) procedure of Breiman and Friedman (1985).

We discuss in **Chapter 8** some further applications of the additive model to a few nonstandard situations, for example, models for survival, case-control and ordered categorical response data. **Chapter 9** explores some special topics including the modelling of interactions, resistant fitting and model selection.

To explore some practical issues that arise when using additive models in a data analysis, **Chapter 10** provides detailed analyses of two real sets of data.

1.2 A word of caution

Because of our enthusiasm, the reader may get the impression that additive modelling is *uniformly better* than linear modelling. Of course this is not so. One has to consider where additive modelling should fit into a data analyst's overall strategy. The additive model is only one extension of the usual component-wise linear model.

We like to think of linear modelling and additive modelling as just two tools in a data analyst's *toolbox*. The question of how best to use these tools is a difficult but very important one. The first case study of Chapter 10 illustrates some of our current approaches.

1.3 How to read this book

If you've read this far and are interested in reading further, Chapters 2, 4 and 6 contain the main ideas as far as smoothing and additive models are concerned. The other chapters may then be selected depending on the specific interests of the reader.

1.4 Data sets used in this book

Here we give a brief description of some of the datasets used in the examples in the book. Listings of two of the shorter datasets appear in Appendix A.

Diabetes data. These data come from a study of the factors affecting patterns of insulin-dependent diabetes mellitus in children. (Sockett *et al.*, 1987). The objective was to investigate the dependence of the level of serum C-peptide on various other factors in order to understand the patterns of residual insulin secretion. The response measurement is the logarithm of C-peptide concentration at diagnosis, and the predictor measurements age and base deficit, a measure of acidity. These two predictors are a subset of those studied in Sockett *et al.* (1987). The data are given in Appendix A.

Heart-attack data. These data are a subset of the Coronary Risk-Factor Study (CORIS) baseline survey, carried out in three rural areas of the Western Cape, South Africa (Rousseauw *et al.*, 1983). The aim of the study was to establish the intensity of ischaemic heart disease risk factors in that high-incidence region. The data represent white males between 15 and 64, and the response variable is the presence or absence of myocardial infarction (MI) at the time of the survey (the overall prevalence of MI is 5.1% in this region). There are 162 cases in our data set, and a sample of 302 controls. We use these data as an illustration of additive models in Chapter 4, where we use the risk factors systolic blood pressure and cholesterol ratio (defined as (total cholesterol-HDL)/HDL). These data are described in more detail in Hastie and Tibshirani (1987a).

Kyphosis data. Data were collected on 83 patients undergoing corrective spinal surgery. The objective was to determine important risk factors for kyphosis, or the forward flexion of the spine of at least 40 degrees from vertical, following surgery. The risk factors are location of the surgery along the spine and age. These data are analysed in some detail in Chapter 10, and listed in Appendix A.

Warm cardioplegia data. Williams *et al.* (1990) studied the effectiveness of a warm cardioplegia induction procedure prior to heart surgery, for young patients with congenital heart disease. The outcome was survived or died; additional risk factors are age, weight, diagnosis, surgeon and operation date. We analyse these data in Chapter 6.

Atmospheric ozone concentration. These data were collected over a one year period in the Los Angeles Basin. The objective was to predict the level of atmospheric ozone concentration using eight meteorological measurements. The data were given to us by Leo Breiman and is analysed in Breiman and Friedman (1985). We analyse these data in a comparative case study in Chapter 10.

1.5 Notation

Here we provide an explanation of some of the notation used in this book. Notation for special quantities such as *MSE* (mean squared error) or *ASR* (average squared residual) can be found in the index.

We represent random variables in uppercase, for example, X_1, Y, and observed values in lowercase, for example, x_{i1}, y_i. A vector of random variables is written in boldface as $\boldsymbol{X} = (X_1, \ldots, X_p)$. We use (column) vector notation to represent a sequence of values, such as $\mathbf{y} = (y_1, \ldots, y_n)^T$, $\mathbf{x} = (x_1, \ldots, x_n)^T$ and $\mathbf{x}_j$ the n observed values for variable j. Collections of vectors are packed into a data matrix such as $\mathbf{X} = (\mathbf{x}_1, \ldots, \mathbf{x}_p)$; the rows of this matrix are denoted by the p-vectors $\mathbf{x}^i$. The trace or sum of the diagonal elements of a square matrix $\mathbf{A}$ is written as $\text{tr}(\mathbf{A})$. Given a vector $\mathbf{a} = (a_1, \ldots, a_m)^T$, $\text{diag}(\mathbf{a})$ represents the $m \times m$ diagonal matrix with ith diagonal element a_i. The notation $\langle \mathbf{u}, \mathbf{v} \rangle$ refers to the inner product of vectors $\mathbf{u}$ and $\mathbf{v}$, and $\|\mathbf{u}\|$ to the norm of $\mathbf{u}$.

A linear predictor refers to an expression of the form $\beta_0 + \sum_1^p X_j \beta_j$, and an additive predictor $\beta_0 + \sum_1^p f_j(X_j)$, where $f_j(X_j)$ denotes an arbitrary function of variable X_j.

We use $\mathbf{f}_j$ to represent the vector of evaluations of the function f_j at the n observed values of X_j, that is $\mathbf{f}_j = \{f_j(x_{1j}), \ldots, f_j(x_{nj})\}^T$.

Expectation is denoted by E and variance by var. The distribution of Y will be sometimes indexed by $\mu = E(Y)$, or by an equivalent parameter $\eta = g(\mu)$, where g is a monotonic function known as the link function.

A log-likelihood is denoted by $l(\boldsymbol{\mu}; \mathbf{y})$ or $l(\boldsymbol{\eta}; \mathbf{y})$ or l for short. In the case of n independent observations this can be expressed as $\sum_{i=1}^n l(\mu_i; y_i)$ where $l(\mu_i; y_i)$ is the log-likelihood contribution of a single observation. A Gaussian (normal) distribution with mean μ and variance σ^2 is denoted by $N(\mu, \sigma^2)$; similarly the multivariate Gaussian is $N(\boldsymbol{\mu}, \boldsymbol{\Sigma})$. A binomial distribution with n draws and success probability p is denoted by $B(n, p)$. The chi-squared distribution with p degrees of freedom is denoted by χ_p^2, and $F_{m,n}$ refers to a F-distribution with m and n degrees of freedom. The notation *df* is an abbreviation for *degrees of freedom*.

When describing examples and case studies, we write variable names such as `age` and `ozone` in typewriter font to distinguish them from the rest of the text.

CHAPTER 2

Smoothing

2.1 What is a smoother?

A smoother is a tool for summarizing the trend of a response measurement Y as a function of one or more predictor measurements $X_1, \ldots, X_p$. It produces an estimate of the trend that is less variable than Y itself; hence the name *smoother*. An important property of a smoother is its *nonparametric* nature: it doesn't assume a rigid form for the dependence of Y on $X_1, \ldots, X_p$. For this reason, a smoother is often referred to as a tool for nonparametric regression. The running mean (or moving average) is a simple example of a smoother, while a regression line is not strictly thought of as a smoother because of its rigid parametric form. We call the estimate produced by a smoother a *smooth*. The single predictor case is the most common, and is called *scatterplot smoothing*.

Smoothers have two main uses. The first use is description. A scatterplot smoother can be used to enhance the visual appearance of the scatterplot of Y vs X, to help our eyes pick out the trend in the plot. Figure 1.1 in Chapter 1 shows a plot of `log(C-peptide)` versus `age` for the diabetes data. It seems that `log(C-peptide)` has a strong dependence on `age` and a scatterplot smoother is helpful in describing this relationship. Figure 2.1 shows a number of scatterplot smooths of these data, each described in turn in this chapter. The second use of a smoother is to estimate the dependence of the mean of Y on the predictors, and thus serve as a building block for the estimation of additive models, discussed in the remainder of the book.

In this chapter we give an overview of some useful smoothers. Most of our discussion concerns scatterplot smoothers; at the end of the chapter we give a brief description of multiple-predictor smoothers.

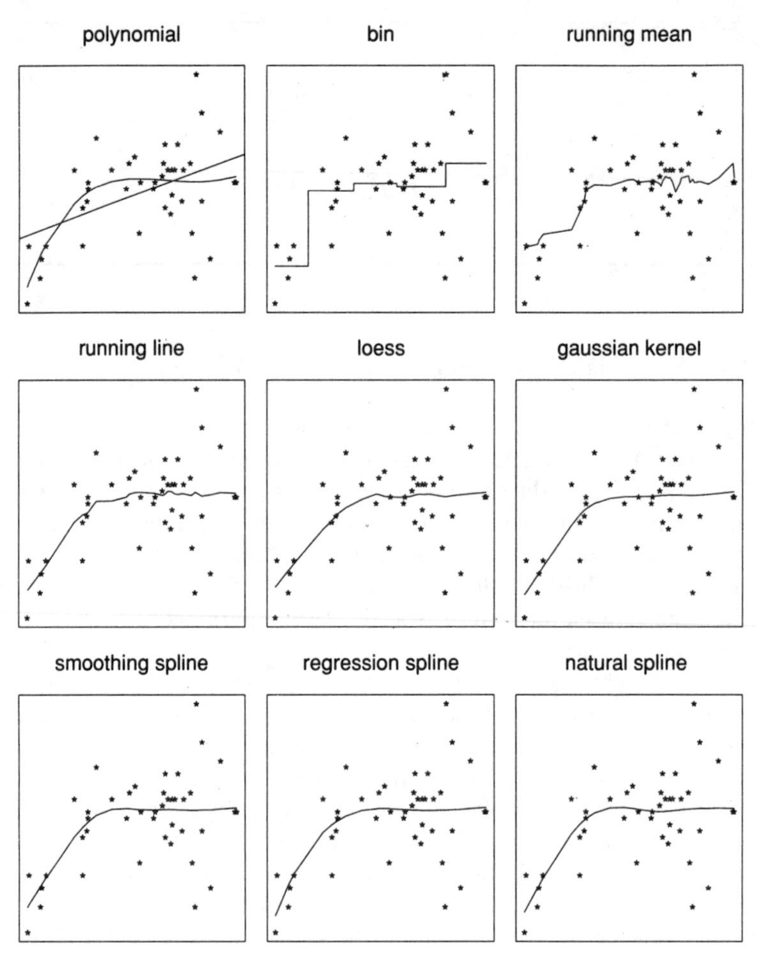

Fig. 2.1. *A variety of smooths applied to the scatterplot of* `log(C-peptide)` *versus* `age`. *In each case the smoothing parameter is chosen so that the degrees of freedom in the fit is about five. The linear fit has only two degrees of freedom. The parametric bin and quartic-polynomial fits have five parameters. The (cubic) regression spline has one interior knot at the median and hence five parameters. The natural cubic regression spline has three interior knots and two endpoint constraints, which also results in five parameters. For the other smooths, the smoothing parameter λ is chosen such that df ≈ 5, where df is a measure of the equivalent degrees of freedom (a concept introduced in Chapter 3).*

The simplest smoother occurs in the case of a *categorical* predictor, for example, `sex` (male, female) or `colour` (red, blue, green etc.) To smooth Y we can simply average the values of Y in each category. This satisfies our requirements for a scatterplot smooth: it captures the trend of Y on X and is smoother than the Y-values themselves. While the reader might not normally think of this as *smoothing*, this simple averaging process is the conceptual basis for smoothing in the most general setting, that of an ordered (noncategorical) predictor. The problem here is that we usually lack replicates at each predictor value. Most smoothers attempt to mimic category averaging through *local averaging*, that is, averaging the Y-values of observations having predictor values close to a target value. The averaging is done in *neighbourhoods* around the target value. There are two main decisions to be made in scatterplot smoothing:

(i) how to average the response values in each neighbourhood, and

(ii) how big to take the neighbourhoods.

The question of how to average within a neighbourhood is really the question of which *brand* of smoother to use, because smoothers differ mainly in their method of averaging. In this chapter we describe a number of different smoothers and compare them informally. Formal recommendations on how to choose among smoothers are difficult to make because few systematic comparisons have been made so far in the literature.

The question of how big to make the neighbourhoods is discussed in the next chapter. The issue underlying this question is very important, however, so we briefly discuss it here. The size of the neighbourhood is typically expressed in terms of an adjustable *smoothing parameter*. Intuitively, large neighbourhoods will produce an estimate with low variance but potentially high bias, and conversely for small neighbourhoods (Fig. 2.2). Thus there is a *fundamental tradeoff between bias and variance*, governed by the smoothing parameter. This issue is exactly analogous to the question of how many predictors to put in a regression equation. In the next chapter we discuss this important issue and the practical question of how to choose the smoothing parameter, based on the data, to trade bias against variance in an optimal way. We also discuss some other important topics, such as linear and nonlinear

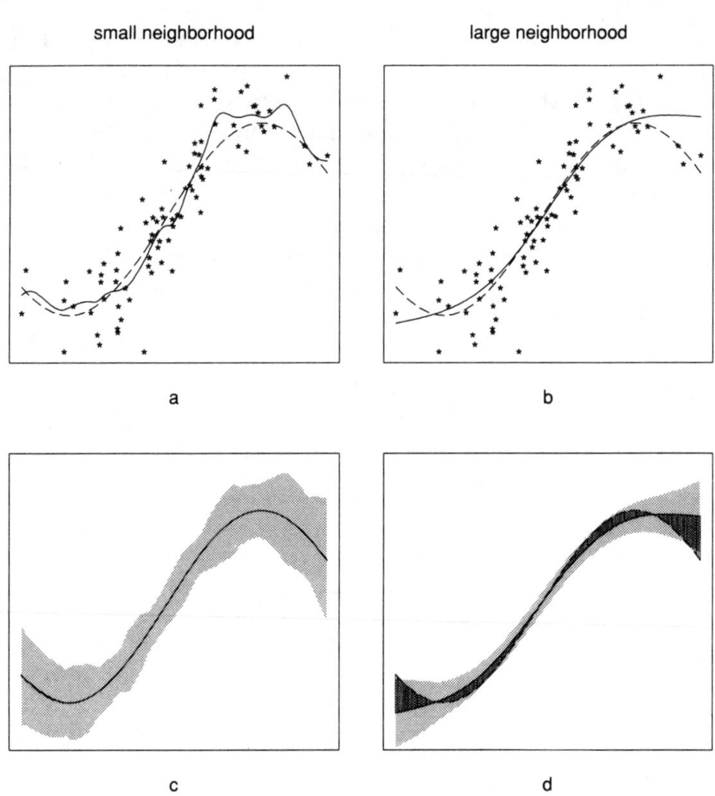

Fig. 2.2. *Bias-variance tradeoff. A sample of size 80 is generated from* $Y = sin(X) + \varepsilon$ *with* $X \sim N(0,1)$ *and* $\varepsilon \sim N(0,1/9)$. *The top two panels represent one realization of the model with a smooth included; panel (a) uses a small neighbourhood and thus does little smoothing, while (b) uses a large neighbourhood and does a lot of smoothing. In terms of approximate degrees of freedom (a concept introduced in Chapter 3), the smoother in (a) uses df* $= 15$, *whereas for (b) df* $= 4$. *The* sin *function is included in both panels. The lower two panels show the bias-variance behaviour of this smoother. The light shaded regions show twice the pointwise standard errors of the fitted smooth values, and thus reflect the variance of the smooths under repeated sampling of 80 pairs from the model. The darker shaded region shows the bias, which is the difference between the generating curve and the average value of the smooth curves under repeated sampling. In panel (c), the bias is virtually nonexistent, while the standard-error bands are quite wide. In panel (d), there is a considerable cost in bias for a moderate decrease in variance.*

smoothers, the incorporation of observation weights, and inference for the fitted smooth.

The amount of smoothing in Fig. 2.1 is calibrated according to *df*, the number of equivalent degrees of freedom. The anxious reader can look ahead to section 2.8 for a brief description, or to section 3.5 for a more detailed definition of *df*.

All the smoothers we shall describe can easily be *robustified* by replacing the averaging or least-squares operation by a more robust procedure. Details can be found in Chapters 3 and 9.

Bibliographic notes on smoothers are given at the end of the next chapter.

2.2 Scatterplot smoothing: definition

Suppose we have response measurements $\mathbf{y} = (y_1, \ldots, y_n)^T$ at design points $\mathbf{x} = (x_1, \ldots, x_n)^T$. We assume that each of $\mathbf{y}$ and $\mathbf{x}$ represent measurements of variables Y and X. In most cases it is useful to think of Y, and sometimes X, as having been generated by some random mechanism, but this is not necessary for the discussion here. In particular we don't need to assume that the pairs (x_i, y_i) are a random sample from some joint distribution. For example, the X-values might be preset dose levels of a drug.

Since Y and X are noncategorical we don't expect to find many replicates at any given value of X. For convenience we assume that the data are sorted by X and for the present discussion that there are no tied X-values, so that $x_1 < \cdots < x_n$. An easy remedy when there are ties is to use weighted smoothers, which we discuss in Chapter 3.

A scatterplot smoother is defined to be a function of $\mathbf{x}$ and $\mathbf{y}$, whose result is a function s with the same domain as the values in $\mathbf{x}$: $s = \mathcal{S}(\mathbf{y} \,|\, \mathbf{x})$. Usually the recipe that defines $s(x_0)$, which is the function $\mathcal{S}(\mathbf{y} \,|\, \mathbf{x})$ evaluated at x_0, is defined for all x_0, but at other times is defined only at $x_1, \ldots, x_n$, the sample values of X. In this latter case some kind of interpolation is necessary in order to obtain estimates at other X-values.

2.3 Parametric regression

A regression line, estimated for example by least-squares, provides an estimate of the dependence of $E(Y)$ on X. It does so, however, by assuming a rigid form for this dependence, and thus it may or may not be appropriate for a given set of data. If the dependence is linear or close to it, the regression line provides a concise and useful summary. For the data in Fig. 2.1, however, the regression line is clearly inappropriate and creates a misleading impression.

In a sense the regression line is an *infinitely smooth* function, and not surprisingly, many of the scatterplot smoothers that we discuss approach the linear regression line as the amount of smoothing is increased. The other extreme, as the amount of smoothing is decreased, is usually some function that interpolates the data.

Other nonlocal parametric fits, such as polynomial regression estimates, share the same pros and cons as the regression line. They are useful if they are appropriate for the data at hand but potentially misleading otherwise. Although parametric fitting certainly isn't the solution to the scatterplot smoothing problem, we still make use of it for comparative purposes and for possible summaries of smooths estimated by other methods. Regression splines are a less rigid form of parametric fitting which are closer in spirit to a smoother; we discuss them later in this chapter.

2.4 Bin smoothers

A bin smoother, also known as a *regressogram*, mimics a categorical smoother by partitioning the predictor values into a number of disjoint and exhaustive regions, then averaging the response in each region. Formally, we choose cutpoints $c_0 < \cdots < c_K$ where $c_0 = -\infty$ and $c_K = \infty$, and define

$$R_k = \{i; c_k \leq x_i < c_{k+1}\}; \qquad k = 0, \ldots, K-1,$$

the indices of the data points in each region. Then $s = \mathcal{S}(\mathbf{y} \mid \mathbf{x})$ is given by $s(x_0) = \text{ave}_{i \in R_k}(y_i)$ if $x_0 \in R_k$. Typically one chooses five regions, for example, and picks the cutoff points so that there is approximately an equal number of points in each region. The bin smooth shown in Fig. 2.1 is constructed in this way and illustrates the limitation of the method. The estimate is not very smooth because it jumps at each cut point. Unless we have reason to believe such discontinuities exist, this estimate is not very appealing. One way to make it more smooth is through the use of overlapping regions, as we'll see in the next section.

2.5 Running-mean and running-line smoothers

Assume that our target value x_0 equals one of the x_js, say x_i. If we have replicates at x_i, we can simply use the average of the Y-values at x_i as our estimate $s(x_i)$. Now we are assuming that we don't have replicates, so instead we can average Y-values corresponding to X-values *close* to x_i. How do we pick points that are close to x_i? A simple way is to choose x_i itself, as well as the k points to the left and k points to the right of x_i that are closest in X-value to x_i. This is called a *symmetric nearest neighbourhood* and we denote the indices of these points by $N^S(x_i)$. We then define the *running mean*

$$s(x_i) = \text{ave}_{j \in N^S(x_i)}(y_j). \tag{2.1}$$

If it is not possible to take k points to the left or right of x_i, we take as many as we can. A formal definition of a symmetric nearest neighbourhood is

$$N^S(x_i) = \{\max(i-k,1),\ldots,i-1,i,i+1,\ldots,\min(i+k,n)\}. \tag{2.2}$$

It is not obvious how to define the symmetric nearest neighbours at target points x_0 other than the x_i in the sample. We can simply interpolate linearly between the fit of the two values of X in the sample adjacent to x_0. Alternatively we can ignore symmetry and take the r closest points to x_0, regardless of which side they are on; this is called a *nearest neighbourhood.* This handles arbitrary values of x_0 in a simple and clean way. The pros and cons of these two types of neighbourhoods are discussed later.

This simple smoother is also called a *moving average*, and is popular for evenly-spaced time-series data. Although it is valuable for theoretical calculation because of its simplicity, in practice it does not work very well. It tends to be so wiggly that it hardly deserves the name *smoother.* Apart from looking unpleasant, it tends to flatten out trends near the endpoints and hence can be

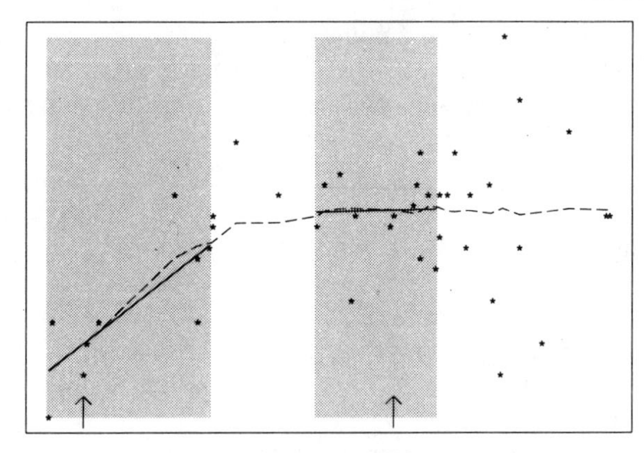

Fig. 2.3. *The running-line fit computes linear fits locally in the symmetric nearest neighbourhoods. The fitted value is the point on the line at the target point of the neighbourhood. Typically in the interior the target point is close to the neighbourhood mean, so a running-line behaves like a running mean. On the boundaries the neighbourhoods become asymmetric and the slope corrects the first-order bias of the running mean.*

severely biased. For example, Fig. 2.1 shows a running mean smooth with $2k+1=11$ or about 25% of the 43 observations.

A simple generalization of the running mean alleviates the bias problem: we compute a least-squares line instead of a mean in each neighbourhood. The *running-line smoother* is defined by

$$s(x_0) = \hat{\alpha}(x_0) + \hat{\beta}(x_0)x_0 \tag{2.3}$$

where $\hat{\alpha}(x_0)$ and $\hat{\beta}(x_0)$ are the least-squares estimates for the data points in $N^S(x_0)$. Figure 2.3 shows the local line computed in two different neighbourhoods: one in the interior, the other near the boundary. As we expect, the fit in the interior is dominated largely by the mean and the slope plays a small role, whereas near the boundary the slope is important for picking up the trend in the asymmetric neighbourhood. A running-line smooth with a neighbourhood size of $2k+1=13$ or 30% for the diabetes data, is shown in Fig. 2.1. It seems to capture the trend in the data quite nicely but is still somewhat jagged.

The parameter k controls the appearance of the running-line smooth. Large values of k tend to produce smoother curves while small values tend to produce more jagged curves. It is more convenient to think not in terms of k but instead in terms of $w = (2k+1)/n$, the proportion of points in each neighbourhood, called the *span*. We denote by $[N^S(x_i)]$ the number of points in $N^S(x_i)$. In the extreme case, if $w = 2$, so that each neighbourhood contains all of the data (note that $w = 1$ won't work because of endpoint effects), the running-line smooth is the least-squares line. On the other hand, if $w = 1/n$, each neighbourhood consists only of one data point and hence the smoother interpolates the data. In the next chapter we discuss the quantitative effects of varying the span and a data-based criterion for choosing it.

The running-line smoother can be computed in $O(n)$ operations (Exercise 2.2), and there is a simple, effective method for span selection. These can be important advantages when it is used as a building block in the iterative algorithms discussed later in this book. On the other hand, it tends to produce curves that are quite jagged so that a second stage of smoothing might be necessary.

One way to improve the appearance of the running-line smooth is through the use of a *weighted* least-squares fit in each neighbourhood. The running-line smoother can produce jagged output because points in a given neighbourhood are given equal (nonzero) weight, while points outside of the neighbourhood are given zero weight. Thus as the neighbourhoods move from left to right, there are discrete changes in the weight given to the leftmost and rightmost points. We can alleviate this problem by giving the highest weight to x_i, and weights smoothly decreasing as we move further away from x_i. Cleveland's (1979) implementation of a locally-weighted running-line smoother, *loess*, is popular and (essentially) works in this way. A full description of locally-weighted running-line smoothers is given later in this chapter; Fig. 2.1 shows one applied to the diabetes data. It is visually smoother than the unweighted running-line smoother. The cost of using smooth weights is a computational one, as the shortcuts available for the unweighted running-line smoother (Exercise 2.2) do not work for locally weighted running-line smoothers. Henceforth we will refer to the unweighted running-line smoother simply as the running-line smoother.

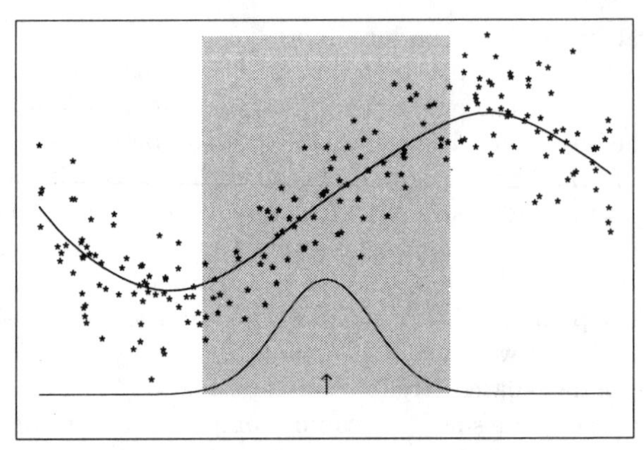

Fig. 2.4. *The Gaussian kernel smoother uses the Gaussian density function to assign weights to neighbouring points. The points within the shaded neighbourhood are those that receive weights that are effectively nonzero for the computation of* $E(Y \mid x_0)$.

2.6 Kernel smoothers

A kernel smoother uses an explicitly defined set of local weights, defined by the *kernel*, to produce the estimate at each target value. Usually a kernel smoother uses weights that decrease in a smooth fashion as one moves away from the target point.

The weight given to the jth point in producing the estimate at x_0 is defined by

$$S_{0j} = \frac{c_0}{\lambda} d\left(\left|\frac{x_0 - x_j}{\lambda}\right|\right) \tag{2.4}$$

where $d(t)$ is an even function decreasing in $|t|$. The parameter λ is the window-width, also known as the bandwidth, and the constant c_0 is usually chosen so that the weights sum to unity, although there are slight variations on this. A natural candidate for d is the standard Gaussian density: this gives the so-called Gaussian kernel smoother (Fig. 2.4). Other popular kernels, with some theoretical justification, are the Epanechnikov kernel,

$$d(t) = \begin{cases} \frac{3}{4}(1 - t^2), & \text{for } |t| \leq 1; \\ 0 & \text{otherwise,} \end{cases}$$

(Epanechnikov, 1969) which minimizes (asymptotic) mean squared error, and the minimum variance kernel,

$$d(t) = \begin{cases} \frac{3}{8}(3 - 5t^2), & \text{for } |t| \leq 1; \\ 0 & \text{otherwise} \end{cases}$$

which minimizes the asymptotic variance of the estimate. Note that the weight given to an observation is a function only of its *metric* distance from x_0, while the weights used by the nearest-neighbour smoothers are typically a function of both *metric* and *rank* distance. Research to date suggests that the choice of the kernel is relatively unimportant compared to the choice of the bandwidth.

Kernel smoothers also exhibit biased endpoint behaviour. Special kernels have been developed to overcome this bias; a simple approach is to use kernel weights in a locally-weighted straight-line fit.

A Gaussian-kernel smooth for the diabetes data is shown in Fig. 2.1, where once again the parameter λ is chosen so that the approximate degrees of freedom of the fitted smooth is five, the same as that of the other smooths in Fig. 2.1.

Computational aspects

One can visualize the action of the kernel smooth as sliding the weight function along the x-axis in short steps, each time computing the weighted mean of y. The smooth is thus similar to a convolution between the kernel and an empirical step function defined on the data. This is indeed the case, although the practical details obscure the resemblance. For example, the kernel is usually truncated at the ends of the data, unless the data itself is periodic. Typically the kernel smooth is computed as

$$s(x_0) = \frac{\sum_{i=1}^{n} d(\frac{x_0 - x_i}{\lambda}) y_i}{\sum_{i=1}^{n} d(\frac{x_0 - x_i}{\lambda})}, \tag{2.5}$$

and so both the numerator and denominator are convolutions. If the x-values are evenly spaced (and preferably a power of two in number), great savings can be made by using the FFT (fast Fourier transform) in performing the calculations. Although evenly spaced data are rarely the case in general regression contexts, reasonable approximations can often reduce the data to a fine grid, where the FFT can be put to work. The details are laid out in Exercise 2.9.

2.7 Running medians and enhancements

A somewhat different approach to smoothing is based on improvements to the simple running mean smoother. First, the running mean is replaced by a running median to make the smoother resistant to outliers in the data. Then the appearance of the estimate is enhanced by applying compound operations known as *Hanning*, *splitting* and *twicing*, in various combinations. We do not discuss these smoothers here because they are mainly useful for evenly spaced data (sequences), especially time series, and they don't typically provide an adequate amount of smoothing for our purposes. The interested reader can pursue the references given in the bibliographic notes. We focus on smoothers for unevenly spaced data (scatterplots), sometimes called *regression smoothers.* In our view, another disadvantage of the enhanced running median smoothers is that they are highly nonlinear functions of the response, and thus it is difficult to assess the amount of fitting that they do. This is an important consideration for the inferential stage of a data analysis. Many of the smoothers that we discuss are *linear* (see section 2.8 below), and this facilitates an approximate assessment of their *degrees of freedom* (section 3.5).

We return to the time series setting in our discussion of the Fourier analysis of smoothers in section 3.7, and a seasonal decomposition procedure in section 8.5.

2.8 Equivalent kernels

The smoothers described in this chapter are defined in quite different ways, with more different ways to come. Their *equivalent kernels* are one way to compare them on common ground. Most of the smoothers studied in this chapter are *linear* in Y, which means that the fit at a point x_0 can be written as $s(x_0) = \sum_{j=1}^{n} S_{0j} y_j$, and the S_{0j} depend on all the x_i and on the smoothing parameter λ. Thus the weight in the fit at x_0 which is associated with the point x_j is S_{0j}, and this sequence of weights is known as the *equivalent kernel* at x_0. A plot of S_{0j} against x_j shows which observations have an influence on the fit at x_0. Looking at Fig. 2.5, we see that the running mean smoother has weights either zero or $1/m$, where m is the number of observations in the neighbourhood.

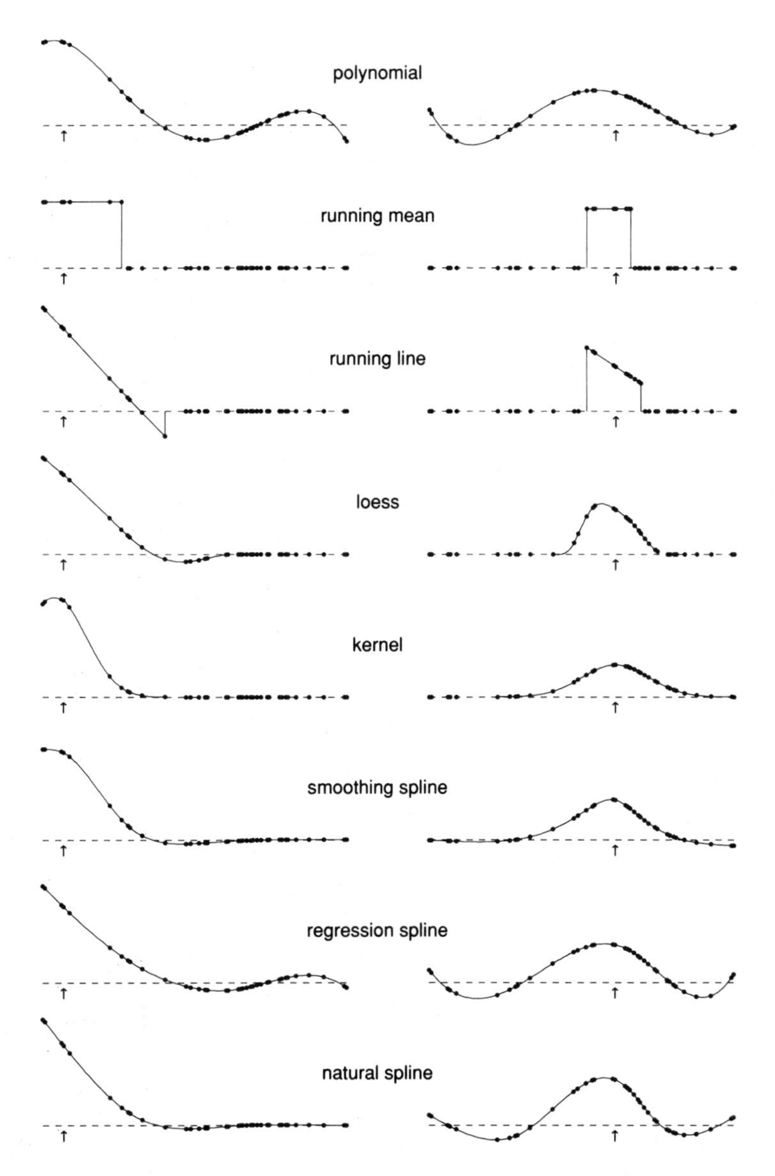

Fig. 2.5. *The equivalent kernels for the smoothers in Fig. 2.1. The bin smoother is omitted, since it resembles the running mean. The arrows indicate the target point in each case. The points are plotted at the pairs* (x_j, S_{0j}).

For each smoother we have computed the equivalent kernel centred at two points: one near the boundary and one at the centre. The weights at the boundary are larger and warn us that end effects may be a problem (as indeed they are).

For the running-line smoother the weights change linearly within the window (Exercise 2.3). For both the running-mean and running-line smoothers the weights drop off abruptly to zero outside the neighbourhood, and account for their jagged appearance. The *loess* smooth on the other hand has a strictly-local neighbourhood yet the weights die down smoothly to zero.

We need to calibrate the smoothers so that they are doing approximately the same amount of smoothing, since the value of the smoothing parameter will clearly widen or narrow the equivalent kernels. We do this using the *equivalent degrees of freedom* or *df*. In the case of projection fits, this is simply the dimension of the projection space; in other cases, such as the kernel smoother, it is the trace of the smoother operator matrix that produces the fitted values at the observed data points. All the smoothers in Fig. 2.5 are calibrated to do about five degrees of freedom worth of smoothing. We discuss this and other definitions of degrees of freedom in more detail in section 3.5.

In the top panel we see the equivalent kernel for the quartic polynomial fit. As we might expect it spreads its influence everywhere. We return to Fig. 2.5 as we encounter the remaining smoothers.

2.9 Regression splines

Polynomial regression has limited appeal due to the global nature of its fit, while in contrast the smoothers discussed so far have an explicit local nature. Regression splines offer a compromise by representing the fit as a *piecewise* polynomial. The regions that define the pieces are separated by a sequence of *knots* or breakpoints, $\xi_1, \ldots, \xi_K$. In addition, it is customary to force the piecewise polynomials to join smoothly at these knots. Although many different configurations are possible (the bin smoother is one), a popular choice consists of piecewise cubic polynomials constrained to be continuous and have continuous first and second derivatives at the knots. Apparently our eyes are skilled at picking up second and lower order discontinuities, but not higher.

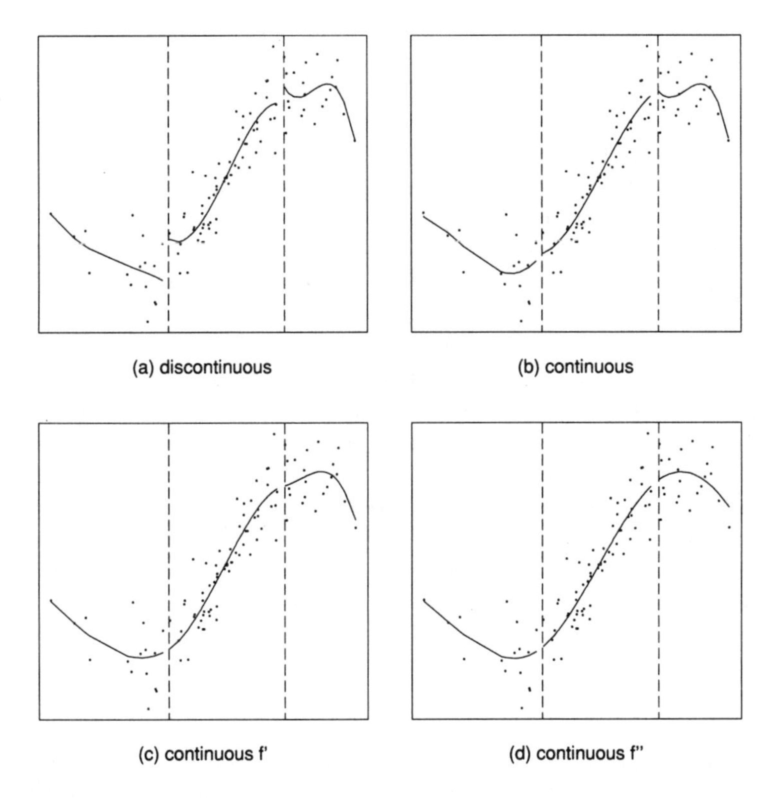

Fig. 2.6. *Piecewise cubic fits to some simulated data. The order of continuity increases from 0 — discontinuous as in figure (a), to 3 — continuous second derivatives as in figure (d). The fitted values are connected within each region, leaving gaps at the knots to make the orders of continuity more visible. The vertical lines indicate the location of the two knots.*

To illustrate this point, Fig. 2.6 shows four piecewise cubic fits to some simulated data, with two interior knots and different orders of continuity. Apart from being smoother, the fit in panel (d) required fewer parameters (six altogether) than the others.

By allowing more knots, the family of curves becomes more flexible. For any given set of knots, the smooth is computed by multiple regression on an appropriate set of basis vectors. These vectors are the basis functions representing the particular family of piecewise cubic polynomials, evaluated at the observed values of the predictor X. This approach is attractive because of its computational and statistical simplicity; for example standard parametric inferential methods can be used to test the importance of any of the parameters.

A variant of polynomial splines are the natural splines; although they are defined for all piecewise polynomials of odd degree, we discuss the natural *cubic* splines. These are cubic splines with the additional constraint that the function is linear beyond the boundary knots. To enforce this condition we have to impose the two constraints in each of the boundary regions: $f''' = f'' = 0$, which reduces the dimension of the space from $K+4$ to K if there are K knots. In practice it is common to supply an additional knot at each extreme of the data, and impose the linearity beyond them. Then with K interior knots (and two boundary knots), the dimension of the space of fits is $K+2$. Natural splines have less flexibility at the boundaries, but this tends to be a plus since the fitted values of regular regression splines have high variance near the boundary. Furthermore, for the same number of parameters, we get two more interior knots.

Computational aspects

The main difficulty when working with regression splines is selecting the number and position of the knots. A very simple approach, referred to as *cardinal splines*, requires a single parameter, the *number* of interior knots. The positions are then chosen uniformly over the range of the data. A slightly more adaptive version places the knots at appropriate quantiles of the predictor variable; e.g., three interior knots placed at the three quartiles. More adventurous schemes use data driven criteria to select the number and positions of the knots. The main challenge is to come up with a sensible procedure, while avoiding a combinatorial nightmare. We describe some specific proposals in Chapter 9.

Another issue is the choice of basis functions for representing the splines (a vector space) for a given set of knots. Suppose these interior knots are denoted by $\xi_1 < \cdots < \xi_K$, and for notational simplicity we augment the set with two boundary knots ξ_0 and ξ_{K+1}. A simple choice of basis functions for piecewise-cubic splines, known as the truncated power series basis, derives from the parametric expression for the smooth

$$s(x) = \beta_0 + \beta_1 x + \beta_2 x^2 + \beta_3 x^3 + \sum_{j=1}^{K} \theta_j (x - \xi_j)_+^3 \qquad (2.6)$$

where the notation a_+ denotes the positive part of a. Evidently (Exercise 2.4) (2.6) has the required properties:

(i) s is cubic polynomial in any subinterval $[\xi_j, \xi_{j+1})$,
(ii) s has two continuous derivatives, and
(iii) s has a third derivative that is a step function with jumps at $\xi_1, \ldots, \xi_K$.

We can write (2.6) as a linear combination of $K + 4$ basis functions $P_j(x)$: $P_1(x) = 1$, $P_2(x) = x$, and so on. Each of these functions must also satisfy the three conditions and be linearly independent in order to qualify as a basis. Even without (2.6) it is clear that we require $K + 4$ parameters to represent piecewise cubics; four for each of $(K + 1)$ cubics, less three per interior knot due to the constraints. To actually smooth some data pairs $\{x_i, y_i\}$, we would construct a regression matrix with $K + 4$ columns, each column corresponding to a function P_j evaluated at the n values of x.

Although (2.6) has algebraic appeal, it is not the recommended form for computing the regression spline. Even though a particular cubic piece is designed to accommodate one interval, it is evaluated at all points to the right of its knot and the numbers usually get large.

The B-spline basis functions provide a numerically superior alternative basis to the truncated power series. Their main feature is that any given basis function $B_j(x)$ is nonzero over a span of at most five distinct knots. In practice this means that their evaluation rarely gets out of hand, and the resulting regression matrix is banded. Of course the B_j are themselves piecewise cubics, and we need $K + 4$ of them if we want to span the space. Their algebraic definition is rather simple (in terms of divided differences), but unhelpful for fully understanding their properties, so we omit it here and refer the reader to de Boor (1978). Figure 2.7 shows the evaluated B-splines used to compute the fits in Fig. 2.6(d).

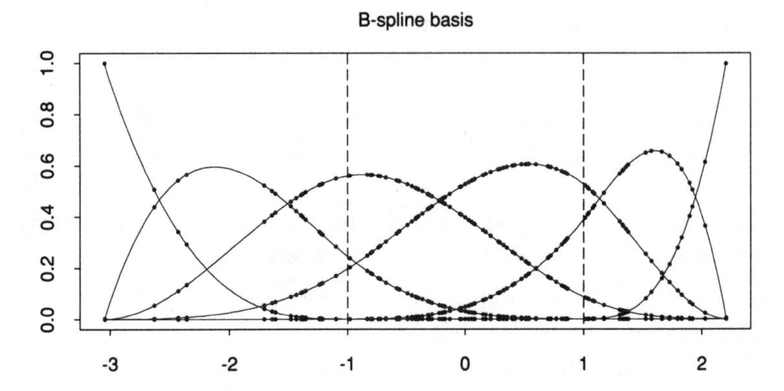

Fig. 2.7. *The six B-spline basis functions for representing the fits in Fig. 2.6(d). The two interior knots are indicated by the vertical broken lines. The points indicate the occurrence of data, and are* $\{x_i, B_j(x_i)\}$ *for* $j = 1, \ldots, 6$.

In summary, regression splines are attractive because of their computational cleanliness, *when the knots are given.* Standard linear model estimation can be applied, which we find very convenient in later chapters when we use regression splines in additive models.

However, the difficulty of choosing the number and position of the knots is a drawback of this approach. When a small number of knots is used, the smoother can show some disturbing nonlocal behaviour. Figure 2.5 includes the equivalent kernels for a cubic regression spline with one interior knot, as well as a natural cubic regression spline with three interior knots. These both use five degrees of freedom in their fits. The right-hand kernel of both, and the left hand kernel of the regression spline do not look much different from those for the quartic polynomial! With more knots this global influence would be damped, but we often don't have that many *df* to spare.

Another problem with regression splines is that the smoothness of the estimate cannot easily be varied continuously as a function of a single smoothing parameter; this is an attractive, if only approximate, property of most of the other smoothers that we discuss.

2.10 Cubic smoothing splines

This smoother is not constructed explicitly like those described so far, but instead emerges as the solution to an optimization problem.

Consider the following problem: among all functions $f(x)$ with two continuous derivatives, find one that minimizes the penalized residual sum of squares

$$\sum_{i=1}^{n}\{y_i - f(x_i)\}^2 + \lambda \int_a^b \{f''(t)\}^2\,dt \qquad (2.7)$$

where λ is a fixed constant, and $a \leq x_1 \leq \cdots \leq x_n \leq b$. This criterion satisfies the requirements for a scatterplot smoother that we mentioned earlier. The first term measures closeness to the data while the second term penalizes curvature in the function. Now it is one thing to state a criterion like (2.7) and quite another to find the optimizing function. Remarkably, it can be shown that (2.7) has an explicit, unique minimizer and that minimizer is a *natural cubic spline* with knots at the unique values of x_i (Exercise 2.6). At face value it seems that the family is overparametrized, since there are as many as $n - 2$ interior knots. This would result in $n + 2$ parameters, although the constraints on each end bring it down to n. We'll see however that the coefficients are estimated in a constrained way as well, and this can bring the *effective* dimension down dramatically.

A cubic smoothing spline fitted to the diabetes data is shown in Fig. 2.1. It looks much like the running-line smooth but is less jagged.

The parameter λ plays the same role as the span in the running-line smooth. Large values of λ produce smoother curves while smaller values produce more wiggly curves. At the one extreme, as $\lambda \to \infty$, the penalty term dominates, forcing $f''(x) = 0$ everywhere, and thus the solution is the least-squares line. At the other extreme, as $\lambda \to 0$, the penalty term becomes unimportant and the solution tends to an interpolating twice-differentiable function. We discuss methods for choosing λ in Chapter 3.

The boundary defined by a and b in (2.7) is arbitrary, as long as it contains the data; the smoothing spline is linear beyond the extreme data points no matter what the values of a and b are. In subsequent chapters we will avoid using integration limits, and assume that each integral is over a range that covers the variable in question.

Computational aspects

Using the fact that the solution to (2.7) is a natural cubic spline with $n-2$ interior knots, we can represent it in terms of a basis for this space of fits. For computational convenience, we will use the unconstrained B-spline basis, and write $s(x) = \sum_1^{n+2} \gamma_j B_j(x)$, where γ_j are coefficients and the B_j are the cubic B-spline basis functions. As written here, s lies in an $(n+2)$-dimensional space; however, the natural splines are a subspace. To solve (2.7) we replace f by s and perform the integration. Defining the $n \times (n+2)$ matrix $\mathbf{B}$ and $(n+2) \times (n+2)$ matrix $\mathbf{\Omega}$ by

$$B_{ij} = B_j(x_i)$$

and

$$\mathbf{\Omega}_{ij} = \int B_i''(x) B_j''(x)\, dx,$$

we can rewrite the criterion (2.7) as

$$(\mathbf{y} - \mathbf{B}\gamma)^T(\mathbf{y} - \mathbf{B}\gamma) + \lambda\gamma^T\mathbf{\Omega}\gamma \tag{2.8}$$

Although at face value it seems that there are no boundary derivative constraints, it turns out that the penalty term automatically imposes them.

Setting the derivative with respect to γ equal to zero gives

$$(\mathbf{B}^T\mathbf{B} + \lambda\mathbf{\Omega})\hat{\gamma} = \mathbf{B}^T\mathbf{y} \tag{2.9}$$

Since the columns of $\mathbf{B}$ are the evaluated B-splines, in order from left to right and evaluated at the *sorted* values of X, and the cubic B-splines have local support, $\mathbf{B}$ is lower 4-banded. Consequently the matrix $\mathbf{M} = (\mathbf{B}^T\mathbf{B} + \lambda\mathbf{\Omega})$ is 4-banded and hence its Cholesky decomposition $\mathbf{M} = \mathbf{L}\mathbf{L}^T$ can be computed easily. One then solves $\mathbf{L}\mathbf{L}^T\hat{\gamma} = \mathbf{B}^T\mathbf{y}$ by back-substitution to give $\hat{\gamma}$ and hence the solution $\hat{s}$ in $O(n)$ operations.

For theoretical purposes, it is convenient to rewrite the solution vector $\mathbf{s}$ (s evaluated at each of the n values of X in the sample) in another form. Let $\mathbf{N}$ be an $n \times n$ nonsingular *natural-spline* basis matrix for representing the solution (Exercise 2.5). Denote by $\hat{\boldsymbol{\beta}}$ the transformed version of $\hat{\gamma}$ corresponding to the change in basis. Then we can write

$$\mathbf{s} = \mathbf{N}\hat{\boldsymbol{\beta}} = \mathbf{N}(\mathbf{N}^T\mathbf{N} + \lambda\boldsymbol{\Omega})^{-1}\mathbf{N}^T\mathbf{y} = (\mathbf{I} + \lambda\mathbf{K})^{-1}\mathbf{y} \qquad (2.10)$$

where $\mathbf{K} = \mathbf{N}^{-T}\boldsymbol{\Omega}\mathbf{N}^{-1}$. In terms of the candidate fitted vector $\mathbf{f}$ and $\mathbf{K}$, the cubic smoothing spline $\mathbf{s}$ minimizes

$$(\mathbf{y} - \mathbf{f})^T(\mathbf{y} - \mathbf{f}) + \lambda\mathbf{f}^T\mathbf{K}\mathbf{f} \qquad (2.11)$$

over all vectors $\mathbf{f}$. It is appropriate to call the term $\mathbf{f}^T\mathbf{K}\mathbf{f}$ a roughness penalty, since it can be shown to be a quadratic form in second differences.

The cubic smoothing spline, because of the implicit way it is defined, doesn't appear to use *local averaging*. However, the equivalent-kernel weights in Fig. 2.5 show that it does possess local behaviour quite similar to kernels or locally-weighted lines. The equivalent kernel is nowhere nonzero, but is close to zero far from the target point. Other examples of equivalent kernels are given in Buja, Hastie and Tibshirani (1989), and an asymptotic form for the equivalent kernel of smoothing splines is derived by Silverman (1984). One might say, then, that a cubic smoothing spline is approximately a *kernel* smoother.

The smoothing parameter λ controls the shape of the kernel or weight function. As one would expect, when λ is decreased the weights tend to concentrate more around the target point, while as λ is increased, they spread out more. We discuss the properties of smoothing splines in more detail in Chapter 5 and section 9.3.6.

2.11 Locally-weighted running-line smoothers

Here we describe in more detail the locally-weighted smoother of Cleveland (1979), currently called *loess* in the S statistical-computing language. Although any polynomial can be fitted locally, we discuss local lines.

A locally-weighted straight-line smooth $s(x_0)$ using k nearest-neighbours is computed in a number of steps:

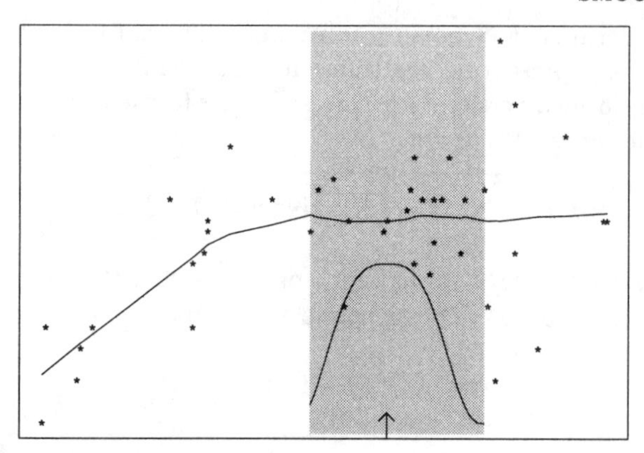

Fig. 2.8. *The loess smoother first computes the λ % nearest-neighbours to the target point, indicated here by the shaded region. A tricube kernel, centered at the target point, becomes zero at the furthest neighbour. The smooth at the target point is the fitted value from the locally-weighted linear fit, with weights supplied by the kernel.*

(i) The k nearest neighbours of x_0 are identified, denoted by $N(x_0)$.

(ii) $\Delta(x_0) = \max_{N(x_0)} |\, x_0 - x_i \,|$ is computed, the distance of the furthest near-neighbour from x_0.

(iii) Weights w_i are assigned to each point in $N(x_0)$, using the *tri-cube* weight function:

$$W\left(\frac{|\, x_0 - x_i \,|}{\Delta(x_0)}\right) \tag{2.12}$$

where

$$W(u) = \begin{cases} (1-u^3)^3, & \text{for } 0 \le u < 1; \\ 0 & \text{otherwise.} \end{cases} \tag{2.13}$$

(iv) $s(x_0)$ is the fitted value at x_0 from the weighted least-squares fit of y to x confined to $N(x_0)$ using the weights computed in (iii).

Figure 2.8 depicts this smoother, using the diabetes data once again. Notice that the kernel is truncated at one end due to the metric asymmetry of the neighbourhood.

Cleveland also discusses the use of a robust regression within each neighbourhood, to protect against outliers. This effectively works by repeatedly smoothing the data, and at each iteration down-weighting points with large residuals. See sections 3.11.6 and 9.2 for more details.

A locally-weighted running-line smooth is shown in Fig. 2.1. The number of nearest-neighbours, usually expressed as a percentage or *span* of the data points, is the smoothing parameter. In Figs 2.1 and 2.8 a span of 21/43 was used, chosen to make the degrees of freedom approximately equal to that of the other smooths.

In principle, nearest neighbourhoods are preferable to symmetric nearest neighbourhoods, because in a neighbourhood with a fixed number of points, the average distance of the points to the target point is less in the nearest neighbourhood (unless the predictors are evenly spaced). In general this should result in less bias. We use symmetric nearest neighbourhoods in the construction of a running-line smoother, however, because of their superior performance at the left and right endpoints. A nearest neighbourhood at the endpoint contains the same number of data points as a neighbourhood in the middle of the data, so that when a least-squares fit is applied there, too much weight is given to points far away from the target point. A symmetric nearest neighbourhood, on the other hand, contains only about half of the number of points as a full neighbourhood, and thus automatically reduces the weight given to far away points. Nearest neighbourhoods work satisfactorily with *loess* at the endpoints, however, because the tri-cube function does the job of down-weighting far away points.

The locally-weighted smoothers are popular, since they enjoy the best of both worlds. They share the ability of near-neighbour smoothers to adapt their bandwidth to the local density of the predictors, while they have the smoothness features of kernel smoothers.

Although in principle it requires $O(n^2)$ operations to compute a locally-weighted smooth, the current implementations compute the fit over a grid of values of x, and use interpolation elsewhere.

2.12 Smoothers for multiple predictors

So far we have discussed smoothers for a single predictor, that is, scatterplot smoothers. What if we have more than one predictor, say $X_1, \ldots, X_p$? Then our problem is one of fitting a p-dimensional surface to Y. The multiple regression of Y on $X_1, \ldots, X_p$ provides a simple, but very limited, estimate of the surface. On the other hand, it is easy conceptually to generalize the running mean, locally-weighted running-line, and kernel smoothers to this setting. The first two smoothers require a definition of *nearest neighbours* of a point in p-space. *Nearest* is determined by a distance measure and for this the most obvious choice is Euclidean distance. Note that the notion of symmetric nearest neighbours is no longer meaningful when $p > 1$. Having defined a neighbourhood, the generalization of the running mean estimates the surface at the target point by averaging the response values in the neighbourhood.

This highlights an important detail: what shape neighbourhood do we use? Nearest neighbourhoods defined in terms of Euclidean distance are typically spherical, but one could imagine more general neighbourhoods defined in terms of a covariance matrix $\mathbf{\Sigma}$ of the predictors. Such a metric would consider points lying on an ellipse centered at the target point to be equidistant from the target. This generalization is not a purely academic issue — it may be important if the covariates are measured in different units or are correlated. In linear regression, the coefficients automatically scale each covariate correctly. Various strategies are used in practice, an obvious one being to standardize the individual variables prior to smoothing. This might not always be the best choice, however. For example, if the underlying surface changes more rapidly with one variable than the other, one may wish to have the neighbourhoods thinner in the direction of the first variable. This can be achieved by differential scaling of the variables.

The kernel smoother is generalized in an analogous way to the running mean. Given two p-vectors $\mathbf{x}^0$ and $\mathbf{x}^i$ in predictor space, the weight given to this ith point for the fit at $\mathbf{x}^0$ is $S_{0i} = (c_0/\lambda)d(\|\mathbf{x}^0 - \mathbf{x}^i\|/\lambda)$ where $d(t)$ is an even function decreasing in $|t|$, and $\|\cdot\|$ is a norm, for example squared distance. The constant c_0 is usually chosen to make the weights sum to unity, so that the smoother reproduces the constant function. The multi-predictor Gaussian kernel uses $d(t)$ equal to the standard Gaussian density.

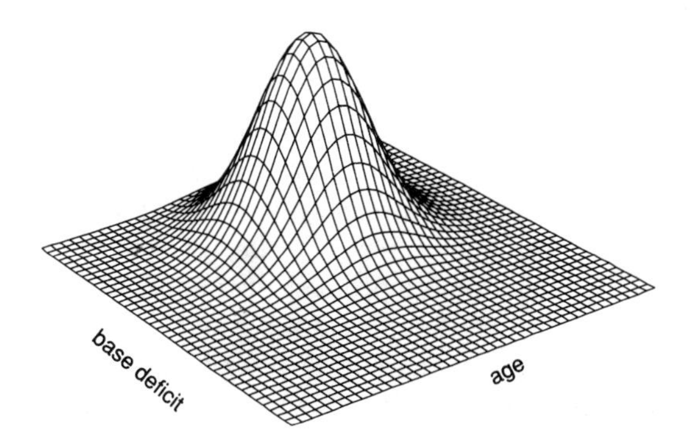

Fig. 2.9. *The weight function for a bivariate Gaussian kernel smoother.*

The resulting weight function is depicted in Fig. 2.9.

The choice of norm is important in this setting, and determines the shape of the neighbourhoods in predictor space. One might argue, for example, that the norm should incorporate covariance information by replacing the squared error norm with the "Mahalanobis" distance

$$\|\mathbf{x}^0 - \mathbf{x}^i\| = (\mathbf{x}^0 - \mathbf{x}^i)^T \boldsymbol{\Sigma}^{-1} (\mathbf{x}^0 - \mathbf{x}^i)$$

where $\boldsymbol{\Sigma}$ is the covariance matrix of the predictors. We do not pursue this issue further.

Locally-weighted lines generalize equally easily to two or more dimensions. Once a multivariate distance is defined, nearest neighbourhoods and tri-cube weights are assigned in exactly the same way, and one computes a local plane rather than a line.

The cubic smoothing spline is more difficult to generalize to two or higher dimensions: the so-called *thin-plate spline* is one example. It derives from generalizing the second derivative penalty for smoothness to a two dimensional Laplacian penalty of the form

$$\int\int\Big\{\Big(\frac{\partial^2 f}{\partial x_1^2}\Big)^2 + \Big(\frac{\partial^2 f}{\partial x_1 \partial x_2}\Big)^2 + \Big(\frac{\partial^2 f}{\partial x_2^2}\Big)^2\Big\}\, dx_1 dx_2.$$

The computations are $O(n^3)$ in number compared to $O(n)$ for univariate splines. Details are not given here; references can be found in the bibliographic notes at the end of Chapter 3.

Another generalization is known as multivariate *tensor product* splines. These are also useful for generalizing univariate regression splines. The basic idea is to construct two-dimensional basis functions by multiplying together one-dimensional basis functions in the respective predictors. We discuss these in more detail for multivariate regression splines in Chapter 9. For the smoothing version of tensor product splines, the effective basis set has dimension n^2 formed by the tensor product of the univariate bases. The fit is a damped regression onto this space of basis functions. See the references at the end of Chapter 3 for details.

We are not devoting much space to multi-predictor smoothers because we don't feel that they are very useful for more than two or three predictors. Indeed, their many shortcomings (e.g. difficulty of interpretation and computation) provide an impetus for studying additive models, the central topic of this book. These comments refer to the generic multivariate smoothers as described here. We are not referring to some of the adaptive multivariate nonparametric regression methods, which might also be termed surface smoothers, and which were designed to overcome some of these objectionable aspects. We discuss some of the problems associated with multi-predictor smoothing in detail at the beginning of Chapter 4. Despite these criticisms, multi-predictor smoothers (especially the bi-predictor) are useful in certain settings, as later examples show.

2.13 Bibliographic notes

In this chapter we describe a number of scatterplot smoothers, including the running-mean and running-line smoothers, cubic regression and smoothing splines, locally-weighted running-line and kernel smoothers. We also touch on surface smoothers. Our discussion focuses on smoothers with which we have some experience, and any imbalance in our introduction simply reflects this. In addition, the emphasis in the book is not on smoothers *per se*, but on smoothers as building blocks in the estimation of additive models.

Some recent theoretical results suggest that for appropriately chosen smoothing parameters, there are not likely to be large differences between locally-weighted running-line , cubic smoothing-splines and kernel smoothers (Silverman, 1984; Muller, 1987). In order to assess the finite-sample operating characteristics of these and other smoothers, a comprehensive Monte Carlo study, akin to the Princeton robustness study, would be very useful. A recent report by Breiman and Peters (1988) has this flavour.

For a detailed bibliography, see the notes at the end of Chapter 3. Here we list some introductory references that we have found useful. Eubank (1988) covered most of the material in various levels of detail, especially smoothing splines. Similarly Härdle (1990) covered nonparametric regression in some detail, with a focus on kernel methods. Silverman's (1985) paper on smoothing splines not only gives a readable introduction but also touches on a variety of applications. Cleveland (1979) introduced the locally-weighted running-line smoother for univariate problems, and Cleveland, Devlin and Grosse (1988) described the multivariate version.

2.14 Further results and exercises 2

2.1 For the diabetes data, compute one of each of the smooths described in this chapter for the variable `base deficit`. Compare qualitatively the resulting estimates in terms of notable features and smoothness.

2.2 *Updating formula for running-line smooth.* Suppose we add a point (x_{j+1}, y_{j+1}) to a neighbourhood containing j points. If the means of X and Y, variance of X, and covariance of X and Y for the first j points are $\bar{x}_j$, $\bar{y}_j$, S_j^x, and S_j^{xy}, show that

$$\bar{x}_{j+1} = (j\bar{x}_j + x_{j+1})/(j+1)$$

$$\bar{y}_{j+1} = (j\bar{y}_j + y_{j+1})/(j+1)$$

$$(j+1)S_{j+1}^x = jS_j^x + \frac{j+1}{j}(x_{j+1} - \bar{x}_{j+1})^2$$

$$(j+1)S_{j+1}^{xy} = jS_j^{xy} + \frac{j+1}{j}(x_{j+1} - \bar{x}_{j+1})(y_{j+1} - \bar{y}_{j+1}). \quad (2.14)$$

Derive the corresponding equations for deleting a point. Together these can be used to update the least-squares slope and intercept, and hence the entire fit can be computed in $O(n)$ operations.

2.3 Derive an expression for S_{ij}, the coefficient of y_j in the expression $s(x_i) = \sum_{j=1}^{n} S_{ij}y_j$ for the running-line fit at x_i, using $N^S(x_i)$ to denote the set of indices of the k symmetric nearest-neighbours to x_i.

2.4 Show that the truncated power series representation

$$s(x) = \beta_0 + \beta_1 x + \beta_2 x^2 + \beta_3 x^3 + \sum_{j=1}^{K} \theta_j (x - \xi_j)_+^3$$

satisfies the three conditions of a cubic spline below (2.6) in the text.

2.5 *Basis for natural splines.* Suppose $\mathbf{B}$ is an $n \times (K+4)$ matrix containing the evaluations of the cubic B-spline basis functions with K interior knots evaluated at the n values of X. Let $\mathbf{C}$ be the $2 \times (K+4)$ matrix containing the second derivatives of the basis functions at the boundary points x_1 and x_n. Show how to derive $\mathbf{N}$ from $\mathbf{B}$, an $n \times (K+2)$ basis matrix for the natural cubic splines with the same interior knots and boundary knots at the extremes of X.

2.6 *Derivation of smoothing splines; Reinsch (1967).* Consider the following optimization problem: minimize

$$\int_a^b \{f''(x)\}^2\, dx \quad \text{subject to} \quad \sum_{i=1}^{n} \{y_i - f(x_i)\}^2 \le \sigma \qquad (2.15)$$

over all functions $f(x)$ with continuous first and integrable second derivatives in the interval $[a, b]$ for which $a \le x_1 \le \cdots \le x_n \le b$. To solve this problem, consider the Lagrangian functional

$$F(f) = \int_a^b \{f''(x)\}^2\, dx + \rho \left[\sum_{i=1}^{n} \{y_i - f(x_i)\}^2 - \sigma + z^2 \right] \qquad (2.16)$$

where z is an auxiliary variable.

(i) To find the minimum of $F(f)$, compute $\partial F\{f + \delta h(x)\}/\partial\delta$ where $h(x)$ is any twice differentiable function (this is a standard technique in the calculus of variations). Integrate by parts the resulting expression (twice) and conclude that the solution is a piecewise cubic polynomial with knots at the x_i. Show that the solution and its first two derivatives are continuous at the knots, but that the third derivative may be discontinuous there. Show also that the second derivative is zero outside of the range of the x_i.

(ii) Compute $\partial F(f)/\partial\rho$ and $\partial F(f)/\partial\sigma$ and conclude that either $\hat{\rho} = 0$ implying that the least-squares line is a solution to (2.15), or $\hat{\rho} \neq 0$, in which case the solution is nonlinear.

(iii) Establish an equivalence between the (2.15) and the penalized least-squares problem (2.7) given in the text.

2.7 Suppose each observation has associated with it a weight w_i.

(i) Derive the appropriate cubic smoothing spline using a weighted residual sum of squares.

(ii) Suppose there are ties in the x-values. Suggest a way to overcome this for smoothing splines.

2.8 *Semi-parametric regression; Green, Jennison and Seheult (1985).* Suppose we have a set of n observations of p predictors arranged in the $n \times p$ matrix $\mathbf{X}$, an additional covariate $\mathbf{z}$, and a response vector $\mathbf{y}$. We wish to fit the model $y_i = \mathbf{x}^i\boldsymbol{\beta} + f(z_i) + \varepsilon_i$ by penalized least-squares. Construct an appropriate penalized residual sum of squares, and show that the minimizers must satisfy the following pair of *estimating* equations:

$$\boldsymbol{\beta} = (\mathbf{X}^{\mathbf{T}}\mathbf{X})^{-1}\mathbf{X}^{\mathbf{T}}(\mathbf{y} - \mathbf{f})$$
$$f = \mathcal{S}(\mathbf{y} - \mathbf{X}\boldsymbol{\beta} \,|\, \mathbf{z})$$

where $\mathcal{S}$ is an appropriate smoothing-spline operator, and $\mathbf{f}$ represents the function f evaluated at the n values z_i.

2.9 *Efficient kernel smoothing; Silverman (1982), Härdle (1986)* Consider the Nadaraya-Watson form of the kernel smooth given by equation (2.5):

$$s(x_0) = \frac{\sum_{i=1}^{n} d(\frac{x_0 - x_i}{\lambda}) y_i}{\sum_{i=1}^{n} d(\frac{x_0 - x_i}{\lambda})}.$$

We wish to use the fast Fourier transform (FFT) to compute both the numerator and denominator. With that end in mind we define a fine grid that covers the range of X. Describe how you would approximate the data on the grid and use the FFT to perform the computations. Pay particular attention to:

(i) what happens at the boundaries, and
(ii) a method for overcoming discontinuities due to the approximation.

2.10 Extend the methods described in the previous exercise in order to compute a kernel-weighted running-line fit.

2.11 Suppose $\mathbf{S}$ is an $n \times n$ smoother matrix for smoothing against the n unique predictor values $\mathbf{x} = (x_1, \ldots, x_n)^T$. That is, if $s(x_i) = \sum_{j=1}^{n} S_{ij} y_j$, then $\mathbf{S}$ has ijth entry S_{ij}. For all of the smoothers that we have described, $\mathbf{S1} = \mathbf{1}$, where $\mathbf{1}$ is a column vector of ones, and for most $\mathbf{Sx} = \mathbf{x}$. For which smoothers is the second statement true, and what are the implications of these two results? For smoothing splines the symmetric statements are also true: $\mathbf{S}^T\mathbf{1} = \mathbf{1}$ and $\mathbf{S}^T\mathbf{x} = \mathbf{x}$, since $\mathbf{S}$ is symmetric in this case. Is this true for any of the other smoothers? What are the implications of these conditions? How would you modify $\mathbf{S}$ for the other smoothers in order that the first or both of these latter conditions are satisfied. Try some examples and see what happens to the local support properties of the equivalent kernels.

2.12 Given data points $(x_1, y_1), \ldots, (x_n, y_n)$ (with the x_i unique and increasing), the function minimizing

$$\int_{x_1}^{x_n} \{f''(t)\}^2 \, dt$$

subject to $f(x_i) = y_i, \quad i = 1, \ldots, n$, is a natural cubic spline with knots at the values of x_i. This is known as the solution to the *interpolation* problem. Using this fact, provide an heuristic argument to show that any solution to the smoothing problem (2.7) must be a cubic spline.

CHAPTER 3

Smoothing in detail

3.1 Introduction

In the previous chapter we introduce a number of techniques for smoothing a response variable on one or more predictor variables. In this chapter we delve more deeply into the smoothing problem, investigating a number of topics of both practical and theoretical importance. This chapter can be skipped by the reader eager to get to the heart of the book (Chapter 4), to be returned to later.

For ease of presentation, single predictor (scatterplot) smoothing is the subject of study, although most of the discussion generalizes to multiple predictor smoothing. Among the topics covered are the fundamental trade-off between bias and variance, and how one can choose the smoothing parameter to balance this trade-off; degrees of freedom of a smoother, weighted smoothing, resistant smoothing, and inference for the fitted smooth. Some asymptotic results for smoothers are discussed briefly.

3.2 A formal model for scatterplot smoothing

In the previous chapter we did not assume any formal relationship between the response Y and the predictor X. In order to study the smoothing problem further, and to lay the groundwork for additive models, we do so now. We assume that

$$Y = f(X) + \varepsilon \tag{3.1}$$

where $E(\varepsilon) = 0$ and $\text{var}(\varepsilon) = \sigma^2$ (the assumption of constant variance will be relaxed when we discuss weighted smoothers). We also assume that the errors ε are independent. Model (3.1) says that $E(Y \mid X = x) = f(x)$ and in this formal framework, the goal of a scatterplot smoother is to estimate the function f. To emphasize this, we denote the fitted functions by $\hat{f}$ rather than the s we used in the previous chapter. How can the running-mean, locally-weighted running-lines and cubic smoothing splines be seen as estimates of $E(Y \mid X = x)$? This is clearest for the first two, since they are constructed by *averaging* Y-values corresponding to X values close to a target value x_0. The average involves values of $f(x)$ close to $f(x_0)$; since $E(\varepsilon) = 0$, this implies that $E\{\hat{f}(x_0)\} \approx f(x_0)$. This argument is made more formal in the next section. As for a cubic smoothing spline (hereafter referred to simply as a smoothing spline), it can be shown that as $n \to \infty$ and the smoothing parameter $\lambda \to 0$, under certain regularity conditions, $\hat{f}(x) \to f(x)$. This says as we get more and more data, the smoothing-spline estimate will converge to the true regression function $E(Y \mid X = x)$.

3.3 The bias-variance trade-off

In scatterplot smoothing there is a fundamental trade-off between the bias and variance of the estimate, and this trade-off is governed by the smoothing parameter. We illustrate this phenomenon in Fig. 2.2, and we go into more detail here.

The trade-off is most easily seen in the case of the running-mean smoother. The fitted running-mean smooth can be written as

$$\hat{f}_k(x_i) = \sum_{j \in N_k^S(x_i)} \frac{y_j}{2k+1}. \tag{3.2}$$

This has expectation and variance

$$\begin{aligned} E\{\hat{f}_k(x_i)\} &= \sum_{j \in N_k^S(x_i)} f(x_j)/(2k+1) \\ \text{var}\{\hat{f}_k(x_i)\} &= \frac{\sigma^2}{2k+1}. \end{aligned} \tag{3.3}$$

For ease of notation we assume that x_i is near the *middle* of the data so that $N_k^S(x_i)$ contains the full $2k+1$ points. Hence we see that *increasing k decreases the variance* but tends to *increase the bias* because the expectation $\sum_{j \in N_k^S(x_i)} f(x_j)/(2k+1)$ involves more terms with $f(\cdot)$ values different from $f(x_i)$. Similarly, *decreasing* k *increases the variance* but tends to *decrease the bias.* This is the trade-off between bias and variance, much like that encountered when adding or deleting terms from a linear regression model. The same qualitative behaviour occurs for the smoothing spline, with the bias decreasing and variance increasing as $\lambda \to 0$, and the converse holding as $\lambda \to \infty$.

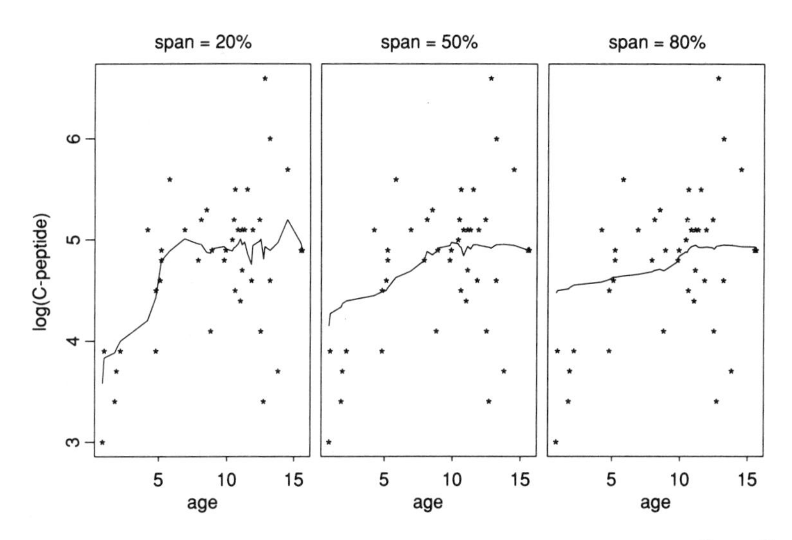

Fig. 3.1. *Running mean smooth of diabetes data, for spans 20%, 50% and 80%.*

Figure 3.1 shows the running-mean smooths using spans of 20%, 50%, and 80% for the diabetes data. Notice how the larger spans produce smoother but flatter curves.

We can get an approximate expression for the bias in terms of the curvature of f. We expand f in a Taylor series

$$f(x_j) = f(x_i) + (x_j - x_i)f'(x_i) + (x_j - x_i)^2 f''(x_i)/2 + R. \quad (3.4)$$

Assuming that the data are equally spaced with $x_{i+1} - x_i = \Delta$, and ignoring the remainder term R, some simple algebra shows that the bias in $\hat{f}_k$ is

$$E\{\hat{f}_k(x_i) - f(x_i)\} \approx \frac{k(k+1)}{6} f''(x_i)\Delta^2. \quad (3.5)$$

Hence the bias increases with k^2 and the second derivative of the function.

An optimal choice of span trades the bias of the estimate against its variance. One reasonable criterion that combines both bias and variance is mean squared error:

$$E\{\hat{f}_k(x_i) - f(x_i)\}^2 = \text{var}\{\hat{f}_k(x_i)\} + [E\hat{f}_k(x_i) - f(x_i)]^2. \quad (3.6)$$

Using the expressions (3.3) and (3.5) for the bias and variance of $\hat{f}_k(x_i)$, the k minimizing the mean squared error at x_i is seen to be approximately

$$k_{opt} = \left\{ \frac{9\sigma^2}{2\Delta^4\{f''(x_i)\}^2} \right\}^{1/5} \quad (3.7)$$

(Exercise 3.1). This is not useful in practice, however, because we usually do not know $f''(x_i)$.

3.4 Automatic selection of smoothing parameters

Most of the smoothers that we discuss in Chapter 2 have a single smoothing parameter. We use the symbol λ to denote this smoothing parameter, even though for different smoothers λ is defined differently. Whenever appropriate, we indicate the particular value of the smoothing parameter used to compute the fit $\hat{f}_\lambda$, as well as other quantities that depend on λ such as $S_{ij}(\lambda)$, the ijth element of the smoother matrix.

3.4.1 *Cross-validation*

In choosing the smoothing parameter, we need not try to minimize the mean squared error at each x_i, but instead we focus on a *global* measure such as *average mean-squared error*

$$MSE(\lambda) = \frac{1}{n}\sum_{i=1}^{n} E\{\hat{f}_\lambda(x_i) - f(x_i)\}^2. \quad (3.8)$$

Another quantity that differs from *MSE* only by a constant function of σ^2 is the *average predictive squared error*

$$PSE(\lambda) = \frac{1}{n}\sum_{i=1}^{n} E\{Y_i^* - \hat{f}_\lambda(x_i)\}^2 \quad (3.9)$$

where Y_i^* is a new observation at x_i, that is $Y_i^* = f(x_i) + \varepsilon_i^*$ where ε_i^* is independent of the ε_is. It is easy to show that $PSE = MSE + \sigma^2$, and it turns out that the quantities discussed below are really estimates of PSE rather than MSE. Notice that we are conditioning on the observed values of X in these summary measures. An alternative (and often more appropriate strategy) is to minimize the expected squared prediction errors averaged over the true distribution of X. This requires some extra knowledge or assumptions, so for the remainder of the chapter we use PSE as defined.

Cross-validation works by leaving points (x_i, y_i) out one at a time and estimating the smooth at x_i based on the remaining $n-1$ points. This is an attempt to mimic the use of training- and test-samples for prediction. One then constructs the *cross-validation sum of squares*

$$CV(\lambda) = \frac{1}{n}\sum_{i=1}^{n}\{y_i - \hat{f}_\lambda^{-i}(x_i)\}^2 \tag{3.10}$$

where $\hat{f}_\lambda^{-i}(x_i)$ indicates the fit at x_i, computed by leaving out the ith data point. We use CV for span selection as follows: $CV(\lambda)$ is computed for a number of values of λ over a suitable range and then the minimizing $\hat{\lambda}$ is selected.

This procedure is loosely justified by the fact that

$$E\{CV(\lambda)\} \approx PSE(\lambda), \tag{3.11}$$

although a stronger justification requires evidence that the minimizer of $CV(\lambda)$ is close in some sense to the minimizer of $PSE(\lambda)$. To see (3.11) we note that

$$\begin{aligned} E\{Y_i - \hat{f}_\lambda^{-i}(x_i)\}^2 &= E\{Y_i - f(x_i) + f(x_i) - \hat{f}_\lambda^{-i}(x_i)\}^2 \\ &= \sigma^2 + E\{f(x_i) - \hat{f}_\lambda^{-i}(x_i)\}^2 \end{aligned}$$

using the important fact that the cross-product term $E\{y_i - f(x_i)\}\{f(x_i) - \hat{f}_\lambda^{-i}(x_i)\}$ is zero because $\hat{f}_\lambda^{-i}(x_i)$ doesn't involve y_i. Similarly, $E\{Y_i^* - \hat{f}_\lambda(x_i)\}^2\} = \sigma^2 + E\{f(x_i) - \hat{f}_\lambda(x_i)\}^2$. Then assuming $\hat{f}_\lambda^{-i}(x_i) \approx \hat{f}_\lambda(x_i)$, we have that $E\{CV(\lambda)\} \approx PSE(\lambda)$. We will examine this approximation later for the special case of *linear* smoothers. Note that the naive estimate of PSE, the average squared residual

$$ASR(\lambda) = \frac{1}{n}\sum_{i=1}^{n}\{y_i - \hat{f}_\lambda(x_i)\}^2 \tag{3.12}$$

is not a good estimate of $PSE(\lambda)$. We would expect this intuitively because minimizing ASR over the smoothing parameter leads to $\hat{f}_\lambda(x_i) = y_i$, that is, an interpolating estimate. Later in our analysis of linear smoothers we shed more light on the failure of ASR.

Figure 3.2 displays three methods for span selection for a simulated data set of a quadratic function with noise added. The data are shown in the top left; the ASR curve for a running-mean smooth of various spans is displayed with the true PSE curve (solid lines) in the top right figure. The bottom left figure shows the CV curve along with the true PSE. Notice how the ASR curve is roughly increasing, while the CV curve is roughly concave. This particular realization has two minima, one at 10%, and a global minimum at 30%; the minimum of the PSE curve occurs approximately at span= 20%. The bottom right figure shows the C_p statistic, discussed below. Notice that C_p and CV look qualitatively the same; in particular, their minima occur at the same locations, a feature that persisted in other simulations from this model.

These pictures are qualitatively representative of a number of simulations that we tried from this model, and show a certain amount of variability. Some but not all of this variability is due to the crude smoother that is used. Of course, of more importance is the location of the minimum $\hat{\lambda}$ of each curve, and its behaviour as an estimate of the location of the value of the smoothing parameter that minimizes PSE. We return to this issue in section 3.4.5.

3.4.2 *The bias-variance trade-off for linear smoothers*

A smoother is linear if $\mathcal{S}(a\mathbf{y}_1 + b\mathbf{y}_2 \,|\, \mathbf{x}) = a\mathcal{S}(\mathbf{y}_1 \,|\, \mathbf{x}) + b\mathcal{S}(\mathbf{y}_2 \,|\, \mathbf{x})$ for any constants a and b. If we focus on the fit at the observed points $x_1, \ldots, x_n$, a linear smoother can be written as

$$\hat{\mathbf{f}} = \mathbf{S}\mathbf{y} \tag{3.13}$$

where $\mathbf{S} = \{S_{ij}\}$ is an $n \times n$ matrix that we call a *smoother matrix*.

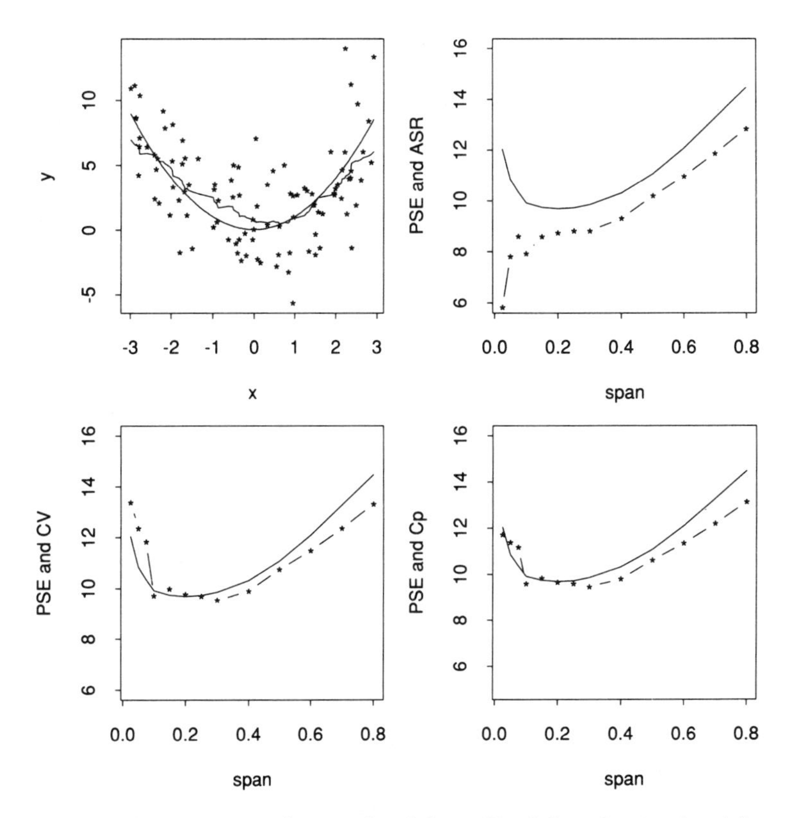

Fig. 3.2. *Span selection for simulated data. Top left is the simulated data from a quadratic function of* X *plus noise. The generating quadratic is included, as well as a running-mean smooth with a span of* 30%. *In the other figures, the solid line is the mean squared prediction error (PSE) of the estimate as a function of the span. The broken lines show the average squared residual (ASR), the average squared cross-validated residual (CV) and* C_p *estimates. The latter two both have a minimum near* 30%, *while the PSE curve has a minimum near* 20%. *The points on these curves indicate the actual values of the three criteria; the rest of thc curvc is filled in by interpolation.*

The running-mean, running-line, smoothing spline, kernel, locally-weighted running-line and regression spline smoothers are all linear smoothers. In Fig. 2.5 we plot their equivalent kernels; the rows of **S** are the equivalent kernels for producing fits at each of the observed data sites x_i (see section 2.8).

Not all smoothers are linear, and later in the chapter we see examples of nonlinear smoothers.

Consider a linear smooth $\hat{\mathbf{f}}_\lambda = \mathbf{S}_\lambda \mathbf{y}$ and let $\mathbf{b}_\lambda = \mathbf{f} - E(\mathbf{S}_\lambda \mathbf{y}) = \mathbf{f} - \mathbf{S}_\lambda \mathbf{f}$ denote the *bias* vector, where $f_i = f(x_i)$. Then the following formulae are easy to derive:

$$
\begin{aligned}
MSE(\lambda) &= \frac{1}{n}\sum_{i=1}^{n} \mathrm{var}(\hat{f}_{\lambda i}) + \frac{1}{n}\sum_{i=1}^{n} b_{\lambda i}^2 \\
&= \frac{\mathrm{tr}(\mathbf{S}_\lambda \mathbf{S}_\lambda^T)}{n}\sigma^2 + \frac{\mathbf{b}_\lambda^T \mathbf{b}_\lambda}{n} \qquad (3.14) \\
PSE(\lambda) &= \left\{1 + \frac{\mathrm{tr}(\mathbf{S}_\lambda \mathbf{S}_\lambda^T)}{n}\right\}\sigma^2 + \frac{\mathbf{b}_\lambda^T \mathbf{b}_\lambda}{n}.
\end{aligned}
$$

The first term in each of $MSE(\lambda)$ and $PSE(\lambda)$ measures variance while the second term measures squared bias. If we increase the amount of smoothing, we expect the bias to increase while the variance decreases, and conversely when the amount of smoothing is decreased. This turns out to be generally true, because as the amount of smoothing increases, $\mathrm{tr}(\mathbf{S}_\lambda \mathbf{S}_\lambda^T)$ tends to decrease, while the elements of $\mathbf{b}_\lambda$ tend to increase, and conversely. As a matter of fact, $\mathrm{tr}(\mathbf{S}_\lambda \mathbf{S}_\lambda^T)$ is one of the quantities that we use to calibrate the amount of smoothing performed by a linear smoother. Note that in the case of least-squares regression, $\mathbf{S}_\lambda$ is idempotent so that $\mathrm{tr}(\mathbf{S}_\lambda \mathbf{S}_\lambda^T) = \mathrm{tr}(\mathbf{S}_\lambda) = \mathrm{rank}(\mathbf{S}_\lambda)$ which equals the number of linearly independent predictors in the model.

3.4.3 *Cross-validation for linear smoothers*

The cross-validation sum of squares CV involves the quantity $\hat{f}_\lambda^{-i}(x_i)$, the *fit at x_i with the ith point removed,* also known as the jackknifed fit at x_i. We haven't said clearly what we mean by this. If we start with a linear smoother based on n data points, in order to define $\hat{f}_\lambda^{-i}(x_i)$ we must define the corresponding smoother for $n-1$ points.

There is a simple, unambiguous way to define $\hat{f}_\lambda^{-i}(x_i)$ given only the smoother matrix $\mathbf{S}_\lambda$. First note that any reasonable smoother is constant preserving, that is $\mathbf{S}_\lambda \mathbf{1} = \mathbf{1}$, where $\mathbf{1}$ is an n-vector of ones. Thinking of the elements of each row of $\mathbf{S}_\lambda$ as weights, this implies that the sum of the weights in each row is one. We define $\hat{f}_\lambda^{-i}(x_i)$ to be the fit obtained by setting the weight on the ith observation to zero and increasing the remaining weights so that they sum to one. A formal definition is

$$\hat{f}_\lambda^{-i}(x_i) = \sum_{\substack{j=1\\ j\neq i}}^{n} \frac{S_{ij}(\lambda)}{1 - S_{ii}(\lambda)} y_j. \tag{3.15}$$

It is easily checked that (3.15) implies the important relationship

$$\hat{f}_\lambda^{-i}(x_i) = \sum_{\substack{j=1\\ j\neq i}}^{n} S_{ij}(\lambda) y_j + S_{ii}(\lambda)\hat{f}_\lambda^{-i}(x_i). \tag{3.16}$$

Relation (3.16) holds for linear regression and greatly simplifies the computation of deletion diagnostics. In that case (3.16) says that if we add a new point that lies exactly on the regression surface, then that point doesn't change the fitted regression.

The smoothing spline, by its nature, is defined for all x_0 and n, and thus $\hat{f}_\lambda^{-i}(x_i)$ is already defined for such a smoother. Happily, it turns out that (3.16) holds for a cubic smoothing spline as well. To see this, suppose that $\hat{f}_\lambda^{-i}$ minimizes the penalized least-squares criterion

$$\sum_{\substack{j=1\\ j\neq i}}^{n} \{y_j - g(x_j)\}^2 + \lambda \int \{g''(x)\}^2 \, dx \tag{3.17}$$

for sample size $n-1$ and suppose we add the point $\{x_i, \hat{f}_\lambda^{-i}(x_i)\}$ to our data set. Then $\hat{f}_\lambda^{-i}$ results in the same value of (3.17) and must still minimize (3.17) for sample size n, for if some other cubic spline produced a smaller value of (3.17), it would also produce a smaller value of (3.17) over the original $n-1$ points.

The relation (3.16) implies that

$$y_i - \hat{f}_\lambda^{-i}(x_i) = \frac{y_i - \hat{f}_\lambda(x_i)}{1 - S_{ii}(\lambda)} \tag{3.18}$$

and thus the fit $\hat{f}_\lambda^{-i}(x_i)$ can be computed from $\hat{f}_\lambda(x_i)$ and $S_{ii}(\lambda)$; there is no need to actually remove the ith point and recompute the smooth.

With result (3.18) in hand, the cross-validation sum of squares can be rewritten as

$$CV(\lambda) = \frac{1}{n}\sum_{i=1}^{n}\left\{\frac{y_i - \hat{f}_\lambda(x_i)}{1 - S_{ii}(\lambda)}\right\}^2. \tag{3.19}$$

Using the approximations $S_{ii} \approx \{\mathbf{S}\mathbf{S}^T\}_{ii}$ and $1/(1 - S_{ii})^2 \approx 1 + 2S_{ii}$, we obtain (Exercise 3.2)

$$E\{CV(\lambda)\} \approx PSE(\lambda) + \frac{2}{n}\sum_{i=1}^{n} S_{ii}(\lambda) b_i^2(\lambda). \tag{3.20}$$

Thus $CV(\lambda)$ adjusts $ASR(\lambda)$ so that in expectation the variance term is correct but in doing so induces an error of $2S_{ii}(\lambda)$ into each of the bias contributions.

3.4.4 *The C_p statistic*

A more direct way of constructing an estimate of PSE is to correct the average squared residual ASR. It is easy to show that

$$E\{ASR(\lambda)\} = \left\{1 - \frac{\operatorname{tr}(2\mathbf{S}_\lambda - \mathbf{S}_\lambda \mathbf{S}_\lambda^T)}{n}\right\}\sigma^2 + \frac{\mathbf{b}_\lambda^T \mathbf{b}_\lambda}{n} \tag{3.21}$$

(Exercise 3.3). From this we see that $E\{ASR(\lambda)\}$ differs from $PSE(\lambda)$ by $2\operatorname{tr}(\mathbf{S}_\lambda)\sigma^2/n$. If we knew σ^2 we could simply add $2\operatorname{tr}(\mathbf{S}_\lambda)\sigma^2/n$ to $ASR(\lambda)$ and it would then have the correct expectation. Of course σ^2 is rarely known; since we want an estimate with little bias, we might use

$$\hat{\sigma}^2 = RSS(\lambda^*)/\{n - \operatorname{tr}(2\mathbf{S}_{\lambda^*} - \mathbf{S}_{\lambda^*}\mathbf{S}_{\lambda^*}{}^T)\}$$

where $RSS(\lambda^*)$ is the residual sum of squares from a smooth $\mathbf{S}_{\lambda^*}\mathbf{y}$ that does relatively little smoothing. Other (related) estimates of σ^2 have been proposed, based on differences of adjacent Y values. The result is a form of the so-called *Mallows's C_p* statistic

$$C_p(\lambda) = ASR(\lambda) + 2\operatorname{tr}(\mathbf{S}_\lambda)\hat{\sigma}^2/n$$

originally proposed as a covariate-selection criterion for linear regression models, and sometimes known as C_ℓ in this context. The bottom right panel in Fig. 3.2 shows the C_p statistic (broken line) along with the true mean squared error (solid line) for the simulated data in that figure. The C_p curve has its minimum at 30%, somewhat higher than the optimal 20%.

3.4.5 *Variations on cross-validation and a comparison with C_p*

Computation of CV is easy if the diagonal elements of the smoother matrix can be readily obtained. For example, for a locally-weighted running-line smoother, the elements of $\mathbf{S}$ are simply computed as

$$S_{ij} = \frac{w_j^i}{\sum_{j \in N(x_i)} w_j^i} + \frac{(x_i - \bar{x}_w^i)(x_j - \bar{x}_w^i) w_j^i}{\sum_{j \in N(x_i)} w_j^i (x_j - \bar{x}_w^i)^2},$$

where w_j^i represents the weight that the jth point receives in the neighbourhood of the ith target point, and $\bar{x}_w^i$ is the weighted mean of the x_j in $N(x_i)$.

Until recently it was not known how to compute the diagonal elements S_{ii} for a smoothing spline in $O(n)$ operations, and this led to two variations of cross-validation. *Generalized cross-validation (GCV)* replaces S_{ii} by its average value $\text{tr}(\mathbf{S})/n$, which is easier to compute (Exercise 3.10):

$$GCV(\lambda) = \frac{1}{n} \sum_{i=1}^{n} \left\{ \frac{y_i - \hat{f}_\lambda(x_i)}{1 - \text{tr}(\mathbf{S}_\lambda)/n} \right\}^2. \tag{3.22}$$

Currently, algorithms are available to compute S_{ii} for the cubic smoothing spline in $O(n)$ operations, so the original motivation for generalized cross-validation is no longer valid. Generalized cross-validation can be mathematically justified in that asymptotically it minimizes mean squared error for estimation of f. We shall not provide details but instead refer the reader to O'Sullivan(1985).

Figure 3.3 shows the results of a small simulation that demonstrates CV and GCV. Fifty samples (each of size 50) were simulated from the quadratic model in Fig. 3.2, holding the X-values fixed at 50 random sites. For each the CV and GCV curves are computed using a smoothing spline, and the minima are located and indicated by vertical lines. The curves are plotted as a function of $\text{tr}(\mathbf{S}_\lambda)$ which is more interpretable than λ itself. The thicker curves are the true *PSE* curves for the model. We can learn several interesting features from these plots. The CV and GCV curves behave similarly, and have considerable variance about the true *PSE* curve. Although the location of their minima is of more importance, these also show high variability. If we interpret $\text{tr}(\mathbf{S}_\lambda)$ as degrees of freedom, these vary between just above two (the minimum attainable) and 12, rather excessive for a three parameter model.

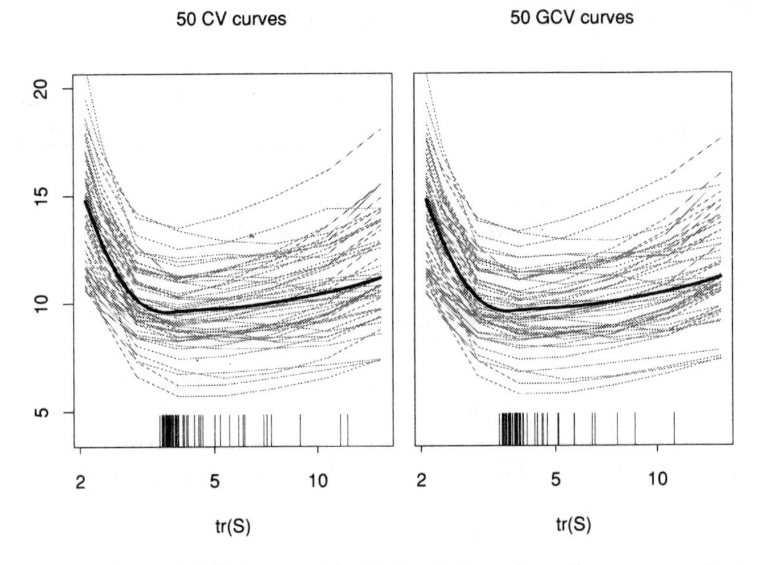

Fig. 3.3. *CV and GCV curves for 50 simulations from the same quadratic model as in Fig. 3.2. Here the smoother is a cubic smoothing spline. Each light curve represents a CV/GCV curve for one of the simulations. Each vertical line at the base of the plot corresponds to a minimum of a curve. We have plotted the curves as a function of* $\mathrm{tr}(\mathbf{S}_\lambda)$ *(and on the log scale) rather than* λ *itself, since this is a more meaningful calibration. The solid curve is the true PSE curve.*

Our experience and that of others has indicated that GCV tends to undersmooth, and in these situations the GCV curve typically has multiple minima. Undersmoothing is particularly prevalent in small datasets, where short trends in the plot of Y against X are interpreted as high-frequency structure.

We have seen that generalized cross-validation can be viewed as an approximation to cross-validation. Here we show a simple way to compare C_p to GCV. Using the approximation $(1-x)^{-2} \approx 1+2x$ we obtain

$$GCV(\lambda) \approx \frac{1}{n}\sum_{i=1}^{n}\{y_i - \hat{f}_\lambda(x_i)\}^2 + 2\mathrm{tr}(\mathbf{S}_\lambda)\frac{1}{n}\sum_{i=1}^{n}\{y_i - \hat{f}_\lambda(x_i)\}^2.$$

Note that the right hand side is the same as the C_p statistic, except that the estimate $\frac{1}{n}\sum_{i=1}^{n}\{y_i - \hat{f}_\lambda(x_i)\}^2$ is used for $\hat{\sigma}^2$, while C_p uses a separate estimate based on a low bias smoother.

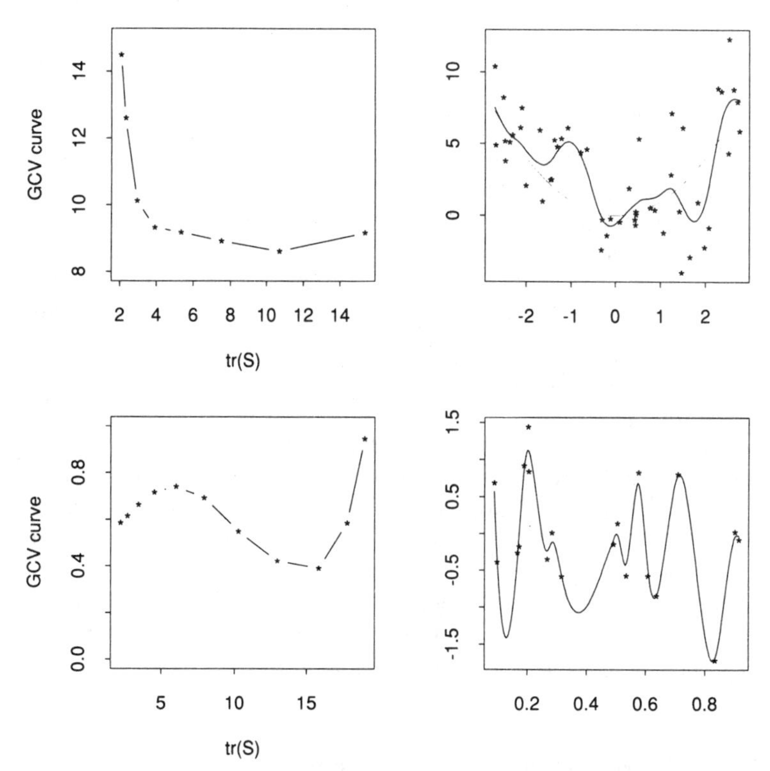

Fig. 3.4. *Two cases where GCV (and CV and C_p) fail. The top row is simulation 17 from Fig. 3.3 having 50 observations. Simulation 17 had the largest* $\mathrm{tr}(\mathbf{S})$ *amongst all 50 simulations in Fig. 3.2, when either CV, GCV or C_p were used as the selection criterion. The left panel shows the GCV curve, and the right panel shows the data, the model and the fitted smoothing spline. The second row is a similar example, where the 20 observations are white noise. The GCV function has two minima, and the global minimum produces a curve that virtually interpolates the data.*

This discussion suggests that for linear smoothers, there is little to choose between C_p and cross-validation. However, if our smoother is nonlinear, cross-validation is the only simple method available for smoothing parameter selection. Expansions such as (3.21) do not necessarily exist for nonlinear smoothers, and hence the C_p statistic doesn't make sense in this case. Of course the theoretical justification for cross-validation is not solid for nonlinear smoothers.

3.4.6 *Discussion*

Although cross-validation and the other automatic methods for selecting a smoothing parameter seem well founded, their performance in practice is sometimes questionable. Figure 3.3 shows a lot of variation despite the simplicity of the model and the moderate error variance as exhibited in Fig. 3.2. Figure 3.4 isolates a bad case and shows a similar example where 20 observations of white noise are smoothed using a smoothing spline.

We tend to rely more on graphical methods for selecting the smoothing parameter, and use the degrees of freedom measures described in the next section to guide us in selecting reasonable values. Furthermore, we will see that automatic methods are less reliable and far more expensive to implement for additive models, where we need to select several smoothing parameters simultaneously.

Fortunately other practitioners have discovered these problems as well, for there has been a flurry of recent work in this area; see, for example, Härdle, Hall, and Marron (1988).

3.5 Degrees of freedom of a smoother

We have already made use of the effective number of parameters or degrees of freedom (*df*) of a smoother in order to make different smoothers comparable with respect to the amount of fitting they do (Fig. 2.1). In fact, it is reasonable to select the value of a smoothing parameter simply by specifying the *df* for the smooth.

Given a linear smoother operator $\mathbf{S}_\lambda$, we define the degrees of freedom *df* to be simply $df = \text{tr}(\mathbf{S}_\lambda)$. Thus *df* is the sum of the eigenvalues of $\mathbf{S}_\lambda$, and gives an indication of the amount of fitting that $\mathbf{S}_\lambda$ does.

The left panel in Fig. 3.5 shows the degrees of freedom for a running-line smoother for the simulated data of Fig. 3.2, as a function of the span. Notice that the degrees of freedom decreases as the span increases, with a moderate span of 50% corresponding to about four degrees of freedom. Thus the number of degrees of

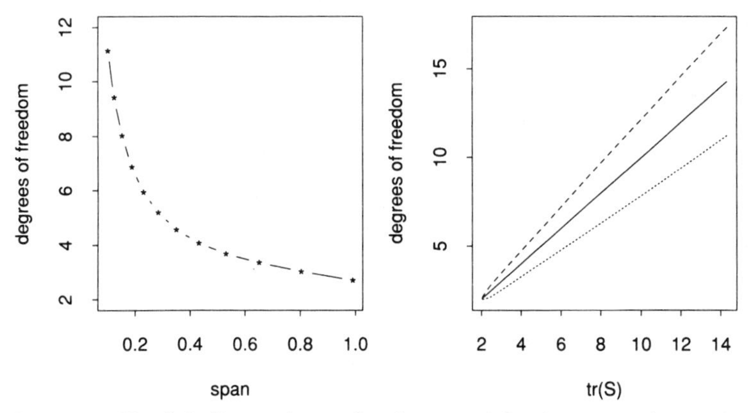

Fig. 3.5. *The left figure shows the degrees of freedom as a function of the span of a running-line smoother, for the simulated data of Fig. 3.2. The right figure shows* $\text{tr}(\mathbf{SS}^T)$ *(dots) and* $\text{tr}(2\mathbf{S} - \mathbf{SS}^T)$ *(dashes) versus* $\text{tr}(\mathbf{S})$ *for a cubic smoothing spline applied to the same data. The* 45% *line (solid) is included for reference.*

freedom used in such a fit is approximately four. This number is a function of the span and the predictor values in the data set, and is not a function of the response Y. Typically the smoothing parameter is the major determinant of the degrees of freedom, while the predictor configuration has little effect. Although a span of 100% appears to imply a linear regression and thus two *df*, we are using symmetric neighbourhoods, so the fits at the endpoints use only 50% of the data.

There are two other popular definitions of degrees of freedom for a linear smoother besides $\text{tr}(\mathbf{S}_\lambda)$, namely $n - \text{tr}(2\mathbf{S}_\lambda - \mathbf{S}_\lambda\mathbf{S}_\lambda^T)$ and $\text{tr}(\mathbf{S}_\lambda\mathbf{S}_\lambda^T)$. These definitions, as well as $\text{tr}(\mathbf{S}_\lambda)$, can be motivated by analogy with the linear-regression model and are useful for different purposes. In addition, they can be extended to nonlinear smoothers, although they may then depend on the distribution of Y.

To be more specific, assume that $Y_i = f(x_i) + \varepsilon_i$, where $f(x_i)$ is the true function and the errors ε_i are independent and identically distributed with mean zero and variance σ^2. The degrees of freedom $df = \text{tr}(\mathbf{S}_\lambda)$ can also be motivated as the C_p correction for the average squared residual. In particular, the C_p statistic corrects *ASR* to make it unbiased for *PSE* by adding a quantity $2\text{tr}(\mathbf{S}_\lambda)\hat{\sigma}^2/n$, where $\hat{\sigma}^2$ is an estimate of σ^2. The definition

$df = \text{tr}(\mathbf{S}_\lambda)$ is also popular in the smoothing-spline literature where $\mathbf{S}\sigma^2$ emerges as the posterior covariance of $\mathbf{f}$, after appropriate Bayesian assumptions are made (section 3.6, Exercise 3.14).

From (3.21) we deduce that the residual sum of squares $RSS(\lambda) = \sum_{i=1}^n \{y_i - \hat{f}_\lambda(x_i)\}^2$ has expectation

$$E\{RSS(\lambda)\} = \left\{n - \text{tr}(2\mathbf{S}_\lambda - \mathbf{S}_\lambda \mathbf{S}_\lambda^T)\right\}\sigma^2 + \mathbf{b}_\lambda^T \mathbf{b}_\lambda.$$

Thus we define the degrees of freedom for error as

$$df^{\text{ err}}(\lambda) = n - \text{tr}(2\mathbf{S}_\lambda - \mathbf{S}_\lambda \mathbf{S}_\lambda^T)$$

since in the linear regression case this is $n - p$, where p is the number of parameters. If we are smoothing noise ($\mathbf{f} = \mathbf{0}$), then $n - df^{\text{ err}}(\lambda)$ is the expected drop in $RSS(\lambda)$ due to overfitting; that is, $n - df^{\text{ err}}(\lambda) = E\{\sum_{i=1}^n y_i^2 - RSS(\lambda)\}/\sigma^2$. This latter definition is useful for general (nonlinear) smoothers. In section 3.9 we show how $df^{\text{ err}}$ arises in the assessment of the variance estimate $\hat{\sigma}^2$ and for comparing models.

Lastly, recall that in a linear model $\sum_{i=1}^n \text{var}\{\hat{f}_\lambda(x_i)\} = p\sigma^2$. For a linear smoother $\hat{\mathbf{f}}_\lambda = \mathbf{S}_\lambda \mathbf{y}$, and the sum of the variances is $\text{tr}(\mathbf{S}_\lambda \mathbf{S}_\lambda^T)\sigma^2$; hence we define the degrees of freedom for variance as

$$df^{\text{ var}}(\lambda) = \text{tr}(\mathbf{S}_\lambda \mathbf{S}_\lambda^T).$$

For a general (nonlinear) smoother, the definition is $df^{\text{ var}}(\lambda) = \sum_{i=1}^n \text{var}\{\hat{f}_\lambda(x_i)\}/\sigma^2$.

Note that if $\mathbf{S}$ is a symmetric projection matrix, then $\text{tr}(\mathbf{S}_\lambda)$, $\text{tr}(2\mathbf{S}_\lambda - \mathbf{S}_\lambda \mathbf{S}_\lambda^T)$, and $\text{tr}(\mathbf{S}_\lambda \mathbf{S}_\lambda^T)$ coincide. This is the case for linear and polynomial regression smoothers, as well as regression splines. For general nonprojection smoothers, they are not equal. Interestingly, however, they are equal for the (unweighted) running-line smoother since each row of a running-line smoother matrix is a row of a projection matrix, and thus $\sum_j S_{ij}^2 = S_{ii}$ for all i.

The right panel of Fig. 3.5 shows the quantities $\text{tr}(\mathbf{S}_\lambda \mathbf{S}_\lambda^T)$ and $\text{tr}(2\mathbf{S}_\lambda - \mathbf{S}_\lambda \mathbf{S}_\lambda^T)$ plotted against $\text{tr}(\mathbf{S}_\lambda)$ for a smoothing spline based on the same simulated data. Notice that there are substantial differences when $\text{tr}(\mathbf{S}_\lambda)$ is large, say greater than six. For smoothing splines one can show (Exercise 3.7):

(i) $\text{tr}(\mathbf{S}_\lambda \mathbf{S}_\lambda^T) \le \text{tr}(\mathbf{S}_\lambda) \le \text{tr}(2\mathbf{S}_\lambda - \mathbf{S}_\lambda \mathbf{S}_\lambda^T)$ and

(ii) all three of these functions are decreasing in λ.

It appears from Fig. 3.5 that $\text{tr}(\mathbf{S}_\lambda \mathbf{S}_\lambda^T)$ and $\text{tr}(2\mathbf{S}_\lambda - \mathbf{S}_\lambda \mathbf{S}_\lambda^T)$ vary linearly with $\text{tr}(\mathbf{S}_\lambda)$. Of the three definitions, $df = \text{tr}(\mathbf{S}_\lambda)$ is the easiest to compute, since it requires only the diagonal of $\mathbf{S}_\lambda$. For many of the smoothers this is available at a cost of $O(n)$ operations, while $\text{tr}(\mathbf{S}_\lambda \mathbf{S}_\lambda)^T$ costs $O(n^2)$ operations. In Appendix B we develop approximations to df^{err} using df based on this relationship. The smoothers in Figs 2.1 and 2.5 in Chapter 2 are calibrated using $df = 5$.

3.6 A Bayesian model for smoothing

The cubic smoothing spline can be derived from a number of Bayesian models for smoothing. The infinite-dimensional model specifies that the underlying function f is a sum of a random linear function and an integrated Wiener process. When combined with a Gaussian model for the data, the conditional expectation of f given the data, or *posterior mean* of f, can be shown to be a fitted smoothing spline with an appropriate value for λ. A full discussion of this would be beyond the mathematical scope of this book; references to the work of Wahba and others on this topic can be found in the bibliography.

A fairly simple finite-dimensional model, outlined by Silverman (1985), can be formulated as follows. Recall from section 2.10 that a smoothing spline can be written in terms of the natural B-spline basis functions $N_i(t)$ whose definition is given in that section. Define $n \times n$ matrices $\mathbf{N}$ and $\mathbf{\Omega}$ by $N_{ij} = N_j(x_i)$ and $\Omega_{ij} = \int N_i''(x) N_j''(x)\,dx$. Then if we assume a Bayesian model in which

(i) the data have a Gaussian distribution with mean $\mathbf{N}\gamma$ and variance $\sigma^2\mathbf{I}$, and
(ii) γ has a multivariate Gaussian prior distribution with mean $\mathbf{0}$ and variance $\tau\mathbf{\Omega}^-$ where $\tau = \sigma^2/\lambda$,

it follows that the posterior distribution of γ is multivariate Gaussian with mean $E(\gamma \,|\, \mathbf{y}) = \hat{\gamma} = (\mathbf{N}^T\mathbf{N} + \lambda\mathbf{\Omega})^{-1}\mathbf{N}^T\mathbf{y}$ and covariance matrix $\text{cov}(\gamma \,|\, \mathbf{y}) = (\mathbf{\Omega}/\tau + \sigma^{-2}\mathbf{N}^T\mathbf{N})^{-1}$. Notice that the posterior mean is identical to $\hat{\boldsymbol{\beta}}$ in equation (2.10) in Chapter 2.

Given the posterior distribution of γ, we can make posterior inferences about all aspects of the cubic spline $\sum_{i=1}^{n} N_i(x)\gamma_i$ parametrized by γ. In particular, we can compute the (posterior) mean and variance of $\mathbf{f} = \mathbf{N}\gamma$, the vector of sample evaluations. The posterior mean is of course $\hat{\mathbf{f}} = \mathbf{S}\mathbf{y}$, where $\mathbf{S}$ is the smoothing-spline matrix, and it is not difficult to show that the posterior covariance is $\mathbf{S}\sigma^2$ (Exercise 3.14). Notice that an increase in the prior covariance of γ corresponds to a smaller smoothing parameter, or less smoothing.

A criticism of the above approach is that the prior is somewhat artificial, since one wouldn't normally have much prior information about γ. A more direct approach is to put the prior on $\mathbf{f}$ itself, the vector of evaluations. The prior for $\mathbf{f}$ corresponding to that for γ above has mean zero and covariance $\tau\mathbf{N}^T\mathbf{\Omega}^-\mathbf{N} = \mathbf{K}^-$, where $\mathbf{K}$ is defined in section 2.10 and $\mathbf{K}^-$ is any generalized inverse of $\mathbf{K}$ (see, for example, Rao, 1973, 1b.5).

How reasonable is it to use $\mathbf{K}^-$ as a prior covariance? The eigenstructure of $\mathbf{K}^-$ can shed some light on this question. $\mathbf{K}$ has eigenvalues zero for constant and linear functions of $\mathbf{x}$, and increasingly positive eigenvalues for higher order functions whose zero-crossing behaviour resembles that of polynomials (next section, and Exercises 3.5 and 3.6). Hence $\mathbf{K}^-$ is concentrated on a subspace orthogonal to functions constant and linear in X, and has decreasing positive eigenvalues for increasingly higher order functions. Since direct inversion of $\mathbf{K}$ leads to eigenvalues $+\infty$ corresponding to its null space, it is called *partially improper*. It gives infinite variance to linear combinations of $\mathbf{1}$ and $\mathbf{x}$ (as does the prior for γ). In a sense, then, one is assuming complete prior uncertainty about the linear and constant functions, and decreasing uncertainty about the higher order functions.

This Bayesian model is closer to the infinite dimensional one alluded to above; in fact, we can view the prior distribution as being obtained by sampling the continuous stochastic process (Cox, 1989). It is interesting to note that these two Bayesian approaches (sampled and continuous) give the same point estimate (mean of the posterior) but different posterior bands.

3.7 Eigenanalysis of a smoother and spectral smoothing

For a smoother with symmetric smoother matrix $\mathbf{S}$, the eigendecomposition of $\mathbf{S}$ can be used to describe it behaviour. This is much like the use of a *transfer function* to describe a linear filter for time series, and we make this connection precise below.

Let $\{\mathbf{u}_1, \mathbf{u}_2, \ldots \mathbf{u}_n\}$ be an orthonormal basis of eigenvectors of $\mathbf{S}$ with eigenvalues $\theta_1 \geq \theta_2 \cdots \geq \theta_n$:

$$\mathbf{S}\mathbf{u}_j = \theta_j \mathbf{u}_j, \qquad j = 1, 2, \cdots, n \tag{3.23}$$

or

$$\begin{aligned} \mathbf{S} &= \mathbf{U}\mathbf{D}_\theta \mathbf{U}^t \\ &= \sum_{j=1}^{n} \theta_j \mathbf{u}_j \mathbf{u}_j^T. \end{aligned} \tag{3.24}$$

The cubic smoothing spline is an important example of a symmetric smoother, and its eigenvectors resemble polynomials of increasing degree (Demmler and Reinsch, 1975; Utreras, 1979; Eubank, 1985). In particular, it is easy to show that the first two eigenvalues are unity, with eigenvectors which correspond to linear functions of the predictor on which the smoother is based. The other eigenvalues are all strictly between zero and one (Exercise 3.5). The action of the smoother is now transparent: if presented with a response $\mathbf{y} = \mathbf{u}_j$, it shrinks it by an amount θ_j as in (3.23); for an arbitrary $\mathbf{y}$, it shrinks the component of $\mathbf{y}$ along $\mathbf{u}_j$ by θ_j as in (3.24). Figure 3.6 shows the eigenvalues of a cubic smoothing spline matrix, based on a particular set of X values, and using two different values for the smoothing parameter.

The right panel of Fig. 3.6 shows the third to sixth eigenvectors. Notice that the amount of shrinking appears to increase with increasing order of the eigenvector. As the amount of smoothing is increased (fewer degrees of freedom), the curve decreases to zero more quickly. Demmler and Reinsch (1975) give some theoretical support for these empirical findings. They show that for $k \geq 3$, the number of sign changes in the kth eigenvector of a cubic spline smoother is $k-1$. They also derive asymptotic approximations for the eigenvalues which show that they decrease fairly rapidly with increasing order (Exercise 3.6).

The bin smoother, least squares line, polynomial regression, and regression splines are other symmetric smoothers that we

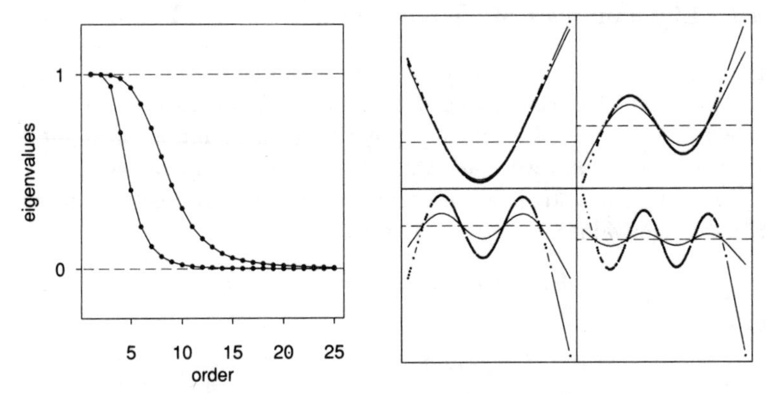

Fig. 3.6. *Left: The first 25 eigenvalues of a smoothing-spline matrix based on 128 unevenly spaced observations. The first two eigenvalues are exactly one, and all are nonnegative. The two curves correspond to two different smoothing parameters; the larger smoothing parameter (more smoothing) results in a curve that dies off more quickly. Right: The third to sixth eigenvectors of the smoothing-spline matrices. In each case, the eigenvector* $\mathbf{u}_j$ *is plotted against* $\mathbf{x}$*, and as such is viewed as a function of* x*. The damped functions represent the smoothed versions of these functions, using the larger smoothing parameter.*

have discussed. They are all in fact orthogonal projections onto different spaces of fits. Thus their eigenvalues are 0 or 1 only, with corresponding eigenspaces consisting of the spaces of residuals and fits respectively. The smoother or projection matrices are the familiar hat matrices of one-way analysis of variance and simple and multiple regression.

What if $\mathbf{S}$ is not symmetric? The running-mean and locally-weighted running-lines are simple examples of smoothers that are not. In this case the eigendecomposition is no longer as useful because the eigenvalues and vectors may be complex, and algebraic and geometric multiplicities may differ. One can, however, turn to the singular-value decomposition of $\mathbf{S}$, which is always real-valued. For a running-line smoother, the first two (left and right) singular vectors tend to be approximately linear, while the remaining ones resemble orthogonal polynomials of increasing degree (not shown). Note that the singular values can be slightly greater than one: for example, when used with the same predictor as in Fig. 3.6, the largest singular value of a running-line smoother of span 0.5 is 1.07.

The analysis of a linear scatterplot smoother through an eigenanalysis of the corresponding smoother matrix is closely related to the study of the transfer function of a linear filter for time series. This analogy can add insight, so we provide a brief summary here. Consider a time series $\{y_t : t = 0, \pm 1, \pm 2, \ldots\}$. Suppose we apply a linear smoother (or *digital filter*) to the series, defined by

$$\hat{y}_t = \sum_{j=a}^{b} c_j y_{t-j}.$$

Note that the weights used to obtain $\hat{y}_t$ are just a shifted version of those used to obtain $\hat{y}_{t-1}$; the smooth is a *convolution* of the series and the filter. This is not generally true for a linear scatterplot smoother. The weights are also known as the *impulse response function*, since if the input has just a single nonzero value of one at time t (an impulse), the output consists only of the weights centered with c_0 at t.

Suppose we have an input series of the form $r\cos(\omega t + \phi)$ sampled at $t = 0, \pm 1, \pm 2, \ldots$. It is easy to check that the output is

$$\mathrm{Re}\Big[\sum_{j=a}^{b} c_j \exp(-i\omega j) r \exp\{i(\omega t + \phi)\}\Big], \tag{3.25}$$

where i is the complex number $\sqrt{-1}$, and Re denotes real part. The function $G(\omega) = \sum_{j=a}^{b} c_j \exp(-i\omega j)$ is called the *transfer function* of the linear filter, and its squared magnitude $\left| G(\omega)^2 \right|$ is called the *power transfer function*.

In general, the action of a linear filter consists of both a damping of the input series and a phase shift, evident upon closer examination of (3.25). Usually, however, the filter is symmetric ($c_{-j} = c_j$ and $a = -b$) and there is no phase shift. Then $G(\omega)$ is real and the output is $G(\omega) r \cos(\omega t + \phi)$. Thus $G(\omega)$ is the eigenvalue of the filter corresponding to the eigenfunction $r\cos(\omega t + \phi)$; it measures the amount by which the filter damps an input cosinusoid of frequency ω.

The transfer function is a convenient tool both for describing the action of a filter, and for designing one. The eigenvalue sequence in Fig. 3.6 resembles the transfer function of a typical *low pass* filter; the higher the frequency, the more damping takes place. We pursue this topic a bit further in sections 8.5 and 9.3.6.

3.8 Variance of a smooth and confidence bands

As before we assume $Y_i = f(x_i) + \varepsilon$ where $E(\varepsilon) = 0$, $\text{var}(\varepsilon) = \sigma^2$. We once again hide the smoothing parameter λ and assume it is fixed at some prespecified value.

3.8.1 *Pointwise standard-error bands*

The covariance matrix of the fitted vector $\hat{\mathbf{f}} = \mathbf{Sy}$ is simply

$$\text{cov}(\hat{\mathbf{f}}) = \mathbf{SS}^T\sigma^2, \tag{3.26}$$

and given an estimate of σ^2 this can be used to form pointwise standard-error bands. Under Gaussian errors and negligible bias, these bands also represent pointwise confidence intervals.

Figure 3.7 gives an example. A smoothing spline with six degrees of freedom (solid line) is computed for the simulated data of Fig. 3.2. Pointwise standard-error bands are computed using ± 2 times the square root of the diagonal of $\mathbf{SS}^T\sigma^2$ (dashes). If the bias is not negligible (something that is very difficult to check), the bands provide a 95% pointwise confidence interval for $\mathbf{g} = \mathbf{Sf}$ rather than for $\mathbf{f}$.

Wahba (1983) and Nychka (1988) provide evidence that the posterior confidence bands derived from the Bayesian model for smoothing splines have good sampling properties, even when a number of abscissa values are considered simultaneously. As detailed in the previous section, the posterior distribution for $\mathbf{f}$ is $N(\hat{\mathbf{f}}, \sigma^2\mathbf{S})$ where $\mathbf{S}$ is the smoothing-spline matrix. The covariance $\sigma^2\mathbf{S}$ may be compared to the covariance $\sigma^2\mathbf{SS}^T$ used in constructing the standard-error bands above; in particular, $\mathbf{S} \geq \mathbf{SS}^T$. The difference is that $\mathbf{S}$ has a component of bias in it. In detail, the mean square error matrix for $\hat{\mathbf{f}}$ is

$$E(\hat{\mathbf{f}} - \mathbf{f})(\hat{\mathbf{f}} - \mathbf{f})^T = \sigma^2\mathbf{SS}^T + \mathbf{bb}^T$$

where the bias vector is $\mathbf{b} = E\hat{\mathbf{f}} - \mathbf{f}$. This bias term depends on the unknown $\mathbf{f}$, but if we average it with respect to its prior $\mathbf{f} \sim N(\mathbf{0}, \tau\mathbf{K}^-)$, we get $E_f\mathbf{bb}^T = \sigma^2(\mathbf{I} - \mathbf{S})\mathbf{K}^-(\mathbf{I} - \mathbf{S})^T = \sigma^2(\mathbf{S} - \mathbf{SS}^T)$. From a frequentist point of view, the Bayesian covariance represents an average mean-squared error with regard to the prior.

Consider again the setup of Fig. 3.5. From the right panel, we note that the differences between $\mathbf{S}$ and $\mathbf{S}\mathbf{S}^T$ are likely to be greater for larger degrees of freedom. For six degrees of freedom, $\text{tr}(\mathbf{S})$ and $\text{tr}(\mathbf{S}\mathbf{S}^T)$ differ by less than 2, so that the diagonal elements of $\mathbf{S}$ and $\mathbf{S}\mathbf{S}^T$ should differ by only about 2/50 on the average.

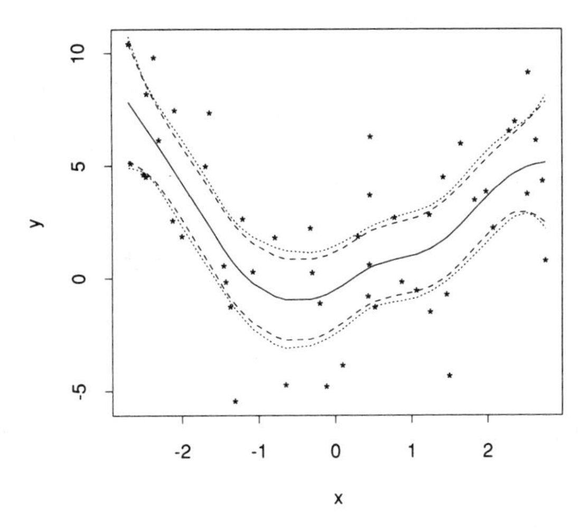

Fig. 3.7. *± 2 standard-error bands for simulated data of Fig. 3.2. The solid line is the cubic smoothing spline with six degrees of freedom. The dots use the square root of the diagonal of $\mathbf{S}\sigma^2$ while the dashes use the square root of the diagonal of $\mathbf{S}\mathbf{S}^T\sigma^2$.*

Figure 3.7 confirms this. The dots show the bands constructed using ± 2 times the square root of the diagonal of $\sigma^2\mathbf{S}$. Little difference can be seen in the two sets of bands.

3.8.2 *Global confidence sets*

Suppose we wish to make inferences about more than one function value at the same time; for example, we may wish to check if a straight line fits in a confidence band and hence provides a satisfactory simplification of our function estimate. Pointwise standard-error bands are not appropriate for this purpose, and we would prefer some kind of global confidence band. A common approach is to widen the pointwise bands to account for the implicit multiple comparisons, and thus make them "global" in some sense. An example given in section 6.4.2 revealed to us the flaw in this reasoning, and set us thinking about the problem. In that example the pointwise standard-error bands contain a straight line, but the increase in the log-likelihood due to replacing the smooth with a straight line is highly significant. Hence a confidence band obtained by enlarging the pointwise band must also contain the straight line.

What is wrong with this reasoning? First of all, remember that a confidence set for the n values for the true underlying function is actually a set in n-dimensional space. A global confidence band, like the one described above, represents the projection or "shadow" of such a set along each direction. There are two points to note:

(i) A confidence band is limited in the amount of information it provides because it gives no indication of the functional shape of the members of the n-dimensional confidence set. For example, functions in the confidence set might all have bends at or near a certain abscissa value, a feature not necessarily enforced by a confidence band.

(ii) The projection of the n-dimensional global confidence set into a confidence band need *not* be wider than a pointwise standard-error band.

For the remainder of this section we explore a technique for generating curves from a global confidence set. This technique attempts to uncover some of the shape information mentioned in point (i) and will be used to illustrate point (ii).

Consider the Gaussian-error model

$$Y_i = f(x_i) + \sigma Z_i; \quad Z_i \sim N(0,1).$$

Our approach is to derive a confidence set for $\mathbf{g} = \mathbf{Sf}$. The standard likelihood-ratio method for constructing a confidence set for $\mathbf{g}$ is to use the approximate studentized pivotal $(\hat{\mathbf{f}} - \mathbf{g})^T(\mathbf{SS}^T\sigma^2)^{-1}(\hat{\mathbf{f}} - \mathbf{g})$ which has a χ^2_n distribution. Since σ^2 is unknown we compute the estimate $\hat{\sigma}^2 = (\mathbf{y} - \hat{\mathbf{f}})^T(\mathbf{y} - \hat{\mathbf{f}})/\{n - 2\text{tr}(\mathbf{S}) + \text{tr}(\mathbf{SS}^T)\}$, and use the approximate pivotal

$$\nu(\mathbf{g}) = (\hat{\mathbf{f}} - \mathbf{g})^T(\mathbf{SS}^T\hat{\sigma}^2)^{-1}(\hat{\mathbf{f}} - \mathbf{g}). \tag{3.27}$$

Suppose that $\nu(\mathbf{g})$ has distribution G, which is not a χ^2_n distribution due to the estimation of σ^2. We describe two approximations to G below; one based on the F distribution, the other using the bootstrap. For the moment assume that G is given. The approximate pivotal leads to a $1-\alpha$ level confidence set for $\mathbf{g}$:

$$C(\mathbf{g}) = \{\mathbf{g}; \nu(\mathbf{g}) \leq G_{1-\alpha}\}. \tag{3.28}$$

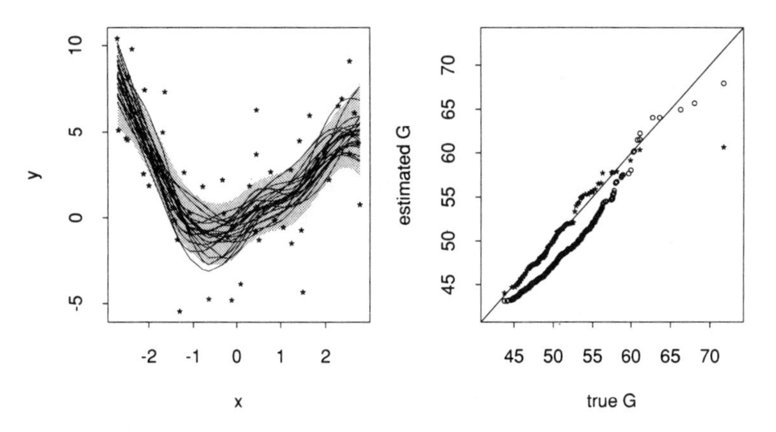

Fig. 3.8. *The left panel shows 20 functions in the simultaneous confidence region* $C(\mathbf{g})$, *along with pointwise confidence intervals (shaded region) from Fig. 3.7. The right panel shows a Q-Q plot of the bootstrap distribution of the approximate pivotal* $\nu(\mathbf{g})$ *versus its exact distribution* G *for this simulation. The circles represent the Q-Q plot of the approximation* $\hat{G} = df^{\text{err}} + df^{\text{var}} F_{df^{\text{var}}, df^{\text{err}}}$; *this shows systematic departures from* G.

How can we display the confidence set $C(\mathbf{g})$? There are too many members in the set to make a useful display of all of them. An alternative is to plot a subset of the functions that lie in $C(\mathbf{g})$. We proceed by sampling a function $\mathbf{g}$ from a convenient distribution and then checking if $\mathbf{g} \in C(\mathbf{g})$. An obvious choice for the sampling distribution is $\mathbf{g} \sim N(\hat{\mathbf{f}}, \hat{\sigma}^2\mathbf{S}\mathbf{S}^T)$. Thus a function is chosen for plotting with probability proportional to its likelihood. The left panel of Fig. 3.8 shows this method applied to the setting of Fig. 3.7. Twenty curves lying in $C(\mathbf{g})$ are shown (with G approximated by the bootstrap estimate G^* described below), along with the shaded region indicating the pointwise standard-error envelope of Fig. 3.7. Notice that the curves lie mostly in the pointwise envelope, counter to the notion that global bands should be wider than pointwise ones. To understand their characteristic behaviour we can examine them individually. For example the curves that tend to be high on the left appear to be lower near the middle.

Note that the sampling of a function $\mathbf{g}$ from $N(\hat{\mathbf{f}}, \mathbf{S}\mathbf{S}^T\hat{\sigma}^2)$ and checking whether $\mathbf{g} \in C(\mathbf{g})$ is similar to sampling a function from the Bayesian posterior $N(\hat{\mathbf{f}}, \mathbf{S}\sigma^2)$ and checking if its posterior probability is at least $1 - \alpha$. The latter corresponds to the sampling of functions from the highest posterior density region with probability proportional to their posterior density.

We can also contrast this approach to the pointwise standard error bands described earlier. For the purpose of this comparison let's assume that σ^2 is known and thus the distribution G is χ^2_n. Then we can obtain the pointwise bands by

(i) Setting the off-diagonal elements of $(\mathbf{S}\mathbf{S}^T)^{-1}$ to zero.
(ii) Inverting each of the n pivotal quantities separately around its χ^2_1 distribution (rather than inverting the entire pivotal around its χ^2_n distribution).

Point (ii) causes the pointwise bands to be narrower than the projection of the global confidence sets, as our intuition about multiple comparisons suggests. However (i) has the opposite effect, because $(\mathbf{S}\mathbf{S}^T)^{-1}$ is symmetric and positive definite and hence $\{(\mathbf{S}\mathbf{S}^T)^{-1}\}_{ii} > \{(\mathbf{S}\mathbf{S}^T)_{ii}\}^{-1}$. The example of Fig. 3.8 shows that the two effects can cancel one another.

We have displayed all the curves sampled from (3.28), and the picture quickly fills up. An alternative approach (suggested by J. Rice) is to display only the *interesting* members of the set, such as the extreme members.

Finally we describe methods for approximation of the distribution G. Our first approximation to G is based on the F distribution. If the model is correct and the smoother is a projection with $\mathbf{S}\mathbf{y}$ unbiased for $\mathbf{f}$, one can show that $G = n - p + pF_{p,n-p}$ exactly. For the nonprojection case, one can argue that $G \approx \hat{G} = df^{\text{err}} + df^{\text{var}} F_{df^{\text{var}}, df^{\text{err}}}$ (Exercise 3.12) with the approximation improving the closer $\mathbf{S}$ is to a projection. Figure 3.8 shows a Q-Q plot (small dots) of $\hat{G}$ against G for this simulation model. This true G was obtained by simulating 1000 realizations of the model. The approximation $\hat{G}$ is not very good, having a shorter right tail than G. A χ^2_n approximation is poorer still.

The Gaussian assumption used above may be relaxed through use of the bootstrap method to estimate the distribution of $\nu(\mathbf{g})$. A bootstrap sample $\mathbf{y}^*$ is generated via

$$\mathbf{y}^* = \hat{\mathbf{f}} + \mathbf{r}^*$$

where $\mathbf{r}^*$ consists of n residuals sampled with replacement from the model residuals $\mathbf{y} - \hat{\mathbf{f}}$. A new smooth $\hat{\mathbf{f}}^* = \mathbf{S}\mathbf{y}^*$ and a corresponding variance estimate $\hat{\sigma}^{*2}$ are computed. Finally, we obtain the bootstrap analogue of $\nu(\mathbf{g})$

$$\nu^*(\hat{\mathbf{g}}) = (\hat{\mathbf{f}}^* - \hat{\mathbf{g}})^T(\mathbf{S}\mathbf{S}^T\hat{\sigma}^{*2})^{-1}(\hat{\mathbf{f}}^* - \hat{\mathbf{g}}).$$

This procedure is repeated many times, giving an empirical distribution of values of $\nu^*(\hat{\mathbf{g}})$, which is the bootstrap estimate G^*.

The right panel of Fig. 3.8 includes a Q-Q plot of 100 values of $\nu^*(\hat{\mathbf{g}})$ versus G. The bootstrap approximation is very good, except in the tail, where it is too short. One really needs more than 100 bootstrap samples, however, to estimate extreme quantiles with any accuracy. This bootstrap approximation is used in the construction of the global confidence band in the left panel of Fig. 3.8.

Note that specific features of $\mathbf{g}$ may be assessed in a similar manner. One could construct a statistic $\kappa(\mathbf{g})$ that captures the feature (perhaps a bend at a certain point), and construct its bootstrap distribution. The observed $\kappa(\hat{\mathbf{f}})$ is then compared to the bootstrap distribution for significance. This is an area of current research.

3.9 Approximate F-tests

In this section we present some heuristic procedures for inference concerning σ^2 and predictor effects, derived by analogy with linear regression. Exact distributional results are not yet available for these tests, but a modest number of simulations by us and others suggest that the approximations given are useful at least as rough guides. The degrees of freedom γ used here is $n - df^{\text{err}} = \text{tr}(2\mathbf{S} - \mathbf{S}\mathbf{S}^T)$.

Given an estimated smooth $\hat{\mathbf{f}} = \mathbf{S}\mathbf{y}$, if we assume that the bias for f in (3.21) is zero, an unbiased estimate of σ^2 is given by

$$\hat{\sigma}^2 = \sum_{i=1}^{n}(y_i - \hat{y}_i)^2/(n-\gamma). \tag{3.29}$$

As a first approximation, inference for σ^2 can be based on $(n-\gamma)\hat{\sigma}^2/\sigma^2 \sim \chi^2_{(n-\gamma)}$. A better approximation can be obtained through a two-moment correction, that is, finding a multiple of $\hat{\sigma}^2/\sigma^2$ that has its mean and variance in a 1:2 ratio. If we assume Y is Gaussian, the mean and variance of $(n-\gamma)\hat{\sigma}^2/\sigma^2$ are $\delta_1 = \text{tr}(\mathbf{I}-\mathbf{S})^T(\mathbf{I}-\mathbf{S})$ and $\delta_2 = 2\text{tr}\{(\mathbf{I}-\mathbf{S})^T(\mathbf{I}-\mathbf{S})\}^2$ and the two-moment approximation works out to

$$\frac{\delta_1}{\delta_2}(n-\gamma)\frac{\hat{\sigma}^2}{\sigma^2} \sim \chi^2_{\frac{\delta_1^2}{\delta_2}(n-\gamma)}. \tag{3.30}$$

If we use this approximation in the construction of pointwise standard-error bands (section 3.8), it leads to the use of the percentiles of the t-distribution with δ_1^2/δ_2 degrees of freedom, rather than the percentiles of the standard Gaussian.

Now suppose that we wish to compare two smooths $\hat{\mathbf{f}}_1 = \mathbf{S}_1\mathbf{y}$ and $\hat{\mathbf{f}}_2 = \mathbf{S}_2\mathbf{y}$. For example the smooth $\hat{\mathbf{f}}_2$ might be rougher than $\hat{\mathbf{f}}_1$, and we wish to test if it picks up significant bias. A standard case that often arises is when $\hat{\mathbf{f}}_1$ is linear, in which case we want to test if the linearity is real. We must assume that $\hat{\mathbf{f}}_2$ is unbiased, and $\hat{\mathbf{f}}_1$ is unbiased under the null hypothesis. Letting RSS_1 and RSS_2 be the residual sum of squares for each of the smooths, and γ_j be the degrees of freedom $\text{tr}(2\mathbf{S}_j - \mathbf{S}_j\mathbf{S}_j^T)$, $j = 1, 2$, the first approximation uses

$$\frac{(RSS_1 - RSS_2)/(\gamma_2 - \gamma_1)}{RSS_2/(n-\gamma_2)} \sim F_{\gamma_2-\gamma_1, n-\gamma_2}. \tag{3.31}$$

This can be improved by applying a two-moment correction to both the numerator and denominator. Defining $\mathbf{R}_j = (\mathbf{I}-\mathbf{S}_j)^T(\mathbf{I}-\mathbf{S}_j)$, $j = 1, 2$; $\nu_1 = \text{tr}(\mathbf{R}_1 - \mathbf{R}_2)$, $\nu_2 = \text{tr}(\mathbf{R}_1 - \mathbf{R}_2)^2$, $\delta_1 = \text{tr}(\mathbf{R}_2)$, and $\delta_2 = \text{tr}(\mathbf{R}_2^2)$, the two moment approximation uses a reference

F distribution $F_{\nu_1^2/\nu_2, \delta_1^2/\delta_2}$ with parameters $(\nu_1/\nu_2)(\gamma_2 - \gamma_1)$ and $(\delta_1/\delta_2)(n - \gamma_2)$ for the statistic in (3.31).

Notice that if the correction factors δ_1/δ_2 and ν_1/ν_2 derived above are equal to one, we obtain the first approximations. Some limited experience with these corrections suggests that the first correction factor is usually close to one, but the ν_1/ν_2 factor can be substantially different from one. Since ν_2 and hence ν_1/ν_2 are difficult to compute (unlike γ_1 and γ_2), it is desirable to investigate how much the first approximation errs and in what direction. For running-line smoothers ν_1/ν_2 tends to lie between 0.65 and 0.75, implying that F-tests based on the first approximation will be somewhat liberal. In particular, tests with nominal size 95% have actual size between 96% and 98%. The error goes in the other direction for smoothing splines, with ν_1/ν_2 typically falling in the range from 1.4 to 2.0. Hence nominal 95% tests have an actual size between 90% and 93%. In this case, one can prove that $\nu_1/\nu_1 \leq 1$ and $\lambda_1/\lambda_2 \leq 1$ and approximate calculation of these factors is feasible (Exercise 3.13). There has not been much detailed theoretical work on these approximations, but the empirical evidence is very promising.

Finally, it is interesting to compare the F-test method of model assessment to the smoothing parameter selection methods, namely cross-validation and C_p, discussed earlier. The comparison is most easily made in terms of the C_p statistic. In choosing between two models with residual sums of squares RSS_1 and RSS_2, and degrees of freedom γ_1 and γ_2, we choose model 2 if the C_p statistic is less than that for model 1. That is, if $RSS_2 + 2\gamma_2\hat{\sigma}^2 < RSS_1 + 2\gamma_1\hat{\sigma}^2$, or

$$\frac{(RSS_1 - RSS_2)/(\gamma_2 - \gamma_1)}{\hat{\sigma}^2} > 2.$$

Hence as long as the denominator above is not very different from $\hat{\sigma}^2$ used in the C_p statistic, we see that the two methods differ solely in the cutoff point used to reject model 1 in favour of model 2. The F-test uses $F_{\gamma_2-\gamma_1, n-\gamma_2}$ while C_p uses 2. Now assuming n is large, we can replace $F_{\gamma_2-\gamma_1, n-\gamma_2}$ by $\chi^2_{\gamma_2-\gamma_1}/(\gamma_2-\gamma_1)$. For $k = 1, 3, 5, 7, 9, \ldots$ $\text{pr}(\chi^2_k/k > 2)$ is about 0.16, 0.11, 0.08, 0.05, 0.04,.... So if the 5% level is the gold standard, C_p is too liberal for low values of $\gamma_2 - \gamma_1$, is on target at seven, and is increasingly conservative thereafter.

3.10 Asymptotic behaviour of smoothers

There has been considerable theoretical study of smoothers, concerning topics such as consistency, asymptotic bias and variance, and the optimal choice of smoothing parameter. The results of such work, although asymptotic in nature, can help direct the choice and use of scatterplot smoothers for real data. We briefly summarize some of these results here.

Consider a running-mean based on r nearest neighbours, as described in section 2.5. Denote by r_n the number of near-neighbours that are used for estimation at a fixed point x for sample size n. Then in order that $\hat{f}_{r_n}(x)$ be consistent for the true function value $f(x)$ as $n \to \infty$, it is intuitively clear that both r_n must approach infinity and r_n must be asymptotically small relative to n so that the bias in $\hat{f}_{r_n}(x)$ vanishes. In fact, one can show that if $r_n \to \infty$ and $r_n/n \to 0$, then under suitable regularity conditions $\hat{f}_{r_n}(x) \to f(x)$ in probability. Now this only specifies a range for the asymptotic size of r_n; the optimal form for r_n can be worked out from the bias and variance of $\hat{f}_{r_n}(x)$. It turns out that the r-nearest-neighbour smoother has asymptotic bias and variance proportional to $(1/8)p(x)^{-3}(r_n/n)^2$ and $2/r_n$ respectively, where $p(x)$ is the marginal density of X. From this it is easily derived that the r_n minimizing the mean squared error of $\hat{f}_{r_n}(x)$ is $r_n \sim n^{4/5}$ (the symbol $\sim$ means *asymptotically equivalent to*). Note that this agrees with our rough calculation in (3.5): if the Xs are equally spaced on the unit interval then $\Delta = n^{-1}$ and we get $k_{opt} \sim n^{4/5}$.

Similar results can be derived for a kernel smoother with bandwidth h_n for sample size n. Under suitable regularity conditions, the kernel estimate is consistent if $n \to \infty, h_n \to 0$, and $nh_n \to \infty$. The asymptotic bias and variance are proportional to $h_n^2/2p(x)$ and $1/nh_np(x)$, and the optimal bandwidth choice is $h_n \sim n^{-1/5}$. If the bandwidth is allowed to vary locally with X, the optimal choice is proportional to $n^{-1/5}p(x)^{-1/5}$.

If one keeps track of the constants of proportionality, the results for nearest-neighbour and kernel smoothers can be seen to asymptotically coincide if $h_n = r_n/2np(x)$. One way to interpret this result is that a nearest-neighbour smoother is like a variable kernel smoother, that is, the bandwidth changes as a function of $p(x)$.

A smoothing spline can be shown to be asymptotically equivalent to a certain variable kernel smoother. The kernel is symmetric with exponentially decaying tails and negative sidelobes. The bandwidth $h_n \sim \lambda_n^{1/4} n^{-1/4} p(x)^{-1/4}$, λ_n being the smoothing parameter for sample size n. Note that the dependence on the local density is of the power $-1/4$, closer to the optimal dependence of $-1/5$ (see above) than that of the nearest-neighbour smoother (-1).

The sequence b_n is called the optimal rate of convergence for a regression curve if:

(i) the mean squared error of estimation is at least cb_n for some constant c and all estimators of that regression curve, and

(ii) the mean squared error of some estimator of the function is no greater than $c_1 b_n$ (for some constant c_1), that is, the rate is achievable.

The following result has been obtained concerning rates of convergence. If the regression function is p times differentiable and the pth derivative is Hölder continuous, the predictors are d dimensional and the kth derivative is to be estimated, then the optimal rate of convergence is

$$n^{-2(p-k)/(2p+d)}.$$

For scatterplot smoothing, $d = 1$ and $k = 0$, and thus the optimal rate of convergence is $n^{-2p/(2p+1)}$. Not surprisingly, the smoother the underlying function, the faster the rate of convergence.

Finally, we list some asymptotic results for automatic smoothing parameter selection. Let $\hat{\lambda}$ be the minimizer of $CV(\lambda)$ or $GCV(\lambda)$, let $\hat{\lambda}_{opt}$ minimize mean squared error for the given data, and let λ_{opt} minimize the true mean squared error. Then it has been shown under regularity conditions that $\hat{\lambda}$ converges to $\hat{\lambda}_{opt}$ and $\hat{\lambda}_{opt}$ converges to λ_{opt}, both at the relatively slow rate of $n^{-3/10}$. On the other hand, the mean squared errors of the functions associated with these estimators converge at the faster rate n^{-1}. This difference in rates is thought to be a reflection of the flatness of the criteria near their minima. More references on asymptotic results for smoothers are given in the bibliography.

3.11 Special topics

3.11.1 *Nonlinear smoothers*

Some smoothers are nonlinear, that is, the fit $\hat{\mathbf{f}}$ cannot be written as $\hat{\mathbf{f}} = \mathbf{S}\mathbf{y}$ for any $\mathbf{S}$ independent of $\mathbf{y}$. A simple example is the running median smoother (mentioned briefly in section 2.7), which is the same as the running mean except that the fitted value is the median of the responses in the neighbourhood. Now for random variables X and Y, $median(X+Y) \neq median(X)+median(Y)$, so by definition this smoother is nonlinear. Running medians have the advantage that they are not very sensitive to outliers but produce curves far too jagged to be useful by themselves as scatterplot smoothers. It makes some sense to apply a running median first to a set of data, before applying some other smoother, in order to make the estimation procedure more robust. However, the running median induces correlation in the errors, so that one has to be careful if one uses a method for automatic smoothing parameter selection in a subsequent step.

We have already seen other examples of nonlinear smoothers. All the smoothers we encountered in Chapter 2 are nonlinear if the selection of the smoothing parameter is based on the data $\mathbf{y}$. Similarly robust smoothers that use residuals to reweight the observations are nonlinear.

An interesting example of a nonlinear smoother is the variable span smoother called *supersmoother*. This smoother is an enhancement of the running-line smoother, the difference being that it chooses a (possibly) different span at each X-value. It does so in order to adapt to the changes (across X) in the curvature of the underlying function and the variance of Y. In regions where the curvature-to-variance ratio is high, a small span is appropriate, while in low curvature-to-variance regions a large span is called for. The supersmoother tries to achieve this effect as follows. Three windows are passed over the data, of small, medium and large spans. For each span, the squared cross-validated residual is computed at each point and smoothed as a function of X. Then the span producing the smallest smoothed squared-residual at x_i is chosen. Finally, the optimal span values are smoothed against X so that the spans don't change too quickly as X varies. In simulations, the supersmoother does seem able to adapt the span appropriately, although some price is paid in increased variance of the estimate. Despite its complex nature, there is an $O(n)$ algorithm for supersmoother that makes repeated use of updating formulae. Of course the same idea can be applied to any direct smoothing method such as kernels and locally-weighted lines.

3.11.2 *Kriging*

There is an equivalence between smoothing splines, and *kriging*, a popular prediction method in geostatistics. The kriging framework is also interesting because it illustrates the stochastic function approach to smoothing.

We suppose that the response Y is composed of a stochastic trend component $f(x)$ and an uncorrelated noise component ε, that is

$$Y_i = f(x_i) + \varepsilon_i$$

where $E(\varepsilon_i) = 0$, $\text{cov}(\varepsilon) = \sigma^2\mathbf{I}$, $Ef(x) = 0$, $\text{cov}\{f(s), f(t)\} = C(s,t)$. Kriging seeks to predict $f(x)$ by a linear function of $\mathbf{y}$

$$f^*(x) = \mathbf{c}^T\mathbf{y}$$

that minimizes $E\{f^*(x) - f(x)\}^2$. The solution is the linear regression (in L_2) of $f(x)$ on $\mathbf{y}$; namely

$$\hat{\mathbf{c}} = \text{var}(\mathbf{y})^{-1}\,\text{cov}\{f(x), \mathbf{y}\}. \tag{3.32}$$

Here $\text{var}(\mathbf{y})$ is the $n \times n$ covariance matrix of $\mathbf{y}$, and $\text{cov}\{f(x), \mathbf{y}\}$ is an n-vector of covariances between each element of $\mathbf{y}$ and $f(x)$. We can relate this to smoothing splines by considering the prediction of $f(x_1), \ldots, f(x_n)$. Let $\mathbf{C}$ be the $n \times n$ matrix with elements $\text{cov}\{f(x_i), f(x_j)\}$. Then using (3.32), the set of predicted values are

$$\hat{\mathbf{f}} = (\mathbf{I} + \sigma^2\mathbf{C}^{-1})^T\mathbf{y}.$$

This is exactly a smoothing spline if we set $\mathbf{C} = \mathbf{K}^-$, $\mathbf{K}$ being the matrix defined in section 2.10. In other words, the kriging estimate that uses $\mathbf{K}^-$ for the covariance of $\mathbf{f}$ coincides with a smoothing spline. Actually, it is easy to show that the kriging estimate coincides with the smoothing spline everywhere, not just at the observed predictor values.

Notice that the assumed covariance for $\mathbf{f}$ is the same as in the Bayesian model of section 3.6. This simply reflects the equivalence between linear least-squares and maximum likelihood for the Gaussian distribution.

In practice, kriging is often applied in spatial or temporal settings, and the covariance function is estimated from a model that uses the space or time proximity of the measurements and other considerations. The kriging method we discuss above is known as *simple kriging*, that is, kriging in the absence of drift. *Universal kriging* incorporates drift and is discussed in the references given in the bibliographic notes.

3.11.3 *Smoothing and penalized least-squares*

Earlier we derived the fact that the smoothing spline $\hat{\mathbf{f}} = \mathbf{S}\mathbf{y}$, where $\mathbf{S} = (\mathbf{I} + \lambda\mathbf{K})^{-1}$, is the minimizer of the penalized least-squares criterion

$$(\mathbf{y} - \mathbf{f})^T(\mathbf{y} - \mathbf{f}) + \lambda\mathbf{f}^T\mathbf{K}\mathbf{f}. \tag{3.33}$$

Can any linear smoother $\hat{\mathbf{f}} = \mathbf{S}\mathbf{y}$ be viewed as the solution to some penalized least squares problem? The answer is yes if $\mathbf{S}$ is symmetric and nonnegative definite, for then $\hat{\mathbf{f}} = \mathbf{S}\mathbf{y}$ minimizes

$$(\mathbf{y} - \mathbf{f})^T(\mathbf{y} - \mathbf{f}) + \mathbf{f}^T(\mathbf{S}^- - \mathbf{I})\mathbf{f} \tag{3.34}$$

over all $\mathbf{f} \in \mathcal{R}(\mathbf{S})$, (the range of $\mathbf{S}$), and $\mathbf{S}^-$ is any generalized inverse of $\mathbf{S}$. This is easily checked by differentiating (3.34) with respect to $\mathbf{f}$ and also noting that (3.34) is a nonnegative definite quadratic form in $\mathbf{f}$ (Exercise 3.9). In some cases one can interpret (3.34) in terms of down-weighting the components of $\mathbf{f}$ depending on their smoothness (Exercise 3.6).

We find this connection between penalized least-squares and linear smoothing to be useful in later chapters when we discuss the additive model.

3.11.4 *Weighted smoothing*

In some instances we may want to give unequal weight to the observations. This is often the case if our model assumes heterogeneous variances, that is

$$Y_i = f(x_i) + \varepsilon_i \tag{3.35}$$

where $\text{var}(\varepsilon_i) = \sigma_i^2$. Then we may give the ith observation weight $1/\sigma_i^2$, assuming that the σ_i are known. If they are unknown we may use some kind of iterative weighted-smoothing technique. Weighted smoothers are also required in the iterative-weighted least-squares type algorithms for fitting generalized additive models in Chapters 4 and 6.

Each of the smoothers described earlier can include weights in a natural way. Unfortunately, these natural ways differ.

For local-averaging smoothers like the running-mean and -line, we can reasonably define the weighted smooth to be that obtained by using weighted means and lines in the neighbourhoods. For the locally-weighted running-line smoother, which already uses local kernel weights, it makes sense to simply multiply the observation weights and the local kernel weights, and use them in the weighted least-squares fit.

How about an arbitrary linear smoother represented by a smoother matrix $\mathbf{S}$? Suppose we want to incorporate the weights w_i, with $\sum_{i=1}^n w_i = 1$. It seems natural to redefine the smoothing weights as $w_j S_{ij}$, and then renormalize so that the rows add to one. Formally we do this as follows. Let $\mathbf{W} = \text{diag}(\mathbf{w})$, a diagonal matrix with diagonal entries w_i, and let $\mathbf{D} = \text{diag}(\mathbf{Sw})$. Then our weighted smoother is

$$\mathbf{D}^{-1}\mathbf{SW}. \tag{3.36}$$

This is easy to implement: simply smooth the sequence $\{w_i y_i\}$ as well as $\{w_i\}$, and form the pointwise ratio of the results. If we apply this method to the running-mean, we obtain a weighted mean in each neighbourhood. This is also the appropriate method for kernel smoothers. Unfortunately, when applied to the running-line smoother, definition (3.36) does not produce a weighted least-squares fit in each neighbourhood.

Still another way to define weighted smoothing is through penalized least-squares, and this produces the standard definition for weighted smoothing splines. The penalized weighted least-squares criterion

$$(\mathbf{y} - \mathbf{f})^T\mathbf{W}(\mathbf{y} - \mathbf{f}) + \lambda\mathbf{f}^T\mathbf{K}\mathbf{f} \tag{3.37}$$

is minimized by $\hat{\mathbf{f}} = \mathbf{Sy}$ where

$$\mathbf{S} = (\mathbf{W} + \lambda\mathbf{K})^{-1}\mathbf{W}. \tag{3.38}$$

Actual computation of $\hat{\mathbf{f}}$ can be accomplished in much the same way as the unweighted smoothing spline, with little extra complexity in the algorithm.

An even stickier problem is that of smoothing with a general (nondiagonal) weight matrix $\mathbf{W}$. For smoothing splines and other symmetric linear smoothers, the penalized least squares approach can be used; the only difficulty is in the computation. For smoothing splines an $O(n)$ algorithm no longer exists unless $\mathbf{W}$ is banded or has a simple structure. For local-averaging smoothers, we can define the general weighted smooth to be that obtained by using the appropriate partition of $\mathbf{W}$ as the weight matrix in each local fit. Again, a detailed study of these approaches has yet to be done.

3.11.5 *Tied predictor values*

We make the assumption at the beginning of the chapter that the X-values in our data are distinct. Often this is not the case, there are ties, and some special provision must be made.

There is a simple solution that can be used with all the smoothers. The data are divided into sets corresponding to the $m \leq n$ unique values of X. A new data set is created with m observations, each one representing a set. An observation consists of the X-value itself, the weighted mean of the Y-values in the set, and a weight equal to the sum of the observation weights in the set. Then a weighted smoothing procedure is applied to the new data set.

This gives the exact solution to the penalized least-squares criterion of smoothing splines when there are ties, and is also exact for kernel smoothers. It raises an issue not dealt with in the previous section for near-neighbour smoothers: the near-neighbourhood sizes should be defined in terms of total weight rather than number of observations.

3.11.6 *Resistant smoothing*

We have not worried so far about the resistance of scatterplot smoothers to influential points or clumps of points. Intuitively it seems that this ought not be as much of a concern here as it is in linear regression, because of the local nature of the fitting. Consider for example the running-line smoother. A data point outside the neighbourhood of a given point will not affect the fitted smooth at that point. However, this is not a reason to rest easy. For one thing, with span of 50% (not uncommonly large), a point in the middle of the X-range will appear in the neighbourhood of half of the points. Secondly, even if a point is not in a given neighbourhood, it can affect the fit there indirectly by influencing the choice of span, if an automatic method like cross-validation is used.

Thus there is a need to make the smoothing procedures discussed earlier more resistant to influential data points. As in linear regression there are two possible kinds of influential points: outliers in Y-space and outliers in X-space. In linear regression, the first kind are usually handled by bounding the least-squares criterion function corresponding to large residuals, e.g. *M-estimation for regression* (Huber, 1981; Hampel *et al.*, 1986, Chapter 6). Outliers in X-space, also called *leverage points*, are typically dealt with by adding to the M-estimation criterion a factor that down-weights points that are far away from the middle of the predictor cloud (*bounded influence regression*). (See Hampel *et al.*, 1986, Chapter 6, for a summary. Rousseeuw, 1984, offers a different approach.)

Relatively little work has been done on the resistance problem for smoothing; most of the efforts have concentrated on the problem of outliers in Y-space. Heuristic suggestions have included pre-smoothing the data with a highly resistant smoother like a running median before applying the nonresistant smoothing method, or making an initial pass over the data to flag (and discard) points that are potential outliers. More formally, if the nonresistant smoothing method uses least-squares fitting (like the running-line smoother), then it can be replaced by a resistant or robust regression. This is the approach used in Cleveland's (1979) locally-weighted running-line smoother described in Chapter 2.

Still another method evolves if one considers how to make the smoothing spline resistant. By analogy with linear regression, we can replace the least-squares criterion by a tapered least-squares criterion

$$\sum_{i=1}^{n} \rho\left\{\frac{y_i - f(x_i)}{\hat{\sigma}}\right\} + \lambda \int \{f''(x)\}^2\, dx \tag{3.39}$$

where $\rho(\cdot)$ is, for example, Huber's function defined by $\rho(r) = r^2/2$ for $|r| < k$ and $\rho(r) = |r| + k^2/2$ for $|r| > k$, and $\hat{\sigma}$ is a resistant estimate of standard deviation. This criterion leads to an iteratively-reweighted smoothing spline. We defer details of this until Chapter 9, where we discuss resistant estimation of additive models. It turns out that this resistant produce is a special case of the *local-scoring procedure* introduced in the next chapter. Similar *M-estimate* approaches have been suggested for kernel smoothers.

Leverage is not such a problem in smoothing; most methods are automatically protected against high-leverage points since the fits are by definition local. The near-neighbour methods are exceptions, especially if lines are fitted within the neighbourhoods. Locally-weighted line smoothers overcome this by weighting the points within the near-neighbourhood. So outlying points in X-space do not have much influence on the fits at other points. What is usually noticeable is that they tend to dominate the picture by forcing a large plotting range. Most of the data are squashed up over a small interval, and the plotted curve really represents only a small percentage of the data whose fits are poorly determined anyway. One remedy is to smoothly transform the X-values prior to smoothing the (X, Y) data, to bring in the points in the tails of X. Alternatively, one could plot the fitted curve over a subset of the range.

3.12 Bibliographical notes

The literature on scatterplot smoothing goes back to at least 1923 when Whittaker introduced smoothing splines. Other important papers in the development of smoothing splines include Schoenberg (1964), Reinsch (1967), Kimeldorf and Wahba (1970, 1971), Wahba and Wold (1975), de Boor (1978), Wahba (1978), Craven and Wahba (1979), Utreras (1980), Wahba (1983), Rice and Rosenblatt (1983), Silverman (1985), and Eubank (1984, 1985). Reinsch was the first to introduce the penalized least-squares criterion and its solution, as it appears in this chapter. Wahba and Wold (1975) helped popularize smoothing splines in statistics. de Boor refined the computation of the cubic spline and provided algorithms. Craven and Wahba proposed generalized cross-validation, and gave some asymptotic results supporting its use. Generalized cross-validation was pursued further in subsequent papers by Wahba and co-workers, including the work by Utreras. Li (1986) showed asymptotic optimality for C_p and generalized cross-validation in ridge regression with applications to spline smoothing. Kimeldorf and Wahba (1971) gave general minimization theorems in reproducing kernel Hilbert spaces, with interpolating and smoothing splines following as special cases. Kimeldorf and Wahba (1970), and Wahba (in the 1978 and 1983 papers) studied the Bayesian derivation of smoothing splines. Silverman (1985) and discussants covered a broad range of topics including a theoretical comparison of smoothing splines and kernel smoothers, a finite sample Bayesian model (section 3.6), and efficient computation of the cross-validation scores. Rice and Rosenblatt (1983) developed the asymptotics for smoothing splines. Eubank (1984, 1985) studied the smoother matrix for smoothing splines and diagnostics based on it. Wold (1974) discussed regression splines, and Wegman and Wright (1983) gave a review of spline methods in statistics. Eubank's (1988) book is a useful detailed reference for smoothing splines, kernel smoothers and other smoothing methods. Another recent comprehensive text on smoothing is that of Härdle (1990).

Local-averaging smoothers date back at least to Ezekiel (1941), who suggested a smoother similar to the running-mean. Watson (1964) proposed estimating the fitted value at X from nonparametric estimates of the bivariate density of X and Y and the univariate density of X. Kernel smoothers were proposed and developed by Nadaraya (1964), Rosenblatt (1971) and Priestley and Chao (1972), in addition to the afore-mentioned paper by Watson. Gasser and Müller (1979, 1984) and Gasser, Müller and Mammitzsch, (1985) studied consistency, variance and bias of kernel estimates. Devroye and Wagner (1980), and Devroye (1981) gave weak conditions for consistency of kernel smooths. Rice (1984) and Müller (1985) proposed methods for bandwidth selection. Cleveland (1979) introduced the (resistant) locally-weighted running-line smoother, and discussed a number of associated topics, including degrees of freedom or *effective number of parameters*, a notion further developed by Tibshirani and Hastie (1987). This smoother is implemented (a function named *loess*) in the S statistical language (Becker, Chambers and Wilks, 1988). Cleveland, Devlin and Grosse (1988) discussed extensions to multiple predictor smoothing. Devlin (1986) and Cleveland and Devlin (1988) covered many aspects of multi-predictor smoothing; the two-moment approximations of section 3.9 are taken from their paper. Clark (1977) discussed the properties of various scatterplot smoothing techniques, and independently proposed locally-weighted smoothing (Clark, 1980). Friedman and Stuetzle (1982) and Friedman (1984) proposed the supersmoother, or variable span smoother. Stone (1977) provided elegant conditions under which linear smoothers are consistent in the L^r sense. Wong (1983) and Li (1984) studied the consistency of cross-validated near-neighbour smoothers.

Silverman (1984a) derived an asymptotically equivalent kernel for the smoothing spline, also discussed in Silverman (1985). Müller (1987) demonstrated the asymptotic equivalence of kernel and locally-weighted running-line smoothing. Mack (1981) derived some asymptotic results for kernel and nearest-neighbour smoothers, from which the summary in section 3.10 is extracted. Stone (1982) proved the result on optimal rates of convergence given in section 3.10. Asymptotic results on bandwidth selection are given in Wong (1983), Härdle and Marron (1985), and Härdle, Hall, and Marron (1988). Prakasa Rao (1983) is a general text on nonparametric functional estimation.

Buja, Hastie and Tibshirani (1989) discussed a number of aspects of linear smoothers, including degrees of freedom and symmetry, and empirically compared the implicit kernels of a number of smoothers.

Bootstrap confidence intervals are discussed in many papers, including Efron (1982), Efron and Tibshirani (1986) and DiCiccio and Romano (1988).

Kriging was proposed by Matheron (1973). A clear discussion of the relation between kriging and smoothing splines was given by Watson (1984). Stein (1988) studied the effect of misspecifying the covariance function.

A different *state-space* approach to Wahba's (1978) Bayesian model for smoothing splines was taken by Weinert, Byrd and Sidhu (1980) (and references therein), Wecker and Ansley (1983), Ansley and Kohn (1987) and Kohn and Ansley (1989). The last two papers use the Kalman filter to derive efficient computational algorithms for smoothing splines.

There is much work in the time-series literature on smoothing of equally-spaced data (sequences), a related but different problem from scatterplot smoothing. As a matter of fact. the aforementioned early paper of Whittaker on smoothing splines was in a time series setting. We will not try to track the history of smoothing time series, but simply list a few sources that we have found useful. The Fourier-domain approach to smoothing is well summarized in Bloomfield (1976), while Tukey (1977) introduces many new ideas for time domain smoothing, including the smoothers referred to in section 2.7. Mallows (1980) studied the properties of certain nonlinear smoothers, including these "Tukey" smoothers, and Goodall (1990) gave a survey of smoothers in this class. Huber (1979) has a section on robust smoothing.

Cross-validation is an old idea, key references being Stone (1974, 1977) and Allen (1974). The C_p statistic was proposed by Mallows (1973). Efron (1986) compared C_p, CV, GCV and bootstrap estimates of error rates, and argued that GCV is closer to C_p than CV. Bates *et al.* (1987) provided general algorithms for performing GCV optimization for a class of smoothers that includes smoothing splines.

3.13 Further results and exercises 3

3.1 Derive expression (3.7) for the value of k that minimizes mean-squared error at a point x_i.

3.2 Using the approximation $S_{ii} \approx \{\mathbf{S}\mathbf{S}^T\}_{ii}$ and standard Taylor-series arguments, derive the approximation

$$E\{CV(\lambda)\} \approx PSE(\lambda) + \frac{2}{n}\sum_{i=1}^{n} S_{ii}(\lambda) b_i^2(\lambda).$$

3.3 Verify result (3.21) for the expectation of the residual sum of squares.

3.4 Show that equation (3.15) for the jackknifed fit:

$$\hat{f}_\lambda^{-i}(x_i) = \sum_{\substack{j=1 \\ j \neq i}}^{n} \frac{S_{ij}(\lambda)}{1 - S_{ii}(\lambda)} y_j$$

is exact for regression splines and Nadaraya-Watson style kernel smoothers. To what extent is it correct for locally-weighted running-line smoothers?

3.5 Consider the expression given for the cubic smoothing-spline matrix, namely $\mathbf{S} = (\mathbf{I}+\lambda\mathbf{K})^{-1}$. Show that $\mathbf{S}$ has eigenvalues in the interval $(0, 1]$. Further show that the eigenvectors of eigenvalue 1 are linear functions of the covariate from which $\mathbf{S}$ was constructed. [Craven and Wahba, 1979; Buja, Hastie and Tibshirani, 1989]

3.6 If one orders the eigenvectors of a smoothing spline by decreasing value of the corresponding eigenvalue, the eigenvectors get less smooth, in a certain sense. [Demmler and Reinsch, 1975]. Specifically, the number of zero-crossings of the eigenvector increases as the eigenvalue decreases. Hence give an interpretation to (3.34) in terms of the down-weighting of terms as a function of their smoothness. How does this change as λ varies?

3.7 Write the eigenvalues computed in the previous exercise as a function of the eigenvalues of $\mathbf{K}$. Derive expressions for $\text{tr}(\mathbf{S}_\lambda)$ and $\text{tr}(\mathbf{S}_\lambda\mathbf{S}_\lambda^T)$ in terms of these eigenvalues, and characterize the bias of $\mathbf{S}_\lambda$. Prove that $\text{tr}(\mathbf{S}_\lambda\mathbf{S}_\lambda^T) \le \text{tr}(\mathbf{S}_\lambda) \le \text{tr}(2\mathbf{S}_\lambda - \mathbf{S}_\lambda\mathbf{S}_\lambda^T)$, and that all three are monotone in λ.

3.8 *Twicing.* Twicing is a method proposed to reduce the bias of a smoother. Specifically, one first smooths the sequence, and then the residuals from this smooth are smoothed and added to the original fit. Describe the action of twicing in terms of the eigen-decomposition above. Under what conditions does twicing have no effect?

3.9 Verify that $\hat{\mathbf{f}} = \mathbf{S}\mathbf{y}$ minimizes the penalized least-squares criterion (3.34):

$$(\mathbf{y}-\mathbf{f})^T(\mathbf{y}-\mathbf{f}) + \mathbf{f}^T(\mathbf{S}^- - \mathbf{I})\mathbf{f}.$$

3.10 *Computation of the GCV statistic.* For a given value of λ, $GCV(\lambda)$ is given by

$$GCV(\lambda) = \frac{1}{n}\sum_{i=1}^{n}\left\{\frac{y_i - \hat{f}_\lambda(x_i)}{1-\text{tr}(\mathbf{S}_\lambda)/n}\right\}^2.$$

Derive a decomposition for this expression, taking $\mathbf{S}_\lambda$ to be a smoothing spline, that enables $GCV(\lambda)$ to be computed for any λ in a further $O(n)$ computations.

[Golub, Heath and Wahba, 1979]

3.11 Suppose one is interested in making inferences about some linear functional $\psi(g)$, where g is the underlying regression function. For example, $\psi(g)$ might be $g'(x)$ for a fixed value of x. Suggest a method for inference based on the Bayesian framework for smoothing splines.

[Silverman, 1985]

3.12 Show that the quantity $\nu(\mathbf{g})$ defined in equation (3.27) is approximately pivotal, with distribution approximately that of $df^{\text{err}} + df^{\text{var}} F_{df^{\text{var}}, df^{\text{err}}}$ where $F_{p,n-p}$ denotes a random variable with an F distribution on p and $n-p$ degrees of freedom.

3.13 Using the explicit expression for the smoothing-spline matrix, show that the F-test correction factors ν_1/ν_2 and λ_1/λ_2 are no larger than one and hence the test based on (3.31) is conservative.

3.14 For the Bayesian model given in section 3.6, show that if γ_0 is the mean of the posterior distribution of γ, then $\mathbf{B}\gamma_0$ is a smoothing spline corresponding to smoothing parameter λ, and also that the posterior covariance of $\mathbf{B}\gamma$ is $\mathbf{S}\sigma^2$, where $\mathbf{S}$ is the smoothing-spline matrix.

[Silverman, 1985]

CHAPTER 4

Additive models

4.1 Introduction

In this chapter we introduce the additive model for multiple regression data and the backfitting algorithm for its estimation. We also discuss another commonly used additive model, the additive logistic model for binomial response data.

The additive model is a generalization of the usual linear regression model, so it is important to outline the limitations of the linear model and why we might want to generalize it. An arbitrary regression surface would be a natural generalization. However, there are problems with the estimation and interpretation of fully general regression surfaces and these problems lead one to restrict attention to additive models.

4.2 Multiple regression and linear models

For the moment we restrict attention to the standard multiple regression problem. We have n observations on a response (or dependent) variable Y, denoted by $\mathbf{y} = (y_1, \ldots, y_n)^T$ measured at n design vectors $\mathbf{x}^i = (x_{i1}, \ldots, x_{ip})$. The points $\mathbf{x}^i$ may be chosen in advance, or may themselves be measurements of random variables X_j for $j = 1, \ldots, p$, or both. We do not distinguish the two situations.

Our goal is to model the dependence of Y on $X_1, \ldots, X_p$. There are several reasons why we may want to do this. The first is *description*: we want a model to describe the dependence of the response on the predictors so that we can learn more about the process that produces Y. The second goal is *inference*: we want to assess the relative contributions of the each of the predictors in explaining Y. Finally, there is *prediction*: we wish to predict Y for some set of values $X_1, \ldots, X_p$.

For all these purposes, the standard tool for the applied statistician is the multiple linear regression model:

$$Y = \alpha + X_1\beta_1 + \cdots + X_p\beta_p + \varepsilon \tag{4.1}$$

where $E(\varepsilon) = 0$ and $\text{var}(\varepsilon) = \sigma^2$. This model makes a strong assumption about the dependence of $E(Y)$ on $X_1, \ldots, X_p$, namely that the dependence is linear in each of the predictors. If this assumption holds, even roughly, then the linear regression model is extremely useful and convenient because:

(i) it provides a simple description of the data,
(ii) it summarizes the contribution of each predictor with a single coefficient and
(iii) it provides a simple method for predicting new observations.

There are many ways to generalize the linear regression model (Fig. 4.2). One class of candidates is the surface smoothers that we discuss in Chapter 2. These can be thought of as nonparametric estimates of the regression model

$$Y = f(X_1, \ldots, X_p) + \varepsilon \tag{4.2}$$

and are rather intuitively defined. We outline one problem with surface smoothers in Chapter 2, namely choosing the shape of the kernel or neighbourhood that defines *local* in p dimensions. A more serious problem common to all surface smoothers has to do with this same *localness* in high dimensions, and has been appropriately dubbed the *curse of dimensionality* : neighbourhoods with a fixed number of points become less local as the dimensions increase (Bellman, 1961).

The simple example in Fig. 4.1 illustrates this point. Suppose that points are uniformly distributed in a p-dimensional unit cube, and we wish to construct a cube-shaped neighbourhood at the origin to capture (on the average) $100 \times \text{span}\%$ of the data. It is simple to show that the subcube should have side length $\text{span}^{1/p}$. For $p = 1$ and span $= 0.1$ this is simply 0.1. For $p = 10$ it is about 0.8, hardly local any more! This tells us that our concept of *local*, in terms of percentage of the data, fails in high dimensions. We need to use kernels with smaller bandwidths, and thus require

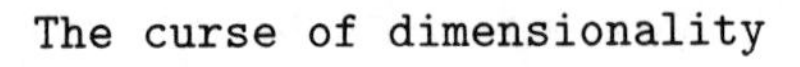

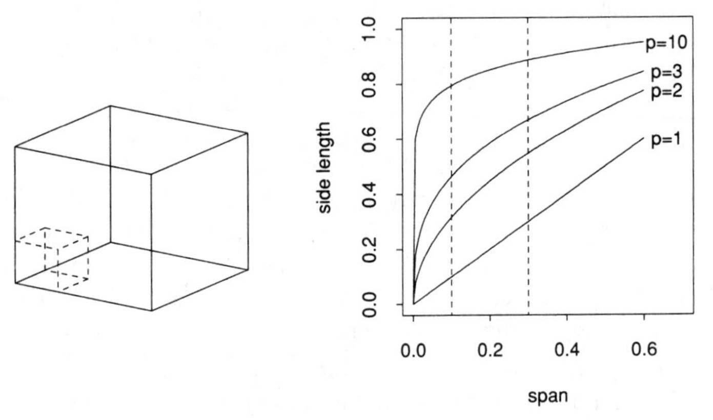

Fig. 4.1. *The large cube represents a uniform distribution in p dimensions, and the subcube represents span% of the volume. The graph on the right shows the side length along any coordinate needed to achieve this span, for different dimensions p.*

far more data in order to get variance-stable estimates in high dimensions. Surface smoothers are nevertheless useful tools for small p (especially $p = 2$ where we can see the fitted surface). For large p they are difficult to interpret.

Several multivariate nonparametric regression techniques have been devised, at least partly in response to the dimensionality problem. Some of these are illustrated in Fig. 4.2.

Recursive-partitioning regression carves the predictor space up into disjoint blocks in a binary style and hence the model can be represented by a binary tree. Each terminal node or leaf of the tree represents a block. The fitted values are constants in each leaf, and the result is a piecewise constant regression surface (Fig. 4.2). The tree is built sequentially, and a new branch replaces a leaf if a split of that region is warranted in terms of the gain in predictive power. The main advantage of the tree approach is that the available data are used judiciously; fine grids are used only in regions that warrant them. Small trees are easy to interpret and explain; for example, "males over 35 with high blood pressure have an average risk of 15%". Large trees, on the other hand, are difficult to represent and interpret. We discuss regression trees in a bit more detail in Chapter 9.

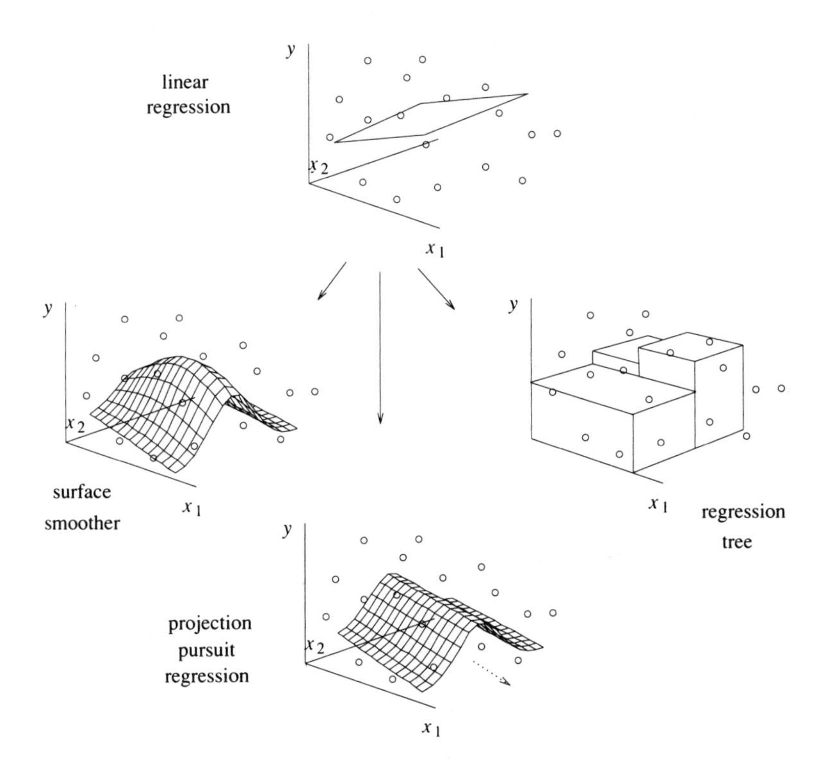

Fig. 4.2. *Some generalizations of multiple linear regression: all are local averages in some sense. Local for the surface smoother means points within a circular neighbourhood, weighted according to their distance from the target. Local for projection pursuit means local in the chosen projection. Local for tree based regression models means falling in the same rectangular region of the predictor space.*

Projection-pursuit regression is a direct attack on the dimensionality issue. It assumes a model of the form

$$Y = \sum_{k=1}^{K} h_k(\boldsymbol{\alpha}_k^T \boldsymbol{X}) + \varepsilon, \tag{4.3}$$

where $\boldsymbol{\alpha}_k^T \boldsymbol{X}$ denotes a one-dimensional projection of the vector $\boldsymbol{X}$, h_k is an arbitrary univariate function of this projection, and

$E(\varepsilon) = 0$, $\text{var}(\varepsilon) = \sigma^2$, and the errors are independent of X. The model builds up the regression surface by estimating these *univariate* regressions along carefully chosen projections defined by the $\boldsymbol{\alpha}_k$. Thus for $K = 1$ and $p = 2$ the regression surface looks like a corrugated sheet and is constant in directions orthogonal to $\boldsymbol{\alpha}_1$ (Fig. 4.2). The directions $\boldsymbol{\alpha}_k$ and number of terms K are chosen to give the best predictive power. The smoothing is always one-dimensional, and consequently no dimensionality problems occur. Of course the neighbourhoods are infinitely large in directions orthogonal to these projections, but the hope is that the surface is relatively constant in these directions. Projection pursuit regression models are parsimonious smooth surface estimators, but are difficult to interpret for K larger than one.

We could go on, but prefer to stop here. Some other multivariate techniques are discussed in Chapter 9. One point that emerges is that, given sufficient data, any of the above models has good predictive power; in fact, under suitable conditions they are all consistent for the true regression surface. They all suffer from being difficult to interpret. Specifically, how do we examine the effect of particular variables once we have fitted a complicated surface? For low dimensional surfaces one can look at slices defined by conditioning on all but one of the variables, but this becomes infeasible in higher dimensions.

4.3 Additive models

The interpretation problem highlights an important feature of the linear model that has made it so popular for statistical inference: the linear model is *additive* in the predictor effects. Once we have fitted the linear model we can examine the predictor effects separately, in the absence of interactions. Additive models retain this important feature: they are additive in the predictor effects. An additive model is defined by

$$Y = \alpha + \sum_{j=1}^{p} f_j(X_j) + \varepsilon \tag{4.4}$$

where as before the errors ε are independent of the X_js, $E(\varepsilon) = 0$ and $\text{var}(\varepsilon) = \sigma^2$. The f_js are arbitrary univariate functions, one for each predictor. For the moment we imagine each of the f_j to be smooth functions, and that somehow they are individually estimated by a scatterplot smoother.

Figure 4.3 illustrates an example of an additive model when there are two predictors, using the diabetes data described earlier. The response is `log(C-peptide)` and the two predictors are `age` and `base deficit`. Both the effects appear to be non linear: `age` has an increasing effect that levels off, and `base deficit` appears quadratic.

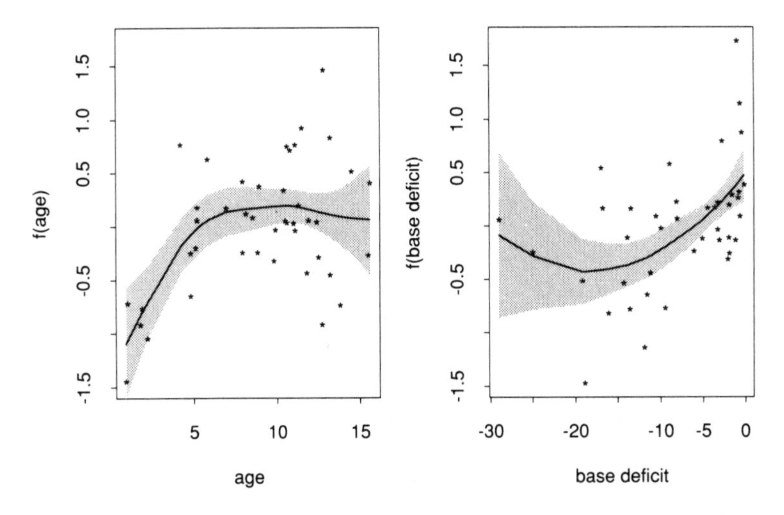

Fig. 4.3. *The additive fit for* `age` *and* `base deficit`. *The shaded region represents twice the pointwise standard errors of the estimated curve. The points plotted are partial residuals: the fitted values for each function plus the overall residuals from the additive model.*

It must be said that even though it is convenient to think of the functions in the additive model as univariate and smooth, neither of these properties are necessary. We shall see that an additive model might have component functions with two or more dimensions, as well as categorical variable terms and their interactions with continuous scale variables. Implicit in (4.4) is the assumption that $E\{f_j(X_j)\} = 0$, since otherwise there will be free constants in each of the functions.

The additive model has an a priori motivation as a data-analytic tool. Since each variable is represented separately in (4.4),

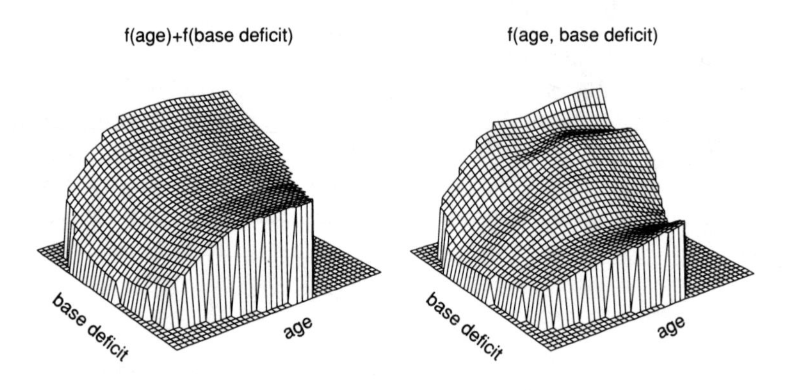

Fig. 4.4. *On the left, the additive surface constructed from functions for* `age` *and* `base deficit` *above. The surface is plotted only over the convex hull of the observed sample points. On the right, the bivariate surface estimated using a bivariate kernel smoother. Both fitted surfaces use approximately the same df of about 10.*

the model retains an important interpretive feature of the linear model: the variation of the fitted response surface holding all but one predictor fixed does not depend on the values of the other predictors. In practice this means that once the additive model is fitted to data, we can plot the p coordinate functions separately to examine the roles of the predictors in modelling the response. Such simplicity does not come free; the additive model is almost always an approximation to the true regression surface, but hopefully a useful one. When fitting a linear regression model, we generally don't believe that the model is correct. Rather we believe that it will be a good first order approximation to the true surface, and that we can uncover the important predictors and their roles using the approximation. Additive models are more general approximations.

Figure 4.4 shows to what extent the additive model for `age` and `base deficit` is an approximation to the bivariate response surface. In this case it appears that a possibly important bump in the regression surface is lost in the additive approximation. On the other hand, the average trend in each variable is well represented.

In some applications it is useful to think of the additive model as a method for estimating simultaneously the appropriate metameter in which to measure the variables. Thereafter the standard linear analysis can be performed on the transformed variables. It usually makes sense to restrict these transformations to be monotone.

The estimated functions from an additive model are the analogues of the coefficients in linear regression. All the pitfalls encountered in interpreting linear regression models apply to additive models, and indeed can be expected to be more severe. We must be careful not to interpret functions for variables that are insignificant, and not have them affect the important ones. One strategy is to select and delete the variables in a stepwise manner, until only important terms are left in the model. Only at this stage should one examine the fitted functions, knowing that each is needed in the construction of the multivariate surface.

We are still concerned with how the functions may interact, whether bends and nonlinear features are important, and whether a simpler class of functions suffices. Standard error curves can be computed for each of the fitted functions, and these go part of the way towards answering these questions. These and related questions are dealt with in Chapters 5, 9 and 10. The remainder of this chapter concentrates on fitting additive models, and seeing how they can be used in more general settings.

4.4 Fitting additive models

There are many ways to approach the formulation and estimation of additive models. Typically they differ in the way the smoothness constraints are imposed on the functions in the model. The following list is by no means complete, but shows in order of increasing generality how semi-parametric additive models can augment the hierarchy of available techniques.

(i) Multiple linear regression, with its vast array of diagnostics and computational tricks, is a standard tool for data analysis.

(ii) More general versions of multiple linear regression add flexibility and can be viewed as parametric nonlinear additive models.

  (a) Some of the variables can be transformed parametrically. Popular choices are log, square-root, inverse and other power transformations.

(b) Each variable can generate a *set* of transformed variables, such as orthogonal polynomials. After the full set has been fitted using multiple linear regression, the fitted function for each variable can be reconstructed.

(c) The estimated constants for categorical variables (entered as dummy variables in the multiple regression) can be viewed as a step function estimate. The main-effects ANOVA model falls into this class. Continuous variables can be divided into categories and treated as categorical variables in the same way.

(iii) The methods above first create a set of *basis functions*, and then perform multiple regression on these basis functions. The dimension of these bases should not grow too large. Regression splines also define a special set of basis functions which are piecewise polynomials centered at pre-chosen knots; B-splines are a popular choice. These can then be used as the variables in a multiple linear regression, resulting in an additive model having one or more terms represented by a piecewise polynomial.

(iv) For one variable, smoothing splines can be seen as a multiple regression onto a large set of such B-spline bases (typically as many as there are data points). The dimensionality is reduced by incorporating into the multiple regression a complicated form of *shrinking*. This has the same flavour, and is similar to, *ridge regression*, a form of biased regression. Later we approximate a smoothing spline by a shrunken regression onto a (reasonably small) set of orthogonal polynomials. This can be done simultaneously for all the variables in the additive model in one large *ridge* regression.

(v) The most general method for estimating additive models allows us to estimate each function by an arbitrary smoother. Some candidates are cubic smoothing splines, locally-weighted running-line, and kernel smoothers. In the special case of smoothing splines, projections or ridge type regressions, we can pose the appropriate penalized least-squares criterion and simply minimize it. The next chapter deals with this problem, which turns out to have nice mathematical properties. The system of equations that needs to be solved, however, is large ($np \times np$ for p variables and n observations). The *backfitting algorithm* is a general algorithm that enables one to fit an additive model using *any* regression-type fitting mechanisms. It is an iterative fitting procedure, and this is the price one pays for the added generality.

Conditional expectations provide a simple intuitive motivation for the backfitting algorithm. If the additive model (4.4) is correct, then for any k, $E(Y - \alpha - \sum_{j \neq k} f_j(X_j) \,|\, X_k) = f_k(X_k)$. This immediately suggests an iterative algorithm for computing all the f_j, which we give in terms of data and arbitrary scatterplot smoothers $\mathcal{S}_j$.

Algorithm 4.1 *The backfitting algorithm*

(i) *Initialize:* $\alpha = \text{ave}(y_i),\ f_j = f_j^0, j = 1, \ldots, p$
(ii) *Cycle:* $j = 1, \ldots, p, 1, \ldots, p, \ldots$
$f_j = \mathcal{S}_j(\mathbf{y} - \alpha - \sum_{k \neq j} \mathbf{f}_k \,|\, \mathbf{x}_j)$
(iii) Continue (ii) until the individual functions don't change.

Recall that $\mathcal{S}_j(\mathbf{y} \,|\, \mathbf{x}_j)$ denotes a smooth of the response $\mathbf{y}$ against the predictor $\mathbf{x}_j$, and produces a function. We want the functions to be fitted simultaneously, so the individual smoothing steps make sense. When readjusting f_j, we remove the effects of all the other variables from $\mathbf{y}$ before smoothing this *partial residual* against x_j. Clearly this is only appropriate if all the functions removed are also correct, and therefore the iteration.

We need to provide initial functions to start the algorithm. Without prior knowledge of the functions, a sensible starting point might be the linear regression of $\mathbf{y}$ on the predictors. Often the backfitting algorithm itself is nested within some bigger iteration, in which case the functions from the previous big iteration loop provide starting values.

The *until* in the backfitting algorithm is not entirely an expression of faith. It turns out that a lot can be said about the convergence of the backfitting algorithm when the smoothers are linear operators. It is a Gauss-Seidel algorithm for solving a certain set of estimating equations, and one can prove convergence in many practical situations. When all the smoothers are projection operators, convergence is guaranteed — in fact, the whole algorithm can then be replaced by a global projection and no iteration is needed. Some smoothers, such as smoothing splines, are not projections but possess the properties of projections that are required for convergence. For others, such as locally-weighted running-line smoothers, we have no proofs of convergence. Our experience has been very promising, however, and one has to look hard for counter examples. Details are given in Chapter 5.

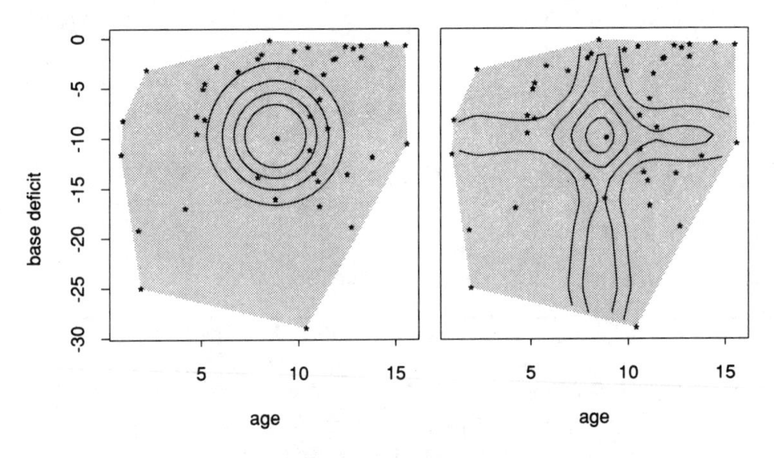

Fig. 4.5. *The two contour plots represent the equivalent kernels for a bivariate kernel surface smoother and an additive model, applied to the diabetes data. The additive model was fitted using univariate kernel smoothers within the backfitting algorithm, calibrated to do roughly the same amount of smoothing as the bivariate kernel.*

To what extent do additive models overcome the *curse of dimensionality*, and what price do they pay? Figure 4.5 compares the equivalent kernel of a bivariate kernel surface smooth to the equivalent kernel for the additive model fit. The equivalent kernels are the linear weights used to compute the fit at a particular point, and as such there are exactly 43 of them, the number of data points in the sample. Since they vary smoothly, we can represent them by a smooth surface, as was done here. As we might have expected, the additive model *smoother* borrows strength from local panels orthogonal to the coordinate directions. In this sense it overcomes the curse of dimensionality, since it is built up of pieces that behave like univariate smoothers. The equivalent kernel is approximately the sum of the two univariate equivalent kernels. It pays a price,

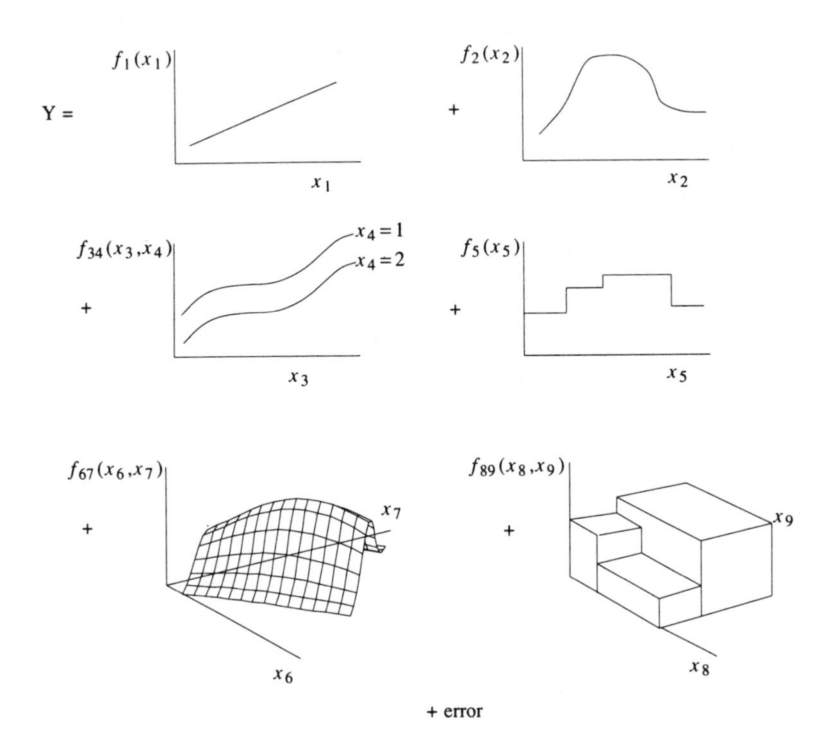

Fig. 4.6. *The additive model can piece together any basic regression techniques as building blocks, including functions of more than one variable.*

since the neighbourhoods are arbitrarily nonlocal in the directions of the coordinate axes.

The algorithm as given above is deliberately vague; the details differ depending on the type of smoother used and the context in which it is used. This modularity is an attractive feature of the backfitting algorithm, since a variety of personal preferences can be accommodated. Figure 4.6 shows some of the possible modules which we discuss in more detail below.

(i) Any of the functions can be fitted linearly, i.e., we set $f_j(x) = x\beta_j$, if this is appropriate. This can most easily be achieved by using the simple linear regression *smoother* for that variable.

(ii) Categorical variables are easily incorporated in the model. If variable x_j has K categories, we can create a K level factor, each level of which is fitted by a constant. It is fitted each time in the backfitting algorithm by a *histogram* type smoother, which fits the partial residual mean in each category.

(iii) Both of the above smoothers are projections and regression splines fall into this class as well. Smarter things can be done computationally with projection-type smoothers. In Chapter 9 we discuss practical implementations of backfitting algorithms that combine all the projection-type smoothers into one big projection that is treated as one operator in the backfitting algorithm.

(iv) One can consider interactions between categorical variables and continuous ones. Suppose we suspect that X_j has a different effect for males and females; instead of a single function f_j for X_j, we can estimate both f_{jM} and f_{jF}. This requires a smoother that can ignore subsets of the data when producing the fit, and any of the smoothers discussed in Chapters 2 and 3 can be modified appropriately.

(v) Interactions between continuous variables can be modelled in a variety of ways. By analogy to linear regression, one can create compound variables such as products of pairs, and fit them linearly or nonparametrically. These might be difficult to interpret, however, and one might resort to categorizing one of them and using the methods in the previous paragraph. Alternatively, one can model the (pairwise) interaction with a general smooth surface. This involves treating the pair as a single variable, and hence using a two-dimensional surface smoother to smooth its partial residual in the backfitting algorithm.

(vi) The last paragraph suggests the most general algorithm, which basically allows each component in the model to be defined by an arbitrary set of variables. Associated with each set is a fitting mechanism which, given a response (partial residual), produces a univariate vector of fitted values as its contribution to the additive model. An example is the regression tree mentioned earlier, which can be used to model collectively a number of categorical variables and their interactions as one term in the additive model.

We do not pursue these ideas further here, but we hope they

have given a taste of what is to come. We now show how the additive model itself can be used as a building block in other data analysis contexts.

4.5 Generalized additive models: logistic regression

The linear model is used for regression in a variety of contexts other than ordinary regression. Common examples include log-linear models, logistic regression, the proportional-hazards model for survival data, models for ordinal categorical responses, and transformation models. The linear model is a convenient but crude first-order approximation to the prediction surface, and in many cases it is adequate. The additive model can be used to generalize all these models in an obvious way. To crystallize ideas we focus on an important example: logistic regression.

In this setting the response variable is dichotomous, such as yes/no or survived/died, and the data analysis is aimed at relating this outcome to the predictors. One quantity of interest is the proportion of either outcome as a function of the predictors. It is convenient to code the response variable Y as zero or one according to the outcome, since this proportion is the mean of Y.

Figure 4.7 is a scatterplot representing some heart attack data described in section 1.5. The ones represent men that have had a heart attack during their lifetime, and the zeros represent those that did not, plotted against their current `systolic blood pressure`. Since Y is binary, $E(Y \mid \texttt{systolic blood pressure})$ represents the proportion of men in the data with a specific `systolic blood pressure` reading who had `coronary heart disease`. Simple smoothing of the 0-1 data provides an estimate of this proportion and is a useful summary of this otherwise uninformative plot. The astute reader will notice rather high proportions of `coronary heart disease` in the plot. These data are in fact a retrospective case-control sample with roughly two controls sampled for each case. In practice this means that one can only estimate the odds-ratio rather than the actual prevalence (Breslow and Day, 1980), and so the prevalence in Fig. 4.7 is exaggerated.

Since the mean is changing with `systolic blood pressure` and therefore the binomial variance is also changing, it would be more efficient to incorporate this variance information when

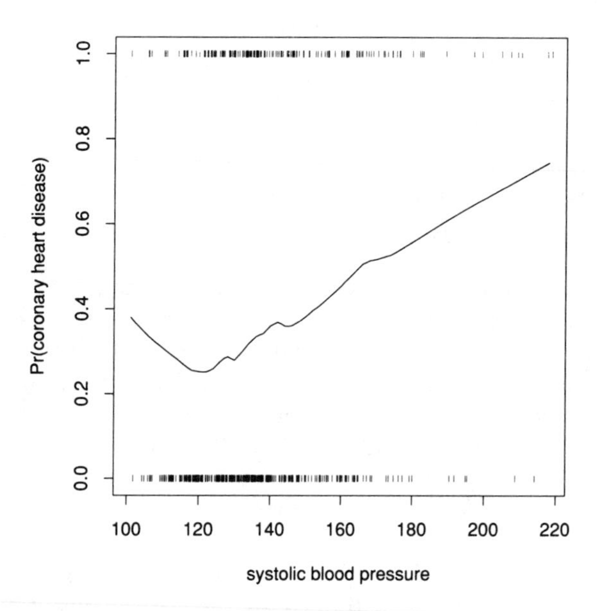

Fig. 4.7. *A plot of the 0-1* `coronary heart disease` *response variable versus* `systolic blood pressure`. *The vertical strokes at zero and one indicate the occurrence of zeros and ones in the response. The scatterplot smooth gives a useful summary of this otherwise uninformative plot.*

smoothing. For example one could compute estimated weights inversely proportional to the binomial variance, and resmooth using a weighted smoother. In this case, and typically in general, this makes little difference, since locally the weights are about constant.

What if there are many predictors, as there are for these data? Do we simply model the proportion additively? In linear modelling of binary data, the most popular approach is *logistic regression* which models the *logit* of the response probability with a linear form

$$\operatorname{logit}\{P(\boldsymbol{X})\} \equiv \log\left\{\frac{P(\boldsymbol{X})}{1-P(\boldsymbol{X})}\right\} = \boldsymbol{X}\boldsymbol{\beta}$$

where $P(\boldsymbol{X}) = \operatorname{pr}(Y = 1 \mid \boldsymbol{X})$. There are several reasons for its popularity, but the most compelling is that the logit model ensures that the proportions $P(\boldsymbol{X})$ lie in $(0, 1)$, without any constraints on the *linear predictor* $\boldsymbol{X}\boldsymbol{\beta}$. This is clear from the inverse relationship $P(\boldsymbol{X}) = \exp(\boldsymbol{X}\boldsymbol{\beta})/\{1 + \exp(\boldsymbol{X}\boldsymbol{\beta})\}$. Our generalization replaces this linear predictor with an additive one:

$$\log\left\{\frac{P(\boldsymbol{X})}{1-P(\boldsymbol{X})}\right\} = \alpha + \sum_{j=1}^{p} f_j(X_j). \tag{4.5}$$

Even for one variable, it makes sense to use this transformation. Scatterplot smoothers, such as locally-linear smoothers, are quite capable of producing fitted values outside $[0,1]$.

We also gain insight from the linear logistic regression methodology on how to estimate the additive version (4.5). Maximum likelihood is the most popular method for estimating the linear logistic model. For the present problem the log-likelihood has the form

$$l(\mathbf{y},\boldsymbol{\beta}) = \sum_{i=1}^{n}\{y_i \log p_i + (1-y_i)\log(1-p_i)\} \tag{4.6}$$

where $p_i = P(\mathbf{x}^i)$. The score equations

$$\frac{\partial l}{\partial \boldsymbol{\beta}} = \sum_{i=1}^{n} \mathbf{x}^i (y_i - p_i) = 0 \tag{4.7}$$

are nonlinear in the parameters $\boldsymbol{\beta}$ and consequently one has to find the solution iteratively.

The Newton-Raphson iterative method can be expressed in an appealing form. Suppose the current estimates are $\hat{\boldsymbol{\beta}}$, yielding estimated probabilities p_i. We form the linearized response $z_i = \hat{\boldsymbol{\beta}}^T\mathbf{x}^i + (y_i - p_i)/\{p_i(1-p_i)\}$ and weights $w_i = p_i(1-p_i)$. A new $\hat{\boldsymbol{\beta}}$ is obtained by weighted linear regression of z_i onto the $\mathbf{x}^i$ with weights w_i. This is repeated till $\hat{\boldsymbol{\beta}}$ converges. A motivation for this procedure is as follows. Suppose that instead of observing y_i, we observed a proportion r_i based on a number of counts. Then we could simply apply the logit transform to the r_i, provided $r_i \in (0,1)$, and use linear regression (perhaps with weights). However, in the extreme case of only one count per observation, or when the observed proportions are one or zero, the logit transform is not defined. The quantity z_i represents the first-order Taylor's series approximation to $\text{logit}(y_i)$ about the current estimate p_i. If $\hat{\boldsymbol{\beta}}$ and hence p_i are fixed, the variance of z_i is $1/\{p_i(1-p_i)\}$, and hence the weights w_i.

This iterative regression algorithm lends itself ideally to the additive model generalization. We define $z_i = \alpha + \sum_{j=1}^{p} f_j(x_{ij}) + (y_i - p_i)/\{p_i(1-p_i)\}$, where the f_j are the current estimates for the additive model components and $p_i = \exp\{\alpha + \sum f_j(x_{ij})\}/[1 + \alpha + \exp\{\sum f_j(x_{ij})\}]$. Weights w_i are defined as before. New functions α and f_j are computed by fitting a weighted additive model to z_i. Of course, this additive model fitting procedure is iterative as well, so there is potentially a lot of computing to be done. Fortunately, the functions from the previous nonlinear step are good starting values for the next step. This procedure is called the *local-scoring* algorithm. The new functions from each local-scoring step are monitored, and the iterations are stopped when their relative change is negligible. More details are provided in Chapter 6.

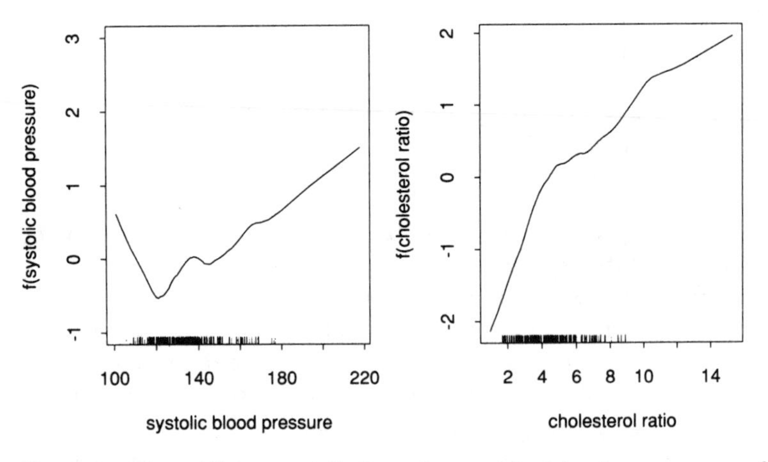

Fig. 4.8. *The additive contribution of* `systolic blood pressure` *and* `cholesterol ratio` *to the logit of the fitted probability. Each curve is plotted on the same scale for ease of comparison. The spikes at the base of each plot represent the frequency of observations; tied values are randomly jittered to avoid overstriking.*

Figure 4.8 shows the joint additive fit of `coronary heart disease` to `systolic blood pressure` and a third variable, the ratio of total cholesterol to HDL cholesterol, which we call `cholesterol ratio` (HDL is the *good* cholesterol, so small values of `cholesterol ratio` are healthy). The curves are estimated using the local-scoring algorithm just described, using a locally-weighted running-line smoother for each variable. The `systolic blood pressure` curve has not changed very much, and this suggests that the two variables contribute independently to the additive model. The investigators in this project were relieved to see the function for `systolic blood pressure`, since this predictor, a known risk factor for heart disease, had appeared to be nonsignificant in a linear analysis. An explanation for the U-shape is given in section 9.5.2.

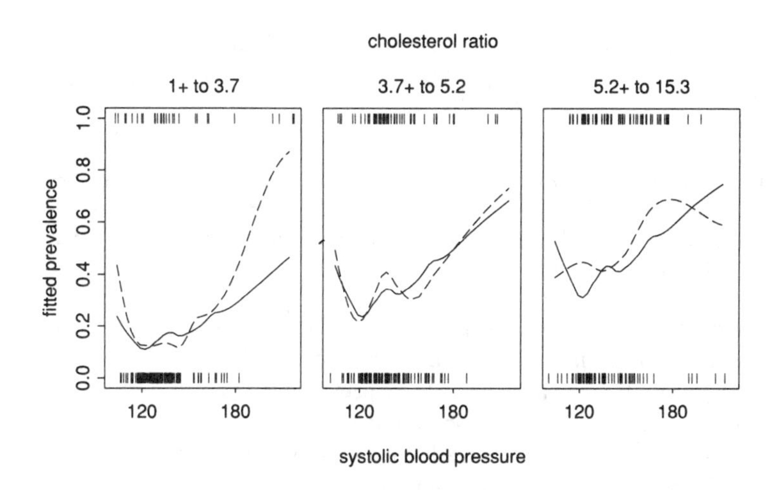

Fig. 4.9. *The conditioned plot is a convenient way of viewing the fitted prevalence surface as a function of both* `systolic blood pressure` *and* `cholesterol ratio`. *The range of* `cholesterol ratio` *is divided into three groups at the tertiles. For each group the surface is plotted as a function of* `systolic blood pressure`, *conditioned on* `cholesterol ratio` *at its median value for the group. The solid curves represent the additive surface on the prevalence scale, computed using the local-scoring algorithm with locally-weighted linear smoothers. The broken curves represent the fitted surface computed using a bivariate locally-weighted linear smoother. The 0-1 data within each group are represented by the vertical bars.*

Notice that we are plotting the functions that contribute to the additive model on the logit scale. For one predictor we can plot either on this scale or on the probability scale, whereas for two variables the separate functions really only make sense on the logit

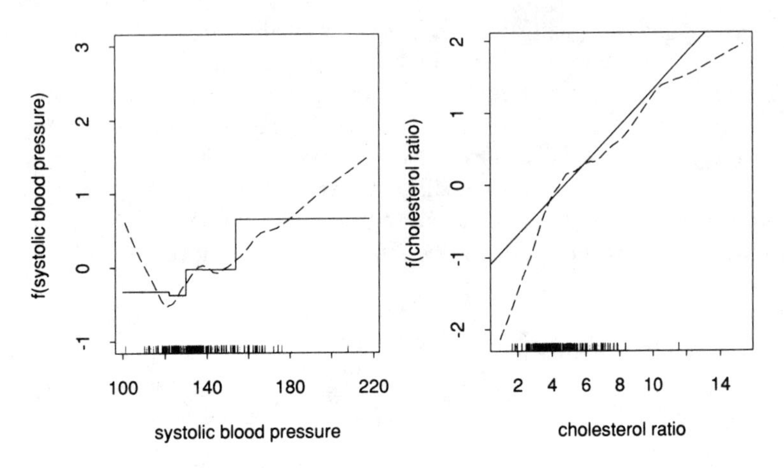

Fig. 4.10. *On the left is a step function fit for* `systolic blood pressure` *with the smooth term previously fit superimposed. The quintiles define the binning intervals. On the right is a linear fit for* `cholesterol ratio` *with the nonparametric fit superimposed.*

scale. It is also helpful to plot the functions on a common scale in order to judge their relative importance visually.

Figure 4.9 compares the fitted prevalence surface derived from the additive model to a prevalence surface computed using the local-scoring algorithm and a locally-weighted linear surface smoother. They seem to have the same general shape for lower values of `cholesterol ratio`, but differ in the higher region.

Figure 4.10 shows an alternative and perhaps more traditional way of finding nonlinearities. We have allowed a step function with 5 steps for the `systolic blood pressure` term, and left `cholesterol ratio` as linear. This can also be written as an additive model $\eta(\boldsymbol{X}) = f_1(X_1) + f_2(X_2)$ with

$$\begin{aligned} f_1(X_1) &= \sum_{k=1}^{5} c_k I(X_1 \in Q_k), \\ f_2(X_2) &= \beta X_2, \end{aligned} \tag{4.8}$$

where the Q_k represent five nonoverlapping intervals defined by the quintiles of X_1. This example gives evidence that the step function is crude, and depending where the cuts are, it can average over areas of nonlinearity. These computations could also be carried out using a backfitting algorithm within the local-scoring procedure. A weighted bin smoother is used for the step function, and simple least-squares for the line. In this case a numerically more efficient approach is actually used, which fits the step function and line jointly in one least-squares step.

One can also test hypotheses about relevant features. For example, we might want to test if the additive model is an adequate approximation to the regression surface. The additive model has a deviance (that is, twice log relative likelihood) of 546.7, with approximate number of parameters equal to 1 (constant) + 3.8 (`systolic blood pressure` curve) + 3.9 (`cholesterol ratio` curve) = 8.7. For the surface the deviance is 527.1 with 14.3 parameters, which is a significant drop of 19.6 for 5.6 degrees of freedom, so it does appear that some interactions are present. Model (4.8) has a deviance of 554.1 with 6 degrees of freedom for fit, an increase of 7.4 with 2.7 degrees of freedom over the nonparametric additive fit.

Iterative reweighted regression is the algorithm used in the GLIM package, an interactive computer program for fitting *generalized linear models.* These are exponential-family regression models which include the logit and probit regression for the binomial, log-linear models for the Poisson, models with gamma errors having constant coefficient of variation, and extensions and variations of these. They also include quasi-likelihood models that don't specify an error distribution but only a mean-variance relationship. The local-scoring algorithm can be applied to the additive analogue of these models, the class of *generalized additive models.* This is the subject of Chapter 6.

Functions are available in the S language for fitting the same class of GLM models, as well as the generalized additive GAM models. These functions use an enhanced version of the formula language used in GLIM, and also have the support of a powerful computing and graphics environment.

Other models, such as the proportional hazards model for survival data and the proportional odds model for ordinal data, do not fit directly into this class and require special treatment. We discuss these in Chapter 8.

4.6 Bibliographic notes

Friedman and Stuetzle (1981) proposed the projection-pursuit regression model, and the backfitting algorithm for its estimation; the same idea was used in the ACE algorithm of Breiman and Friedman (1985). Green, Jennison and Seheult (1985), Denby (1986), and Engle, Granger, Rice and Weiss (1986) studied models with one nonparametric term. These are sometimes called semi-parametric models.

The iteratively-reweighted least-squares approach to exponential family models was developed by Nelder and Wedderburn (1972), and more recently McCullagh and Nelder (1989). GLIM (Baker and Nelder, 1978, copyright, Royal Statistical Society) stands for *Generalized Linear Interactive Modelling*, and is a computer package for fitting generalized linear models. S (Becker, Chambers and Wilks, 1988) is an interactive statistical computing environment; all the graphics in this book were created using S. A thorough account of regression-tree methodology can be found in Breiman, Friedman, Olshen and Stone (1984).

Hastie and Tibshirani (1985, 1986) adapted the additive model technology to generalized linear models, calling them *generalized additive models*. They have written a software package for fitting generalized additive models called GAIM for *Generalized Additive Interactive Modelling*. A more detailed analysis of the coronary heart disease data appeared in Hastie and Tibshirani (1987a and b). Green and Yandell (1985), O'Sullivan, Yandell and Raynor (1986), and Green (1987) discussed generalized semi-parametric models within the framework of penalized likelihood. Landwehr, Pregibon and Shoemaker (1984) discussed smoothing and displaying partial-residual plots in order to expose nonlinearities in logistic-regression models. One can view the local-scoring algorithm as iteratively smoothing partial-residual plots. Mallows (1986) describes augmented partial-residual plots; see also Wong (1987).

The GAIM program and software for the S system are described in Appendix C. Nelder (1989) has written GLIM macros for fitting generalized additive models. Generalized additive models are also featured in the XploRe system, a computing environment for exploratory regression. This system is designed for use on personal computers and is distributed by W. Härdle.

4.7 Further results and exercises 4

4.1 Denote the smoother matrix for smoothing against the sorted variable $\mathbf{x}_j$ by $\mathbf{S}_j$.

(i) What does $\mathbf{S}_j$ look like if $\mathbf{x}_j$ is unsorted? Denote this smoother that operates on unsorted variables by $\mathbf{S}_j^*$.
(ii) At convergence, the backfitting algorithm solves a system of equations. Describe them, using the notation in (i).

4.2 Derive the iteratively-weighted linear least-squares algorithm for maximizing the log-likelihood for Bernoulli data.

4.3 Suppose we wish to augment the log-likelihood above with a quadratic penalty, resulting in the penalized log-likelihood $l(\mathbf{y}, \boldsymbol{\beta}) - \lambda\boldsymbol{\beta}^T\mathbf{K}\boldsymbol{\beta}/2$. This might arise naturally in a Bayesian model with a Gaussian prior for $\boldsymbol{\beta}$. Derive an appropriate iterative algorithm for maximizing this penalized log-likelihood.

4.4 Derive a simple proof that the asymptotic covariance matrix of the regression parameters is given by $(\mathbf{X}^T\mathbf{W}\mathbf{X})^{-1}$ where $\mathbf{X}$ is the $n \times p$ design matrix and $\mathbf{W}$ a diagonal matrix of weights (hint: use the final regression equation from IRLS).

4.5 Generate 50 binary observations according to the model

$$\log[\mathrm{pr}(Y = 1 \mid X = x)/\{1 - \mathrm{pr}(Y = 1 \mid X = x)\}] = x + x^2,$$

where X is uniformly distributed on $[-1, 1]$. Smooth the data with a simple unweighted smoother, and also fit the model $\mathrm{logit}\,\mathrm{pr}(Y = 1 \mid X = x) = f(x)$ as described in this chapter. Compare the results by plotting both the estimated logits and probabilities.

4.6 Derive approximate formulae for the standard errors of the estimates in the previous exercise and add these to the plots above.

4.7 Consider an additive model in the variables `age` and `base deficit` for the diabetes data.

(i) Use backfitting to fit this model with a cubic spline smoother for each predictor.
(ii) Using the same smoothing parameters as in (i), compute the smoothing matrices $\mathbf{S}_1$ and $\mathbf{S}_2$ for the two predictors.

(iii) Solve the backfitting equations $\mathbf{f}_1 = \mathbf{S}_1(\mathbf{y} - \mathbf{f}_2)$, $\mathbf{f}_2 = \mathbf{S}_2(\mathbf{y} - \mathbf{f}_1)$ by matrix methods, and compare the solution to that obtained by backfitting in (i).

4.8 Suppose the two predictors in an additive model are exactly balanced, in that each of the k_1 values for X_1 occur with each of the k_2 values of X_2. Discuss the implications in terms of the backfitting algorithm and the equivalent kernel for the additive fit.

CHAPTER 5

Some theory for additive models

5.1 Introduction

In the previous chapter we introduce the nonparametric additive model and the backfitting procedure, a heuristic method for estimation. In this chapter we provide some theoretical underpinning for these ideas. The technical content and level of this chapter is somewhat higher than the others, so the reader interested only in applications may well decide not to tackle it.

The chapter has two main parts. In the first half the backfitting algorithm is justified as a method for estimating the additive model. Several justifications are provided. We first introduce an L_2 version of backfitting, that is, a backfitting algorithm for square integrable random variables, and show how the intuitive procedure introduced in the last chapter can be viewed as a data analogue of this. A second, different, justification for backfitting comes from a penalized least-squares framework. Unlike the L_2 argument, it makes no appeal to random variables: instead, it applies to finite-sample backfitting procedures that use linear smoothers (recall the discussion of linear smoothers in sections 2.8 and 3.4.2).

An important by-product of these justifications is the set of *estimating equations* that are solved by backfitting. This linear system can be solved noniteratively and in some special cases such a direct solution is more appropriate than backfitting.

We also describe very briefly a more technical abstract derivation using the theory of reproducing-kernel Hilbert-spaces.

In the second half of the chapter we study the existence and uniqueness of the solutions to the additive-model estimating equations and the convergence of the backfitting algorithm. The results pertain to linear smoothers only. One of the interesting notions that arises is *concurvity*, the analogue of collinearity. We also discuss some theory for standard-error bands and degrees of freedom of the estimated smooths, and the relationship of backfitting to the Gram-Schmidt and Gauss-Seidel techniques.

Related theory for generalized additive models is not covered here but is discussed in the next chapter. We don't devote much space to asymptotic issues such as consistency or rates of convergence for additive models; these are briefly mentioned in the bibliographic notes.

5.2 Estimating equations for additive models

The additive model

$$E(Y \mid \boldsymbol{X}) = \sum_{j=1}^{p} f_j(X_j) \tag{5.1}$$

can be estimated by the backfitting algorithm, which we give again below:

Algorithm 5.1 *The backfitting algorithm*

(i) *Initialize:* $f_j = f_j^{(0)}, j = 1, \ldots, p$
(ii) *Cycle:* $j = 1, \ldots, p, 1, \ldots p, \ldots$

$$f_j = \mathcal{S}_j(\mathbf{y} - \sum_{k \neq j} \mathbf{f}_k \mid \mathbf{x}_j)$$

(iii) Continue (ii) until the individual functions don't change.

In the above, the $\mathbf{f}_j$ are the n-vectors $\{f_j(x_{1j}), \ldots, f_j(x_{nj})\}^T$, with x_{ij} in the order of y_i. We have omitted the constant term α in (5.1); we see later that this does not change the resulting estimates.

In order to justify this procedure, we need some way of introducing the smoothness that is provided by the scatterplot smoothers in the algorithm. To put it another way, if we naively tried to minimize

$$\sum_{i=1}^{n} \Big\{ y_i - \sum_{j=1}^{p} f_j(x_{ij}) \Big\}^2 \tag{5.2}$$

then the solution would be any set of functions $(f_j : j = 1, \ldots, p)$ that interpolated the data (assuming for the moment that the X-values are distinct). For example, $f_1(x_{i1}) = y_i \; \forall i$ and $f_j \equiv 0$ for $j > 1$. We discuss several ways of introducing smoothness. One approach is to explicitly add a term to (5.2) that penalizes for lack of smoothness, the *penalized least-squares* approach. Another approach, described next, is to step back and consider random variables instead of data. A Hilbert-space version of the additive model and backfitting can be formulated, with conditional expectation operators playing the role of smoothers. Besides its use here, this formulation is mathematically interesting in its own right. The data version of backfitting is then derived as an empirical version of this Hilbert-space procedure, the smoothness entering when one considers how best to estimate the conditional expectations. A third approach, based on reproducing-kernel Hilbert-spaces, is a more abstract version of the penalized least-squares approach. Later on in the chapter we give yet another derivation, based on a Bayesian stochastic model.

5.2.1 L_2 *function spaces*

Let $\mathcal{H}_j$ for $j = 1, \ldots, p$ denote the Hilbert spaces of measurable functions $\phi_j(X_j)$ with $E\phi_j(X_j) = 0$, $E\phi_j^2(X_j) < \infty$, and inner product $\langle \phi_j(X_j), \phi_j'(X_j) \rangle = E\phi_j(X_j)\phi_j'(X_j)$. In addition, denote by $\mathcal{H}$ the space of arbitrary centered, square integrable functions of $X_1, \ldots, X_p$. We consider the $\mathcal{H}_j$ as subspaces of $\mathcal{H}$ in a canonical way. Furthermore, denote by $\mathcal{H}^{add} \subset \mathcal{H}$ the linear subspace of additive functions: $\mathcal{H}^{add} = \mathcal{H}_1 + \cdots + \mathcal{H}_p$, which is closed under some technical assumptions. These are all subspaces of $\mathcal{H}_{YX}$, the space of centered square integrable functions of Y and $X_1, \ldots, X_p$.

The optimization problem in this population setting is to minimize

$$E\{Y - g(\boldsymbol{X})\}^2 \tag{5.3}$$

over $g(\boldsymbol{X}) = \sum_j f_j(X_j) \in \mathcal{H}^{add}$. Of course, without the additivity restriction, the solution is simply $E(Y \mid \boldsymbol{X})$; we seek the closest additive approximation to this function. Since by assumption $\mathcal{H}^{add}$ is a closed subspace of $\mathcal{H}$ this minimum exists and is unique; the individual functions $f_j(X_j)$, however, may not be uniquely determined. Denote by P_j the conditional expectation operator $E(\cdot \mid X_j)$; as such P_j is an orthogonal projection onto H_j.

The minimizer $g(\boldsymbol{X})$ of (5.3) can be characterized by residuals $Y - g(\boldsymbol{X})$ which are orthogonal to the space of fits: $Y - g(\boldsymbol{X}) \perp \mathcal{H}^{add}$. Since $\mathcal{H}^{add}$ is generated by $\mathcal{H}_j$ ($\subset \mathcal{H}^{add}$), we have equivalently: $Y - g(\boldsymbol{X}) \perp \mathcal{H}_j, \quad \forall j$ or: $P_j\{Y - g(\boldsymbol{X})\} = 0 \quad \forall j$. Component-wise this can be written as

$$\begin{aligned} f_j(X_j) &= P_j\Big\{Y - \sum_{k \neq j} f_k(X_k)\Big\} \\ &= E\Big\{Y - \sum_{k \neq j} f_k(X_k) \,|\, X_j\Big\}. \end{aligned} \tag{5.4}$$

Equivalently, the following system of *estimating equations* is necessary and sufficient for $\mathbf{f} = (f_1, \ldots, f_p)$ to minimize (5.3):

$$\begin{pmatrix} I & P_1 & P_1 & \cdots & P_1 \\ P_2 & I & P_2 & \cdots & P_2 \\ \vdots & \vdots & \vdots & \ddots & \vdots \\ P_p & P_p & P_p & \cdots & I \end{pmatrix} \begin{pmatrix} f_1(X_1) \\ f_2(X_2) \\ \vdots \\ f_p(X_p) \end{pmatrix} = \begin{pmatrix} P_1 Y \\ P_2 Y \\ \vdots \\ P_p Y \end{pmatrix} \tag{5.5}$$

or

$$\mathbf{Pf} = \mathbf{Q}Y,$$

where $\mathbf{P}$ and $\mathbf{Q}$ represent a matrix and vector of operators, respectively, and operator matrix multiplication is defined in the obvious way.

The reader familiar with the Gauss-Seidel method for solving linear systems of equations will recognize backfitting as a formal Gauss-Seidel algorithm for solving the system (5.5). (We say *formal* because the elements in the left matrix of (5.5) are not real numbers or real-valued matrices but conditional expectation operators.) Suppose we wish to solve a linear system of equations $\mathbf{Az} = \mathbf{b}$, with $\mathbf{A} = \{a_{ij}\}$ an $m \times m$ matrix, $\mathbf{b}$ an m-vector and $\mathbf{z}$ the vector of m unknown coefficients. The Gauss-Seidel iterative method solves for each z_i in turn from the relation in the ith row $\sum_{j=1}^{m} a_{ij} z_j = b_i$. This process is repeated for $i = 1, \ldots, m, 1, \ldots, m, \ldots$, using the latest values of each z_j at each step, until convergence.

The connection between backfitting and the Gauss-Seidel method becomes more precise when we consider the corresponding data version of (5.5) using linear smoothers. Recall from section 2.8 that a linear smoother can be written as a *smoother matrix* times the response vector $\mathbf{y}$, that is $\hat{\mathbf{f}} = \mathbf{S}\mathbf{y}$. Examples of linear smoothers include the running-mean, locally-weighted running-line, smoothing splines and kernel smoothers. Consider then a backfitting algorithm that estimates the conditional expectation operator P_j by a linear scatterplot smoother with smoother matrix $\mathbf{S}_j$. Then the data version of the estimating equations (5.5) is the $np \times np$ system

$$\begin{pmatrix} \mathbf{I} & \mathbf{S}_1 & \mathbf{S}_1 & \cdots & \mathbf{S}_1 \\ \mathbf{S}_2 & \mathbf{I} & \mathbf{S}_2 & \cdots & \mathbf{S}_2 \\ \vdots & \vdots & \vdots & \ddots & \vdots \\ \mathbf{S}_p & \mathbf{S}_p & \mathbf{S}_p & \cdots & \mathbf{I} \end{pmatrix} \begin{pmatrix} \mathbf{f}_1 \\ \mathbf{f}_2 \\ \vdots \\ \mathbf{f}_p \end{pmatrix} = \begin{pmatrix} \mathbf{S}_1\mathbf{y} \\ \mathbf{S}_2\mathbf{y} \\ \vdots \\ \mathbf{S}_p\mathbf{y} \end{pmatrix}. \tag{5.6}$$

In short form we write

$$\hat{\mathbf{P}}\mathbf{f} = \hat{\mathbf{Q}}\mathbf{y}.$$

Backfitting is a Gauss-Seidel procedure for solving the above system. The only nonstandard aspect is that we solve for n elements at each step instead of one (block Gauss-Seidel), although since $\hat{\mathbf{P}}$ is block-diagonal this distinction can be dropped.

Now suppose we start with (5.6). Why use an iterative procedure like backfitting to find its solution? Why not use a standard, noniterative method like a QR decomposition? The difficulty is that in general, (5.6) is an $np \times np$ system and since methods like QR require $O(m^3)$ operations to solve an $m \times m$ system, our problem would cost $O\{(np)^3\}$ operations. On the other hand, backfitting exploits the special structure in (5.6), and if the smoothers can be applied in $O(n)$ computations (as is the case for running-lines and smoothing splines), then backfitting requires only $O(np)$ computations. (This assumes that a fixed number of iterations is sufficient for convergence). If, however, the effective dimension of the system (5.6) is really less than np, there may be better methods than backfitting for solving the problem. In particular, if each $\mathbf{S}_j$ is an orthogonal projection and the union of the projection spaces has rank m, then (5.6) is equivalent to an $m \times m$ least-squares problem (Exercise 5.3) and least-squares methods are likely to be preferable if $m << n$. This is the case if the $\mathbf{S}_j$s produce linear or polynomial fits, or if we use regression splines with a small number of knots.

Later in this chapter the properties of the estimating equations (5.6) are studied. We find that there is an intimate connection between the existence of solutions of this system and the convergence of the backfitting procedure for finding these solutions.

5.2.2 *Penalized least-squares*

In this section we provide a different justification for backfitting from that given in the Hilbert-space framework in the previous section. In section 2.10 we derive the cubic smoothing spline as the minimizer of the penalized least-squares criterion

$$\sum_{i=1}^{n}\{y_i - f(x_i)\}^2 + \lambda \int \{f''(x)\}^2 \, dx \tag{5.7}$$

over all twice continuously differentiable functions f. We establish this by using the fact that the solution to (5.7) is a cubic spline, and hence we simplified (5.7) by writing it as a function of $f_i = f(x_i)$, the n evaluations of the minimizing function f. This gave the equivalent form

$$(\mathbf{y}-\mathbf{f})^T(\mathbf{y}-\mathbf{f}) + \lambda \mathbf{f}^T\mathbf{K}\mathbf{f} \tag{5.8}$$

where $\mathbf{K}$ is a certain quadratic penalty matrix. The quantity (5.8) is easily shown to have a minimum given by

$$\hat{\mathbf{f}} = (\mathbf{I} + \lambda\mathbf{K})^{-1}\mathbf{y}. \tag{5.9}$$

We also argued in the opposite direction for other symmetric linear smoothers. That is, given a symmetric linear smoother based on the smoother matrix $\mathbf{S}$, the smooth $\hat{\mathbf{f}} = \mathbf{S}\mathbf{y}$ minimizes

$$(\mathbf{y}-\mathbf{f})^T(\mathbf{y}-\mathbf{f}) + \mathbf{f}^T(\mathbf{S}^- - \mathbf{I})\mathbf{f} \tag{5.10}$$

over all $\mathbf{f} \in \mathcal{R}(\mathbf{S})$ (the range of $\mathbf{S}$), where $\mathbf{S}^-$ is any generalized inverse of $\mathbf{S}$.

In order to extend this idea to the estimation of the additive model, we generalize the criterion (5.7) in an obvious way. We seek to minimize

$$\sum_{i=1}^{n}\Big\{y_i - \sum_{j=1}^{p} f_j(x_{ij})\Big\}^2 + \sum_{j=1}^{p}\lambda_j \int \{f_j''(t)\}^2 \, dt \tag{5.11}$$

over all twice continuously differentiable functions f_j. Before deriving the solution to (5.11), let's take note of some of its features. Notice that each function in (5.11) is penalized by a separate constant λ_j. This in turn determines the smoothness of that function in the solution. Note also that if the λ_js are all zero (no smoothness penalty) the solution to (5.11) is any interpolating set of functions whose evaluations satisfy $\sum_{j=1}^{p} f_j(x_{ij}) = y_i$ for $i = 1, \ldots, n$. On the other hand, if each λ_j goes to infinity, the penalty term goes to infinity unless $f_j''(t) = 0$ for all j, that is, unless each f_j is linear. Hence the problem reduces to standard linear least-squares.

Using a straightforward extension of the arguments used in the single-predictor case, the solution to (5.11) is shown to be a cubic spline in each of the predictors. As before we parametrize by evaluations at the n observations. Thus we may rewrite (5.11) as

$$\left(\mathbf{y} - \sum_{j=1}^{p} \mathbf{f}_j\right)^T \left(\mathbf{y} - \sum_{j=1}^{p} \mathbf{f}_j\right) + \sum_{j=1}^{p} \lambda_j \mathbf{f}_j^T \mathbf{K}_j \mathbf{f}_j \qquad (5.12)$$

where the $\mathbf{K}_j$s are penalty matrices for each predictor, defined analogously to the $\mathbf{K}$ for a single predictor given in section 2.10. Now if we differentiate (5.12) with respect to the function $\mathbf{f}_k$ we obtain $-2(\mathbf{y} - \sum_j \mathbf{f}_j) + 2\lambda_k \mathbf{K}_k \mathbf{f}_k = \mathbf{0}$ or

$$\hat{\mathbf{f}}_k = \left(\mathbf{I} + \lambda_k \mathbf{K}_k\right)^{-1} \left(\mathbf{y} - \sum_{j \neq k} \hat{\mathbf{f}}_j\right). \qquad (5.13)$$

As noted earlier, $(\mathbf{I} + \lambda_k \mathbf{K}_k)^{-1}$ is the smoother matrix for a cubic smoothing spline, and hence (5.13), for $k = 1, \ldots, p$, are just the estimating equations (5.6).

Arguing in the opposite direction, the minimizers of the penalized least-squares criterion

$$\left(\mathbf{y} - \sum_{j=1}^{p} \mathbf{f}_j\right)^T \left(\mathbf{y} - \sum_{j=1}^{p} \mathbf{f}_j\right) + \sum_{j=1}^{p} \mathbf{f}_j^T (\mathbf{S}_j^- - \mathbf{I}) \mathbf{f}_j, \qquad (5.14)$$

over all $\mathbf{f}_j \in \mathcal{R}(\mathbf{S}_j)$, are the solutions to the estimating equations (5.6) (Exercise 5.1).

As we do in the single predictor case (Exercise 3.6), we can interpret each of the penalty terms in (5.12) as a down-weighting of each of the components of $\mathbf{f}_j$, the down-weighting determined by the corresponding eigenvalue of that component and λ_j.

5.2.3 *Reproducing-kernel Hilbert-spaces*

This section describes a more abstract framework for defining and estimating general nonparametric regression models which includes additive models as a special case. We present these results to give the reader a taste of this rich area; the level of mathematics is somewhat higher than the rest of the chapter. The description is close to that of Chen, Gu and Wahba (1989).

A Hilbert space $\mathcal{H}$ of real-valued functions of $t \in \Omega$ is a *reproducing-kernel* Hilbert-space if evaluation is a continuous linear functional. By the Riesz representation theorem, there exist *representers of evaluation* $e_t \in \mathcal{H}$ such that $f(t) = \langle f, e_t \rangle_{\mathcal{H}}$ for $f \in \mathcal{H}$, where $\langle \cdot, \cdot \rangle_{\mathcal{H}}$ denotes the inner-product on $\mathcal{H}$. The consequences of these properties will become clearer as we proceed.

The reproducing kernel itself, $Q(\cdot, \cdot) : \Omega \times \Omega \mapsto \mathbb{R}^1$, is defined by $Q(s,t) = \langle e_s, e_t \rangle_{\mathcal{H}}$, and consequently $e_s = Q(s, \cdot)$, considered as a function of the second argument, with the first held fixed at $s \in \Omega$. We will see that the kernel Q, evaluated at the realizations of t, provides a finite dimensional basis for representing the solution to a class of optimization problems.

Now suppose Ω is a space of vector predictors $\boldsymbol{X} = (X_1, \ldots, X_p)$ and that $\mathcal{H}$ has the decomposition

$$\mathcal{H} = \mathcal{H}_0 + \sum_{k=1}^{q} \mathcal{H}_k,$$

where $\mathcal{H}_0$ is spanned by $\phi_1, \ldots, \phi_M$, and $\mathcal{H}_k$ has the reproducing kernel $Q_k(\cdot, \cdot)$. The space $\mathcal{H}_0$ is the projection component of $\mathcal{H}$, that is, the space of functions that are not to be penalized in the optimization. In the previous section $\mathcal{H}_0$ is the space of functions linear in t.

We are now set up to pose the optimization problem. For a given set of predictors $\mathbf{x}^1, \ldots, \mathbf{x}^n$ (with each $\mathbf{x}^i \in \Omega$), find $f = \sum_{k=0}^{q} f_k$ with $f_k \in \mathcal{H}_k$, $k = 0, \ldots, q$, to minimize

$$\sum_{i=1}^{n} \Big\{ y_i - \sum_{k=0}^{q} f_k(\mathbf{x}^i) \Big\}^2 + \sum_{k=1}^{q} \lambda_k \, \|f_k\|_{\mathcal{H}_k}^2 . \tag{5.15}$$

The first part of the criterion is discrete in nature, and is the reason why reproducing-kernel spaces are natural for these kinds of problems. We do not want small changes in the $\mathbf{x}^i$ to result in vastly different solutions; this is why it is desirable for evaluation to be continuous.

The theory of reproducing kernels guarantees that a minimizer exists, and has the form

$$\begin{aligned} \hat{f}_0(\boldsymbol{X}) &= \sum_{j=1}^{M} \beta_{j0}\phi_j(\boldsymbol{X}) \\ \hat{f}_k(\boldsymbol{X}) &= \sum_{i=1}^{n} \beta_{ik} Q_k(\boldsymbol{X}, \mathbf{x}^i). \end{aligned} \tag{5.16}$$

Furthermore, if the projection onto $\mathcal{H}_0$ is unique, then so is the solution to the larger problem. So even though the problem is posed in an infinite-dimensional space, the minimizing $\hat{f}$ is finite-dimensional, and the Q_k supply bases for representing the solution. The parameters are found by minimizing the finite dimensional quadratic criterion

$$\left\| \mathbf{y} - \mathbf{T}\boldsymbol{\beta}_0 - \sum_{k=1}^{q} \mathbf{Q}_k \boldsymbol{\beta}_k \right\|^2 + \sum_{k=1}^{q} \lambda_k \boldsymbol{\beta}_k^T \mathbf{Q}_k \boldsymbol{\beta}_k \tag{5.17}$$

where $\mathbf{T}$ is the $n \times M$ matrix of evaluations of ϕ_j, with ijth entry $T_{ij} = \phi_j(\mathbf{x}^i)$, and $\mathbf{Q}_k$ is the $n \times n$ evaluated kernel with ijth entry $Q_k(\mathbf{x}^i, \mathbf{x}^j)$. This problem is of dimension at most $qn + M$.

At this point a number of specializations are possible:

(i) If $q = p$ and each of the $\mathcal{H}_k$ are the canonical subspaces of $\mathcal{H}$, then the additive model consists of a sum of univariate functions. Furthermore, by choosing an inner product appropriate for cubic smoothing splines, the problem reduces exactly to (5.11), although the solution is typically represented by a different basis.

(ii) The current specification has $nq + M$ parameters. If attention is restricted to f_0 and $f_+ = \sum_{k=1}^{q} f_k$, with $Q_+ = \sum_{k=1}^{q} Q_k/\lambda_k$, then the dimension of the solution is reduced to $M + n$.

(iii) The general problem as specified by (5.16) can potentially be solved more cheaply by backfitting; the system of estimating equations that characterize the minimum of (5.16) can be written in a form similar to (5.6):

$$\begin{pmatrix} \mathbf{I} & \mathbf{S}_0 & \mathbf{S}_0 & \cdots & \mathbf{S}_0 \\ \mathbf{S}_1 & \mathbf{I} & \mathbf{S}_1 & \cdots & \mathbf{S}_1 \\ \vdots & \vdots & \vdots & \ddots & \vdots \\ \mathbf{S}_q & \mathbf{S}_q & \mathbf{S}_q & \cdots & \mathbf{I} \end{pmatrix} \begin{pmatrix} \mathbf{f}_0 \\ \mathbf{f}_1 \\ \vdots \\ \mathbf{f}_q \end{pmatrix} = \begin{pmatrix} \mathbf{S}_0\mathbf{y} \\ \mathbf{S}_1\mathbf{y} \\ \vdots \\ \mathbf{S}_q\mathbf{y} \end{pmatrix} \tag{5.18}$$

where $\mathbf{S}_0 = \mathbf{T}(\mathbf{T}^T\mathbf{T})^{-1}\mathbf{T}$ and $\mathbf{S}_k = \mathbf{Q}_k(\mathbf{Q}_k + \lambda_i\mathbf{I})^{-1}$. If the computational complexity of the individual operators $\mathbf{S}_k$ is significantly lower than that of the full problem, savings can be made using backfitting-type algorithms.

This is a very brief summary of some powerful machinery; we cite a number of relevant references in the bibliographic section for more details of this approach and for pointers to the large application area.

5.3 Solutions to the estimating equations

5.3.1 *Introduction*

The remainder of the chapter focuses attention on additive models with linear smoothers $\mathcal{S}_1, \ldots, \mathcal{S}_p$, and the algorithms for estimating them.

Most of the smoothers that we have discussed produce function estimates, and so we could discuss issues such as convergence in terms of these functions as well. Instead we restrict attention to the evaluation of these functions at the n realizations of the predictors. We do this mainly for simplicity and clarity, but point out that most of the results cited here, in particular those pertaining to convergence, can be modified to include this more general case. As a consequence, we usually refer to a smoother by its matrix representation $\mathbf{S}$ rather than in the operator form $\mathcal{S}$.

The centerpiece of the discussion is the set of estimating equations (5.6). Before one delves into methods for solving such a system, questions of consistency and degeneracy have to be answered. In other words, we must confirm that the system has at least one solution, and find out whether this solution is unique. We first look at a few special cases which are are easy to work out and illustrate the main issues. Later in the chapter we answer these questions in some generality.

In our discussion we sometimes assume implicitly that the same smoother is used for each of the variables, but this is only for ease of presentation. The results are general in nature and apply to any backfitting procedure in which some linear smoother is used for each of the variables. In fact, there is no need even to assume that each smoother is based on a single predictor: for example a two-dimensional smoother or a least-squares fit on some set of predictors could be included. Even more generally one can think of $\mathbf{S}_1, \ldots, \mathbf{S}_p$ as a set of linear transformations, without reference to predictor variables at all.

Throughout the chapter we assume that the $\mathbf{S}_j$s all reproduce constant functions. Note that this causes a simple kind of non-uniqueness in the backfitting algorithm. Suppose the starting functions are all zero. Then at every stage of the procedure $\hat{\mathbf{f}}_1$ has the same mean as $\mathbf{y}$, but the other $\hat{\mathbf{f}}_j$s have mean $\mathbf{0}$. If the procedure started at $j = 2$, however, the mean of $\mathbf{y}$ would go into $\hat{\mathbf{f}}_2$ instead. A closer look reveals that nonzero starting functions cause a dependence of the final iterates on the values of the starting functions. It is also clear that unless special constraints are built in, the constant in the additive model is not identifiable. This is a special instance of what we call *concurvity*, the analogue of *collinearity* in linear models.

It turns out that such degeneracies do not affect convergence in any important way. In this case a simple fix is possible: assume that $\mathbf{y}$ has been centered to have mean $\mathbf{0}$, and replace $\mathbf{S}_j$ by the matrix that smooths then subtracts off the average of the smooth. It is easy to see that the resultant smoother matrix is $(\mathbf{I} - \mathbf{1}\mathbf{1}^T/n)\mathbf{S}_j$, what we call a *centered* smoother. This ensures that at every stage of the procedure the $\hat{\mathbf{f}}_j$s have mean $\mathbf{0}$.

5.3.2 *Projection smoothers*

The additive model is introduced as a generalization of the linear regression model. What if we use a linear least-squares fit, or any other orthogonal projection, for each predictor? As mentioned in section 5.2.1, the set of estimating equations (5.6) is equivalent to the usual normal equations for linear regression (Exercise 5.3). Thus if $\mathbf{S}_j$ is an orthogonal projection and $\mathcal{L}_{col}(\mathbf{S}_j)$ denotes the subspace spanned by the columns of $\mathbf{S}_j$, we expect the backfitting solution $\hat{\mathbf{y}} = \sum_1^p \hat{\mathbf{f}}_j$ to converge to the projection of $\mathbf{y}$ onto

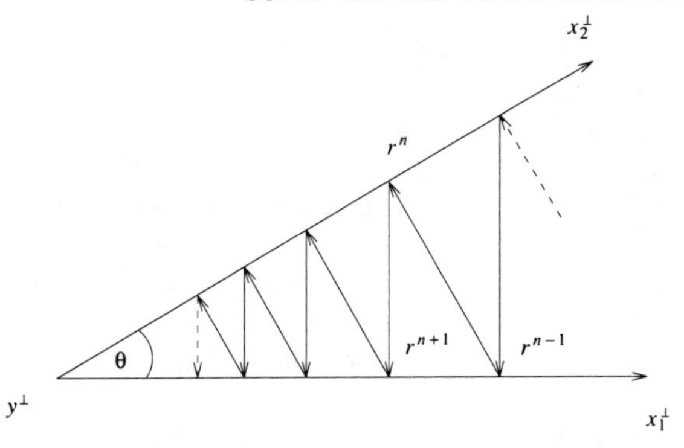

Fig. 5.1. *Backfitting to the least squares linear fit with two covariates. The residual* $\mathbf{r}^{(m)}$ *converges to the least-squares residual* $\mathbf{y}^\perp$ *in a zigzag fashion. In the figure, all vectors are projected onto the* $(\mathbf{x}_1, \mathbf{x}_2)$*-plane.*

$V = \mathcal{L}_{col}(\mathbf{S}_1) \oplus \mathcal{L}_{col}(\mathbf{S}_2) \oplus \ldots \oplus \mathcal{L}_{col}(\mathbf{S}_p)$. Indeed if this wasn't the case we might well question the entire backfitting paradigm. Fortunately, it is fairly easy to show that backfitting does the expected in this special case. We sketch the proof here, leaving the details to the reader (Exercise 5.2). The idea is to show that the residual vector from backfitting converges to the least-squares residual vector, that is, the projection of $\mathbf{y}$ onto the orthogonal complement of V. After one cycle of backfitting, the residual vector from backfitting, say $\mathbf{r}$, is $\mathbf{Cy}$ where

$$\mathbf{C} = (\mathbf{I} - \mathbf{S}_p)(\mathbf{I} - \mathbf{S}_{p-1}) \cdots (\mathbf{I} - \mathbf{S}_1). \tag{5.19}$$

Thus after m cycles the residual is $\mathbf{r}^{(m)} = \mathbf{C}^m\mathbf{y}$. We can split $\mathbf{y}$ into its components in the projection space V and its orthogonal complement, that is $\mathbf{y} = \hat{\mathbf{y}} + \mathbf{y}^\perp$. Now the operator $\mathbf{C}$ *takes residuals* along each predictor in turn, and hence leaves $\mathbf{y}^\perp$ unchanged. Thus

$$\mathbf{r}^{(m)} = \mathbf{C}^m\hat{\mathbf{y}} + \mathbf{y}^\perp. \tag{5.20}$$

The proof is then completed by showing that $\|\mathbf{C}^m\hat{\mathbf{y}}\| \to 0$ and hence $\mathbf{r}^{(m)} \to \mathbf{y}^\perp$ (Exercise 5.2).

Hence we see that the backfitting procedure provides an alternative method for computing least-squares fits. Examination of (5.20) reveals that it works by successively projecting the current residual into the space orthogonal to each $\mathcal{L}_{col}(\mathbf{S}_j)$.

Figure 5.1 gives a picture of this in the two-predictor case. Each $\mathbf{x}_j^{\perp}$ denotes the vector orthogonal to $\mathbf{x}_j$ in the span of $\mathcal{L}_{col}(\mathbf{x}_1, \mathbf{x}_2)$. It shows the backfitting residual $\mathbf{r}^{(m)}$ converging to the least-squares residual $\mathbf{y}^{\perp}$ in a zig-zag fashion. At convergence the backfitting residual vector is orthogonal to each $\mathcal{L}_{col}(\mathbf{S}_j)$ and equals the least-squares residual $\mathbf{y}^{\perp}$. This is a novel but not very practical way of finding the least-squares fit, for it can be very slow if the predictors are correlated. In particular, one can show that for two predictors the difference between the ith iterate and the solution converges to zero geometrically at rate $\cos(\theta)$, where θ is the angle between the two predictor vectors (Exercise 5.4). This is intuitively plausible from Fig. 5.1. There is a close connection between backfitting and the iterative proportional scaling algorithm for fitting log-linear models to contingency tables (Bishop *et al.*, 1975). They both cycle through a system of estimating equations and update one component at a time; convergence is geometric in both cases.

It is interesting to look at two extreme situations that can occur. First, suppose that the predictors are perfectly collinear ($\theta = 0$ in the two predictor case). Then one can easily check that backfitting converges after a single cycle, with $\hat{\mathbf{f}}_1 = \hat{\mathbf{y}}$ and $\hat{\mathbf{f}}_j = \mathbf{0}$ for $j > 1$. More interestingly, suppose that the predictors are mutually uncorrelated ($\theta = 90°$ in the two-predictor case). Then again we have convergence after a single cycle. This brings up the connection of backfitting with the Gram-Schmidt method for solving the least-squares normal equations. Like backfitting, the Gram-Schmidt method works by regressing the current residual onto each predictor in turn. However there is one important difference. After regressing on the jth predictor, the $(j+1)$th through pth predictors are also regressed on the jth predictor and the residual vector from each regression is used in place of the predictor in the remaining steps. This process orthogonalizes the predictors and because of this, the Gram-Schmidt procedure converges after a single cycle. In the orthogonal-projection setting, clearly the Gram-Schmidt method is superior to backfitting and this suggests that a sweeping-out operation be used to improve the backfitting algorithm with general smoothers. This is not useful, however, because the resultant model would no longer be additive in the predictors.

We note also that the above proof of the convergence of the residual vector does not establish that the estimated functions converge to the correct ones. Indeed, if there exists strict collinearity among the predictors it wouldn't be clear to which solution the functions $\hat{\mathbf{f}}_j$ produced by backfitting would converge, if indeed they converge at all. It turns out that backfitting always does converge to a solution representing the projection of $\mathbf{y}$ onto V, but the fixed point depends on the starting functions.

5.3.3 *Semi-parametric models*

Consider an additive model in which all but one term is assumed to be linear — the so called *semi-parametric* model. The backfitting algorithm for estimating such a model can be thought of as having two smoothers: one a projection $\mathbf{S}_1 = \mathbf{X}(\mathbf{X}^T\mathbf{X})^{-1}\mathbf{X}^T$ producing a least-squares fit $\mathbf{X}\hat{\boldsymbol{\beta}}$ on one or more covariates (represented by the full-rank design matrix $\mathbf{X}$), and the other a smoother $\mathbf{S}_2$ producing an estimate $\hat{\mathbf{f}}_2$. The backfitting steps are $\mathbf{f}_1 = \mathbf{S}_1(\mathbf{y} - \mathbf{f}_2) = \mathbf{X}(\mathbf{X}^T\mathbf{X})^{-1}\mathbf{X}^T(\mathbf{y} - \mathbf{f}_2) \equiv \mathbf{X}\boldsymbol{\beta}$, and $\mathbf{f}_2 = \mathbf{S}_2(\mathbf{y} - \mathbf{X}\boldsymbol{\beta})$. It turns out that we can solve for $\hat{\boldsymbol{\beta}}$ and $\hat{\mathbf{f}}_2$ explicitly (Exercise 2.8):

$$\begin{aligned} \hat{\boldsymbol{\beta}} &= \{\mathbf{X}^T(\mathbf{I} - \mathbf{S}_2)\mathbf{X}\}^{-1}\mathbf{X}^T(\mathbf{I} - \mathbf{S}_2)\mathbf{y} \\ \hat{\mathbf{f}}_2 &= \mathbf{S}_2(\mathbf{y} - \mathbf{X}\hat{\boldsymbol{\beta}}) \end{aligned} \tag{5.21}$$

so that iteration is unnecessary. Although $\mathbf{S}_2$ is an $n \times n$ matrix, all we have to do is smooth each of the p columns of $\mathbf{X}$, an operation that can usually be performed in $O(np)$ operations. This provides a computationally simple method for nonparametric analysis of covariance; it is interesting that the smoother matrix $\mathbf{S}_2$ enters as the weight matrix for the regression on $\mathbf{X}$. This manipulation also shows that in this special case, the estimating equations (5.6) are consistent and have a unique solution as long a $\mathbf{X}^T(\mathbf{I} - \mathbf{S}_2)\mathbf{X}$ is invertible. In section 6.7 we discuss this model in more detail.

5.3.4 *Backfitting with two smoothers*

The third special case that we consider is an additive model that involves two linear smoothers. It turns out that one can analyse the properties of the estimating equation solutions and the convergence of backfitting with some fairly elementary calculations, and this exercise sheds light on the main issues that arise in the more difficult p-predictor case.

It is possible to determine general conditions under which the system (5.6) is consistent by checking whether the rank of $\hat{\mathbf{P}}$ is the same as that of the augmented matrix $[\hat{\mathbf{P}}: \hat{\mathbf{Q}}\mathbf{y}]$. However, it is easier to proceed by constructing the solutions to (5.6) through the backfitting procedure.

Recall that the components of each $\mathbf{x}_j$ are in the same order as the components of $\mathbf{y}$. As a technical point, this means that $\mathbf{S}_j$ really means $\mathbf{E}_j^{-1}\mathbf{S}_j\mathbf{E}_j$ where $\mathbf{E}_j$ is the permutation matrix that sorts in the order of $\mathbf{x}_j$ (Exercise 5.19).

Let $\|\mathbf{C}\| = \sup_{\mathbf{a}\neq\mathbf{0}} \|\mathbf{Ca}\| \,/\, \|\mathbf{a}\|$, the 2-norm of the matrix $\mathbf{C}$ (this choice is made for convenience; any matrix norm would do). The system (5.6) can be written

$$\begin{aligned} \mathbf{f}_1 &= \mathbf{S}_1(\mathbf{y} - \mathbf{f}_2) \\ \mathbf{f}_2 &= \mathbf{S}_2(\mathbf{y} - \mathbf{f}_1). \end{aligned} \tag{5.22}$$

Let $\mathbf{f}_1^{(m)}$ and $\mathbf{f}_2^{(m)}$ denote the estimates at the mth stage of the backfitting algorithm, with $m = 0$ denoting the starting functions. Backfitting consists of alternating the steps

$$\begin{aligned} \mathbf{f}_1^{(m)} &= \mathbf{S}_1(\mathbf{y} - \mathbf{f}_2^{(m-1)}) \\ \mathbf{f}_2^{(m)} &= \mathbf{S}_2(\mathbf{y} - \mathbf{f}_1^{(m)}). \end{aligned} \tag{5.23}$$

Using induction one shows that for $m \geq 1$

$$\begin{aligned} \mathbf{f}_1^{(m)} &= \mathbf{y} - \sum_{j=0}^{m-1} (\mathbf{S}_1\mathbf{S}_2)^j(\mathbf{I} - \mathbf{S}_1)\mathbf{y} - (\mathbf{S}_1\mathbf{S}_2)^{m-1}\mathbf{S}_1\mathbf{f}_2^{(0)}, \\ \mathbf{f}_2^{(m)} &= \mathbf{S}_2 \sum_{j=0}^{m-1} (\mathbf{S}_1\mathbf{S}_2)^j(\mathbf{I} - \mathbf{S}_1)\mathbf{y} + \mathbf{S}_2(\mathbf{S}_1\mathbf{S}_2)^{m-1}\mathbf{S}_1\mathbf{f}_2^{(0)}. \end{aligned} \tag{5.24}$$

Then a sufficient condition for $\mathbf{f}_1^{(m)}$ and $\mathbf{f}_2^{(m)}$ to converge is $\|\mathbf{S}_1\mathbf{S}_2\| < 1$. If this is the case, we can solve (5.24) to obtain

$$\begin{aligned} \mathbf{f}_1^{(\infty)} &= \{\mathbf{I} - (\mathbf{I} - \mathbf{S}_1\mathbf{S}_2)^{-1}(\mathbf{I} - \mathbf{S}_1)\}\mathbf{y} \\ \mathbf{f}_2^{(\infty)} &= \mathbf{S}_2(\mathbf{I} - \mathbf{S}_1\mathbf{S}_2)^{-1}(\mathbf{I} - \mathbf{S}_1)\mathbf{y} \\ &= \{\mathbf{I} - (\mathbf{I} - \mathbf{S}_2\mathbf{S}_1)^{-1}(\mathbf{I} - \mathbf{S}_2)\}\mathbf{y}. \end{aligned} \tag{5.25}$$

The fit $\hat{\mathbf{y}}$ is given by

$$\begin{aligned}\hat{\mathbf{y}} &= \mathbf{f}_1^{(\infty)} + \mathbf{f}_2^{(\infty)} \\ &= \{\mathbf{I} - (\mathbf{I} - \mathbf{S}_2)(\mathbf{I} - \mathbf{S}_1\mathbf{S}_2)^{-1}(\mathbf{I} - \mathbf{S}_1)\}\mathbf{y}\end{aligned} \tag{5.26}$$

which is symmetric in $\mathbf{S}_1$ and $\mathbf{S}_2$, as some simple calculations show.

This proves that if $\|\mathbf{S}_1\mathbf{S}_2\| < 1$, the estimating equations are consistent, and have a unique solution. In addition, the final iterates from the backfitting procedure are independent of the starting values and starting order.

Is $\|\mathbf{S}_1\mathbf{S}_2\| < 1$ typically? If the smoothers are not centered, we would have $\mathbf{S}_1\mathbf{S}_2\mathbf{1} = \mathbf{1}$ so that $\|\mathbf{S}_1\mathbf{S}_2\| = 1$, but the centering makes $\mathbf{S}_1\mathbf{S}_2\mathbf{1} = \mathbf{0}$. However, smoothers like the cubic spline smoother have a second unit eigenvalue corresponding to the linear functions. Consider a backfitting algorithm with two covariates $\mathbf{x}_1$ and $\mathbf{x}_2$ using cubic smoothing splines. If the data show strict collinearity through the origin (i.e. $\mathbf{x}_2 = c\mathbf{x}_1$), we still have $\|\mathbf{S}_1\mathbf{S}_2\| = 1$. With higher order splines, for example, quintic smoothing splines which also have unit quadratic eigenvectors, similar situations involving linear and quadratic functions are possible. The condition $\|\mathbf{S}_1\mathbf{S}_2\| = 1$ is an example of *concurvity*, a phenomenon that we study later in the general p-covariate case.

As it turns out, if $\mathbf{S}_1$ and $\mathbf{S}_2$ are symmetric with eigenvalues in $(-1, 1]$, one can prove that the estimating equations are consistent. Furthermore, the backfitting algorithm converges despite the presence of concurvity, and the fitted values $\mathbf{f}_1^{(\infty)} + \mathbf{f}_2^{(\infty)}$ are independent of the starting functions. Concurvity will, however, lead to a dependence of the limits $\mathbf{f}_1^{(\infty)}$ and $\mathbf{f}_2^{(\infty)}$ on the starting guess $\mathbf{f}_2^{(0)}$. That is $\mathbf{f}_+^{(\infty)} = \mathbf{f}_1^{(\infty)} + \mathbf{f}_2^{(\infty)}$ is unique but $\mathbf{f}_1^{(\infty)}$ and $\mathbf{f}_2^{(\infty)}$ are not. Cubic smoothing splines satisfy these conditions; however, the conditions are sufficient and not necessary. Empirical evidence suggests that the results may also hold for smoothers such as locally-weighted lines, for which the smoother matrix is asymmetric and has modulus greater than one.

We can usefully view these results as an extension of those for linear regression with a singular regression-matrix $\mathbf{X}$: the fit $\hat{\boldsymbol{\mu}} = \mathbf{X}\hat{\boldsymbol{\beta}}$ is unique, but $\hat{\boldsymbol{\beta}}$ is not.

It is not surprising that the condition $\|\mathbf{S}_1\mathbf{S}_2\| = 1$ leads to a dependence of the backfitting solutions on the starting functions, for it is immediate in this case that the backfitting solutions themselves are not unique. Since each $\mathbf{S}_j$ is assumed to have eigenvalues in $(-1, 1]$, $\|\mathbf{S}_1\mathbf{S}_2\|$ can only equal 1 if there is some vector $\mathbf{a}$ that is reproduced by both $\mathbf{S}_1$ and $\mathbf{S}_2$. But this will lead to nonuniqueness of the solution to backfitting, for if $\hat{\mathbf{f}}_1$ and $\hat{\mathbf{f}}_2$ are solutions, then so are $\hat{\mathbf{f}}_1 + \mathbf{a}$ and $\hat{\mathbf{f}}_2 - \mathbf{a}$, evident upon examination of (5.22).

5.3.5 *Existence and uniqueness: p-smoothers*

The two-smoother problem has revealed important aspects of the estimating equations and backfitting which can be used to motivate the more general results that we discuss here. In fact, the results in the general case are qualitatively the same as in the simpler setting.

In the two-smoother problem, we are forced to restrict attention to symmetric smoother matrices with eigenvalues in $(-1, 1]$. The exclusion of -1 as an eigenvalue is necessary to avoid oscillatory behaviour in the algorithm. In the p-smoother case, it turns out that we need to assume that each smoother is symmetric and has eigenvalues in $[0, 1]$. We can give no intuitive reason for this stronger condition being necessary here, but note that this is not a practical limitation of the results, because any reasonable symmetric smoother should satisfy this property.

The first issue to be settled is the existence of at least one solution to the estimating equations (5.6). It turns out that if $\mathbf{S}_1, \ldots, \mathbf{S}_p$ are symmetric with eigenvalues in $[0, 1]$, the estimating equations (5.6) have at least one solution for every $\mathbf{y}$. Given that at least one solution exists, is it unique? Nonuniqueness occurs in the two smoother case when $\|\mathbf{S}_1\mathbf{S}_2\| = 1$. This leads us to ask: what interrelation among the $\mathbf{S}_j$s in the p-smoother case will lead to nonuniqueness of the solution to backfitting? The answer lies in the estimating equations $\hat{\mathbf{P}}\mathbf{f} = \hat{\mathbf{Q}}\mathbf{y}$. Suppose that there is a $\mathbf{g}$ such that $\hat{\mathbf{P}}\mathbf{g} = \mathbf{0}$. Then the system $\hat{\mathbf{P}}\mathbf{f} = \hat{\mathbf{Q}}\mathbf{y}$ has an infinite number of solutions because if $\{\mathbf{f}_j^{(\infty)} : j = 1, \ldots, p\}$ is a solution, then so is $\{\mathbf{f}_j^{(\infty)} + c\mathbf{g}_j : j = 1, \ldots, p\}$ for any c.

We think of this phenomenon as the analogue of collinearity, and define the *concurvity space* of the system (5.6) to be the space of functions $\mathbf{g}$ satisfying $\hat{\mathbf{P}}\mathbf{g} = \mathbf{0}$. Concurvity in function space is similarly defined, with regard to the system (5.5).

A quick word on notation. An unsubscripted vector $\mathbf{g}$ denotes the np vector of evaluations $\mathbf{g}^T = (\mathbf{g}_1^T, \ldots, \mathbf{g}_p^T)$, while an additive fit is denoted by $\mathbf{f}_+ = \sum_{j=1}^p \mathbf{f}_j$.

In the two variable case it is easy to show that $\|\mathbf{S}_1\mathbf{S}_2\| < 1$ and $\|\mathbf{S}_2\mathbf{S}_1\| < 1$ if and only the concurvity space is empty. Thus we see that concurvity plays an important role in the behaviour of backfitting. It is natural, then, to try to pin down exactly how concurvity can occur. We can do this, after a bit of preparation. Let $\mathbf{S}_j, \quad j = 1, \ldots, p$ be symmetric smoother matrices with eigenvalues in $[0, 1]$. Let $\mathcal{M}_1(\mathbf{S}_j)$ be the space spanned by the eigenvectors of $\mathbf{S}_j$ with eigenvalue $+1$ (that is, they pass through the smoother unchanged), for $j = 1, \ldots, p$. Then $\hat{\mathbf{P}}\mathbf{g} = \mathbf{0}$ if and only if $\mathbf{g}_j \in \mathcal{M}_1(\mathbf{S}_j)\ \forall j$ and $\mathbf{g}_+ = \mathbf{0}$.

In other words, we have concurvity if and only if the spaces $\mathcal{M}_j(\mathbf{S}_j)$ are linearly dependent; that is, there exist $\mathbf{g}_j \in \mathcal{M}_1(\mathbf{S}_j)$ not all zero satisfying $\mathbf{g}_+ = \mathbf{0}$. Given such a linear degeneracy, any solution $\mathbf{f}_1, \ldots, \mathbf{f}_p$ of $\hat{\mathbf{P}}\mathbf{f} = \hat{\mathbf{Q}}\mathbf{y}$ leads to nonuniqueness in the form of additional solutions $\mathbf{f}_1 + c\mathbf{g}_1, \ldots, \mathbf{f}_p + c\mathbf{g}_p$.

The result above says that concurvity involves only functions in the eigenspaces corresponding to eigenvalue $+1$. In the case of cubic smoothing splines, those eigenspaces correspond to linear functions of each predictor, and thus exact concurvity only exists if the predictors are exactly collinear. However, approximate concurvity is of practical concern, when the predictors are clustered around some lower dimensional manifold. Note that if quintic splines or quadratic regression are used, the eigenspaces $\mathcal{M}_1(\mathbf{S}_j)$ consist of the quadratic functions in the jth variable; hence concurvity may involve truly nonlinear degeneracies between the variables.

5.3.6 *Convergence of backfitting: p-smoothers*

With this definition of concurvity in hand, we can state the main result for the convergence of backfitting. Consider a backfitting algorithm with symmetric smoothers $\mathbf{S}_j, \quad j = 1, \ldots, p$, having eigenvalues in $[0, 1]$. Then if the $\mathbf{S}_j$ do not exhibit concurvity, it can be shown that backfitting converges to the unique solution of (5.6), independent of the starting functions. If there is concurvity, backfitting converges to one of the solutions of (5.6), and the starting functions determine the final solutions.

Note that these results apply to, amongst others, smoothing

splines, regression splines, and simple linear and polynomial regression. They also can be applied to a backfitting algorithm that uses a mixture of these smoothers, for example a cubic smoothing spline for one variable, a simple linear fit for another variable, etc. The smoothers need not even be univariate; the results apply to two or higher-dimensional surface smoothers as well.

5.3.7 *Summary of the main results of the section*

For two smoothers $\mathbf{S}_1$ and $\mathbf{S}_2$:

(i) if $\|\mathbf{S}_1\mathbf{S}_2\| < 1$, then the estimating equations (5.22) have a unique solution (5.25) and the backfitting algorithm converges to this unique solution;

(ii) if $\mathbf{S}_1$ and $\mathbf{S}_2$ are symmetric with eigenvalues in $(-1, 1]$, then the estimating equations (5.22) have at least one solution, and the backfitting algorithm converges to one of the solutions. This solution is dependent on the starting function $\mathbf{f}_2^{(0)}$.

In general for p symmetric smoothers $\mathbf{S}_1, \ldots, \mathbf{S}_p$ with eigenvalues in $[0, 1]$:

(i) The estimating equations (5.6) have at least one solution for every $\mathbf{y}$.

(ii) Let $\mathcal{M}_1(\mathbf{S}_j)$ be the space spanned by the eigenvectors of $\mathbf{S}_j$ with eigenvalue $+1$ (that is, they pass through the smoother unchanged), for $j = 1, \ldots, p$. Then $\hat{\mathbf{P}}\mathbf{g} = \mathbf{0}$ if and only if $\mathbf{g}_j \in \mathcal{M}_1(\mathbf{S}_j)$ $\forall j$ and $\mathbf{g}_+ = \mathbf{0}$. Either of these conditions characterize the *concurvity* space of the estimating equations.

(iii) If the concurvity space is empty, backfitting converges to the unique solution of (5.6), independent of the starting functions.

(iv) If the concurvity space is not empty, backfitting converges to one of the solutions of (5.6), and the starting functions determine the final solutions.

5.4 Special topics

5.4.1 *Weighted additive models*

Consider a weighted penalized least-squares criterion of the form

$$\left(\mathbf{y}-\sum_j \mathbf{f}_j\right)^T \mathbf{W}\left(\mathbf{y}-\sum_j \mathbf{f}_j\right)+\sum_j \lambda_j \mathbf{f}_j^T \mathbf{K}_j \mathbf{f}_j \tag{5.27}$$

where $\mathbf{W}$ is a diagonal matrix of weights, and λ_j is a smoothing parameter and $\mathbf{K}_j$ is a smoothing-spline penalty matrix for the jth predictor. These weights might represent the relative precision of each observation or might arise as part of another iterative procedure, for example the local-scoring procedure described in Chapters 4 and 6. The estimating equations for this problem have the same form as for the unweighted case, except that the smoothers are weighted smoothing splines given by $\mathbf{S}_j = (\mathbf{W} + \lambda_j \mathbf{K}_j)^{-1}\mathbf{W}$. We could generalize all the results presented so far to deal with the weighted case by simply computing norms and inner products in the metric of $\mathbf{W}$. However, it is simpler to map the problem back to the unweighted case, using the transformations $\mathbf{y}' = \mathbf{W}^{1/2}\mathbf{y}$, $\mathbf{f}_j' = \mathbf{W}^{1/2}\mathbf{f}_j$, $\mathbf{K}_j' = \mathbf{W}^{-1/2}\mathbf{K}_j\mathbf{W}^{-1/2}$. Note that $\mathbf{S}_j$ is not symmetric, but $\mathbf{W}^{1/2}\mathbf{S}\mathbf{W}^{-1/2}$ is symmetric with eigenvalues in $[0,1]$, and unit eigenvalues corresponding to linear functions of the jth variable. Thus the convergence results for the unweighted case can be directly applied.

5.4.2 *A modified backfitting algorithm*

In the previous chapter we mention the possibility of modifying the backfitting algorithm to improve its efficiency. The basic idea is as follows. Many smoothers have a *projection* part and a *shrinking* part. For example, a cubic smoothing spline has unit eigenvalues that are constant and linear functions of the predictor (its projection part), and eigenvalues less than one for other eigenvectors. The idea is to combine all of the projection operations for all of the predictors into one large projection, and use only the nonprojection parts of each smoother in an iterative backfitting-type operation.

This modification has several advantages. When a smoothing-spline or running-line smoother is used for several predictors, practical experience has shown that if the predictors are correlated, many iterations may be required to get the correct average slope of the functions. By performing all of the projections in one operation, all of the function slopes are simultaneously estimated. A second advantage is in collinearity/concurvity situations. In a backfitting problem with symmetric smoothers having eigenvalues in $[0,1]$, we have seen in the previous section that the only nonuniqueness (*concurvity*) occurs in the eigenspaces with eigenvalue one. By separating out the estimation of these components of the functions, the nonunique part of the solutions is conveniently allocated to the projection step. This makes it easy to characterize the solutions and eliminates the dependence of the final solutions on the starting functions.

Let us now be more specific. Let $\mathbf{G}_j$ be the matrix that projects onto $\mathcal{M}_1(\mathbf{S}_j)$, the space of eigenvalue one for the jth smoother. Using $\mathbf{G}_j$, we define the modified smoother matrices

$$\tilde{\mathbf{S}}_j = (\mathbf{I} - \mathbf{G}_j)\mathbf{S}_j.$$

Note that $\tilde{\mathbf{S}}_j$ has the effect of subtracting out the component of the smoothed value that lies in $\mathcal{M}_1(\mathbf{S}_j)$. The general form of the modified backfitting algorithm is given below.

Algorithm 5.2 *The modified backfitting algorithm*

(i) Initialize $\tilde{\mathbf{f}}_1, \ldots, \tilde{\mathbf{f}}_p$ and set $\tilde{\mathbf{f}}_+ = \tilde{\mathbf{f}}_1 + \cdots + \tilde{\mathbf{f}}_p$.
(ii) Regress $\mathbf{y} - \tilde{\mathbf{f}}_+$ onto the space $\mathcal{M}_1(\mathbf{S}_1) + \cdots + \mathcal{M}_1(\mathbf{S}_p)$, that is, set $\mathbf{g} = \mathbf{G}(\mathbf{y} - \tilde{\mathbf{f}}_+)$, where $\mathbf{G}$ is the orthogonal projection onto $\mathcal{M}_1(\mathbf{S}_1) + \cdots + \mathcal{M}_1(\mathbf{S}_p)$ in $\mathbb{R}^n$.
(iii) Apply one cycle of backfitting to $\mathbf{y} - \mathbf{g}$ using smoothers $\tilde{\mathbf{S}}_i$; this step yields an updated additive fit $\tilde{\mathbf{f}}_+ = \tilde{\mathbf{f}}_1 + \cdots + \tilde{\mathbf{f}}_p$.
(iv) Repeat steps (ii) and (iii) until convergence. The final estimate for the overall fit is $\mathbf{f}_+ = \mathbf{g} + \tilde{\mathbf{f}}_+$.

Note that it is not sufficient to perform the projection step only once. It must be iterated with the other steps because when $\tilde{\mathbf{f}}_+$ is changed in step (ii), the projection component $\mathbf{g}$ no longer equals $\mathbf{G}(\mathbf{y} - \tilde{\mathbf{f}}_+)$. An alternative to step (iii) is to iterate it to convergence rather than cycling through once; we find that this tends to slow down convergence in terms of the number of smooths performed.

In order justify this procedure, we must not only show that it does converge, but that it converges to the same solution as the original backfitting procedure. It turns out (to follow) that in the case of symmetric smoothers with eigenvalues in $[0,1]$, the modified backfitting procedure does solve the original problem. For other linear smoothers, the modified backfitting procedure might still make sense, but it solves a slightly different problem.

We now state some convergence results about modified backfitting algorithms:

(i) If $\mathbf{S}_j$, $j = 1,\ldots,p$, are symmetric and have eigenvalues in $[0,1]$, then the modified backfitting algorithm converges in the sense that $\mathbf{g}$ and $\tilde{\mathbf{f}}_1,\ldots,\tilde{\mathbf{f}}_p$ converge.

(ii) Suppose the modified backfitting algorithm has converged with smoothers $\tilde{\mathbf{S}}_j$, yielding functions $\tilde{\mathbf{f}}_j$ and $\mathbf{g}_j \in \mathcal{M}_1(\mathbf{S}_j)$. Then the components $\mathbf{f}_j = \mathbf{g}_j + \tilde{\mathbf{f}}_j$ are solutions to the estimating equations with smoothers $\mathbf{S}_j^* = \mathbf{G}_j + (\mathbf{I} - \mathbf{G}_j)\mathbf{S}_j$.

Notice that if $\mathbf{S}_j$ is symmetric, we have $\mathbf{S}_j^* = \mathbf{S}_j$ and thus the solutions to the modified algorithm solve the estimating equations with smoothers $\mathbf{S}_j$.

If the $\mathbf{S}_j$ are symmetric and have eigenvalues in $[0,1]$ then $\tilde{\mathbf{S}}_j = \mathbf{S}_j - \mathbf{G}_j$, and $\left\|\tilde{\mathbf{S}}_j\right\| < 1$. Smoothing splines belong to this class, and hence the algorithm always converges for them. If cubic smoothing splines are used for all predictors, $\mathbf{G}$ is the *hat* matrix corresponding to the least-squares regression on $(\mathbf{1},\mathbf{x}_1,\ldots,\mathbf{x}_p)$. The nonlinear functions $\tilde{\mathbf{f}}_j$ are uniquely determined. Concurvity (collinearity) can show up only in the $\mathbf{G}$ step, where it is dealt with in the standard linear least-squares fashion. At convergence, one may then decompose $\mathbf{g} = \sum \mathbf{g}_j$ and reconstruct final components $\mathbf{f}_j = \mathbf{g}_j + \tilde{\mathbf{f}}_j$. If $\mathbf{S}_j$ is a cubic smoothing spline and if $\mathbf{y}$ is centered initially, then $\mathbf{g}_j = \hat{\beta}_j \cdot \mathbf{x}_j$, where $\hat{\beta}_1,\ldots,\hat{\beta}_p$ are the coefficients from the multiple linear regression of $\mathbf{y} - \tilde{\mathbf{f}}_+$ on $\mathbf{x}_1,\ldots,\mathbf{x}_p$.

5.4.3 *Explicit solutions to the estimating equations*

By manipulating the fixed points of the modified backfitting procedures, an expression for the solutions to the estimating equations (5.6) can be derived (Exercise 5.6). Let $\tilde{\mathbf{A}}_j = (\mathbf{I} - \tilde{\mathbf{S}}_j)^{-1}\tilde{\mathbf{S}}_j$, $\tilde{\mathbf{A}} = \sum_1^p \tilde{\mathbf{A}}_j$, and $\mathbf{B} = (\mathbf{I} + \tilde{\mathbf{A}})^{-1}\tilde{\mathbf{A}}$. Then the solutions are

$\tilde{\mathbf{f}}_+ = (\mathbf{I} - \mathbf{BG})^{-1}\mathbf{B}(\mathbf{I} - \mathbf{G})\mathbf{y}$ and $\mathbf{g} = \mathbf{G}(\mathbf{y} - \tilde{\mathbf{f}}_+)$. These can be combined to obtain

$$\begin{aligned} \mathbf{f}_+ &= \{\mathbf{G} + (\mathbf{I} - \mathbf{G})(\mathbf{I} - \mathbf{BG})^{-1}\mathbf{B}(\mathbf{I} - \mathbf{G})\}\mathbf{y} \\ \tilde{\mathbf{f}}_j &= (\mathbf{I} - \tilde{\mathbf{S}}_j)^{-1}(\mathbf{y} - \mathbf{g} - \tilde{\mathbf{f}}_+) \\ \mathbf{f}_j &= \mathbf{g}_j + \tilde{\mathbf{f}}_j \end{aligned} \tag{5.28}$$

and the individual $\mathbf{g}_j$s are any vectors $\mathbf{g}_j \in \mathcal{M}_1(\mathbf{S}_j)$ such that $\sum_1^p \mathbf{g}_j = \mathbf{g}$. Interestingly, (5.28) reveals that for symmetric smoother matrices with eigenvalues in $[0, 1]$, a direct solution can be obtained in $O(n^3p)$ operations, the number required for computing $(\mathbf{I} - \tilde{\mathbf{S}}_j)^{-1}$ for $j = 1, \ldots, p$. This is less than the $O\{(np)^3\}$ operations that are needed to solve the estimating equations (5.6) in general.

5.4.4 *Standard errors*

From the previous sections we note that each estimated function in the additive fit is the result of a linear mapping or smoother applied to $\mathbf{y}$. This means that the variance formula developed in Chapter 3 can be applied to the additive model. At convergence, we can express $\hat{\mathbf{f}}_j$ as $\mathbf{R}_j\mathbf{y}$ for some $n \times n$ matrix $\mathbf{R}_j$. If the observations have independent and identically distributed errors, then $\text{cov}(\hat{\mathbf{f}}_j) = \mathbf{R}_j\mathbf{R}_j^T\sigma^2$ where $\sigma^2 = \text{var}(Y_i)$. As in the least-squares case, if $\hat{\mathbf{P}}$ in equation (5.6) has singular values close to $\mathbf{0}$, this will be reflected in $\text{cov}(\hat{\mathbf{f}})$ as large variances and covariances.

Direct computation of $\mathbf{R}_j$ is formidable, except in very special cases such as the semi-parametric model. Our best general approach to date is to apply the backfitting procedure to the each of the n unit n-vectors that are the columns of $\mathbf{I}_n$, the $n \times n$ identity matrix. The result of backfitting applied to the ith unit vector produces fitted vectors $\hat{\mathbf{f}}_j^i$, $j = 1, \ldots, p$, where $\hat{\mathbf{f}}_j^i$ is the ith column of $\mathbf{R}_j$. Similarly, $\hat{\mathbf{f}}_+^i$ is the ith column of $\mathbf{R}$. The standard-error bands in Fig. 4.3 of Chapter 4 are constructed using $\pm$ twice the square root of the diagonal elements of $\hat{\sigma}^2\mathbf{R}_j\mathbf{R}_j^T$. Since the backfitting algorithm is $O(kn)$ for $O(n)$ smoothers, this procedure is $O(kn^2)$. Now $k = pmC$, where p is the number of predictors, m is the number of backfitting iterations, and C is the constant for the particular smoother. For smoothing splines, typical numbers might be $k = 5 \times 5 \times 35 = 875$, which is likely to be larger than n, so this task can be tedious in practice. We have nevertheless used it in many examples, although usually not often within any single analysis.

The global confidence set techniques described in Chapter 3 can be extended to apply to additive models; the procedure is similar to the univariate case. Suppose the additive model is correct, i.e., $Y_i = f_+(\mathbf{X}_i) + \varepsilon_i$, and our estimate of $\mathbf{R}\mathbf{f}_+ = \mathbf{g}_+$ is $\hat{\mathbf{f}}_+ = \mathbf{R}\mathbf{y}$. Then an approximate pivotal for $\mathbf{g}_+$ is

$$\nu(\mathbf{g}_+) = (\hat{\mathbf{f}}_+ - \mathbf{g}_+)^T (\mathbf{R}\mathbf{R}^T \hat{\sigma}^2)^{-1} (\hat{\mathbf{f}}_+ - \mathbf{g}_+). \tag{5.29}$$

Assuming that we have an estimate of the dispersion parameter $\hat{\sigma}^2$ and the distribution G of ν, then we can construct a simultaneous $1 - \alpha$ confidence set of all the component functions:

$$C(\mathbf{g}_1, \ldots, \mathbf{g}_p) = \{\mathbf{g}_1, \ldots, \mathbf{g}_p; \nu(\mathbf{g}_+) \leq G_{1-\alpha}\}. \tag{5.30}$$

We will not pursue this topic further here; it is an area of current research and we need to gain more experience with it.

5.4.5 *Degrees of freedom*

Each of the definitions for degrees of freedom given in Chapter 3 has a natural analogue here. The overall degrees of freedom df is simply $\text{tr}(\mathbf{R})$, where $\mathbf{R}$ is the (smoother) matrix that produces $\hat{\mathbf{f}}_+ = \mathbf{R}\mathbf{y}$. In addition, the posterior covariance of $\mathbf{f}_+$, in the Bayesian treatment of the additive model given in section 5.4.6, is proportional to $\mathbf{R}$, for appropriate choice of the prior covariances.

Similarly, the degrees of freedom for error is $df^{\text{err}} = n - \text{tr}(2\mathbf{R} - \mathbf{R}\mathbf{R}^T)$. More usefully, for model comparison, we need a notion of the change in the error degrees of freedom Δdf^{err} due to an individual term. Let $\mathbf{R}_{(j)}$ denote that operator that produces the additive fit with the jth term removed. Then we define df_j^{err}, the degrees of freedom for error due to the jth term:

$$df_j^{\text{err}} = \text{tr}(2\mathbf{R} - \mathbf{R}\mathbf{R}^T) - \text{tr}(2\mathbf{R}_{(j)} - \mathbf{R}_{(j)}\mathbf{R}_{(j)}^T).$$

This is the expected increase in the residual sum of squares (up to a scale factor) if the jth predictor is excluded from the model, assuming its exclusion does not increase the bias. Approximate F and χ^2 tests that make use of df_j^{err} are discussed in section 6.8.

The sum of the variances of the fitted values is a meaningful concept for an additive fit and thus $df^{\text{var}} = \text{tr}(\mathbf{R}\mathbf{R}^T)$. Further, the sum of the variances of the fitted component function $\hat{\mathbf{f}}_j$ is $\sigma^2\text{tr}(\mathbf{R}_j\mathbf{R}_j^T)$; the effect of predictor correlation on this quantity is explored in Exercise 5.17.

None of these definitions are attractive from a computational point of view. In particular, it would be convenient to use $\text{tr}(\mathbf{S}_j)-1$ or even $\text{tr}(2\mathbf{S}_j - \mathbf{S}_j\mathbf{S}_j^T) - 1$ to select the amount of smoothing prior to including the jth predictor in a model, and as an approximation to df_j^{err} for model comparison. We subtract one since there is a redundant constant in $p-1$ of the p terms in the model; in general we subtract the dimension of $\bigcap_j \mathcal{M}_1(\mathbf{S}_j)$. In the extreme case of exact concurvity, it is possible to show that $\text{tr}(2\mathbf{S}_j - \mathbf{S}_j\mathbf{S}_j^T) - 1$ is an upper bound for df_j^{err} (Exercise 5.7); for a balanced additive model (the other extreme; Exercise 5.18) it is equal to df_j^{err}. Buja, Hastie and Tibshirani (1989) carried out some small simulation experiments and found that adding up the individual degrees of freedom gave a good approximation to the true degrees of freedom. The only exceptions occurred when the predictors had extremely high correlation or when a very small smoothing parameter was used.

In the examples in this book, we use the convenient approximation $\text{tr}(\mathbf{S}_j) - 1$ to select the amount of smoothing, while we use the exact quantities df_j^{err} for model comparisons via approximate F or χ^2 tests. Further details are given in section 6.8.

5.4.6 *A Bayesian version of additive models*

Just as in the case of a single smoother, there is a rather simple Bayesian approach to additive models. As in section 3.6, there is a functional (stochastic process) version, and a finite dimensional (sampled) version; we focus on the latter.

The model is

$$\mathbf{y} = \mathbf{f}_1 + \cdots + \mathbf{f}_p + \varepsilon \tag{5.31}$$

with $\mathbf{f}_j \sim N(\mathbf{0}, \sigma^2\mathbf{Q}_j)$ independently for all j and independent of $\varepsilon \sim N(\mathbf{0}, \sigma^2\mathbf{I})$. Note from section 3.6 that each $\mathbf{Q}_j$ corresponds to the inverse of some penalty matrix $\mathbf{K}_j$; also, from section 5.2.3, $\mathbf{Q}_j$

can be identified with the realization of a reproducing kernel. For the moment we assume that the priors are proper and nonsingular.

Straightforward derivations (Exercise 5.9) show the following:

(i) The prior for $\mathbf{f}_+ = \mathbf{f}_1 + \cdots + \mathbf{f}_p$ is $N(\mathbf{0}, \mathbf{Q}_+)$ with $\mathbf{Q}_+ = \sum_j \mathbf{Q}_j$.

(ii) The posterior mean for $\mathbf{f}_+$ is $E(\mathbf{f}_+ \,|\, \mathbf{y}) = \mathbf{Q}_+(\mathbf{I} + \mathbf{Q}_+)^{-1}\mathbf{y}$. Similarly the posterior means for the individual functions are $E(\mathbf{f}_j \,|\, \mathbf{y}) = \mathbf{Q}_j(\mathbf{I} + \mathbf{Q}_+)^{-1}\mathbf{y}$.

(iii) The posterior covariance of $\mathbf{f}_+$ is $\sigma^2\mathbf{Q}_+(\mathbf{I} + \mathbf{Q}_+)^{-1}$, and for $\mathbf{f}_j$ they are $\sigma^2(\mathbf{Q}_j - \mathbf{Q}_j(\mathbf{I} + \mathbf{Q}_+)^{-1}\mathbf{Q}_j)$.

Notice that the solution involves the inversion of an $n \times n$ *unstructured* matrix, which takes $O(n^3)$ operations, unless of course backfitting is used.

Similar equations can be derived for partially improper priors. These give infinite variance to certain components in $\mathbb{R}^n$, which correspond to components in $\mathcal{M}_1(\mathbf{S}_j)$. The simplest way to formulate the problem is along the lines taken in section 5.2.3, where the projection space is explicitly isolated. In addition, smoothing parameters are thought of as prior variances for the $\mathbf{f}_j$s; we have absorbed these into the $\mathbf{Q}_j$s in the above formulation.

5.5 Bibliographic notes

The theory of nonparametric additive modelling is relatively recent. Breiman and Friedman (1985) proposed the ACE algorithm, a procedure more general than the additive model (allowing response transformations, and discussed in Chapter 7) and proved many results on convergence and consistency of backfitting in both Hilbert-space and data settings. Most of this chapter is based on the paper by Buja, Hastie, and Tibshirani (1989) in which some of Breiman and Friedman's convergence results were extended and the concurvity and degrees of freedom results are presented. In some cases, results stronger than those given in this chapter can be derived; see Buja *et al.* (1989) and the discussions. We have traded generality for simplicity and interpretability in this chapter.

Backfitting goes back a long way. Friedman and Stuetzle (1981) defined the term in the context of projection-pursuit regression, while Wecker and Ansley (1983) suggested its use in the context of economic models. Some of the earlier references include Papadakis (1937) for the separation of fertility trends in the analysis of field trials, and Shiskin, Young and Musgrave (1967) in the X-11 system for decomposing time series (Chapter 8). Kohn and Ansley (1989) independently studied the properties of additive models in the stochastic setting, including convergence.

Much of the statistical theory and practice of one and higher-dimensional spline models is due to Grace Wahba and her co-workers. Some relevant references are Kimeldorf and Wahba (1971), Wahba (1978, 1980, 1986), O'Sullivan (1983), Gu and Wahba (1988), and Chen, Gu and Wahba (1989). Section 5.2.3 is based almost entirely on this last reference. Cox (1989) described the Bayesian formulation of additive models, summarized in section 5.4.6.

The notion of concurvity was introduced by Buja, Donnell and Stuetzle (1986) and was also discussed in Buja *et al.* (1989). Bickel, Klaassen, Ritov, and Wellner (1990) studied the theory for semi-parametric models, and in the two predictor projection case, showed convergence of the backfitting functions and computed the rates of convergence.

The semi-parametric approach discussed in section 5.3.3 was considered by Engle, Granger, Rice and Weiss (1986), Denby (1986), Wahba (1986), Green and Yandell (1985), Green (1985), Eubank (1985), Heckman (1986, 1988), Speckman (1988), and Shiau and Wahba (1988). Green (1985), and Green and Yandell (1985) looked at regression and more general models with a single nonparametric term (we discuss their work in Chapter 6). Heckman (1986) proved consistency of the regression estimate in a regression model with a single cubic spline term and showed that the estimates of the regression coefficients and nonparametric function are Bayes estimates under an appropriate diffuse prior, generalizing the work of Wahba (1978). Rice (1986) studied the convergence rates for these partially-splined models. Speckman (1988) compared the bias and variance of estimators for this model, and proposed a new estimate (Exercise 5.13) with asymptotically lower-order bias. This estimate was also suggested by Denby (1986). Shiau and Wahba (1988) did a thorough study of bias and variance for these models, and Heckman (1988) studied minimax estimators.

Stone (1982) studied rates of convergence for additive models, with the functions estimated by polynomials or regression splines. He proved the interesting result that the optimal rate of convergence for an estimate of the additive model is the same as that for a single function, discussed in section 3.10. Thus an increase in the dimension p does not decrease the rate of convergence, as it does if one is estimating a general (nonadditive) p-dimensional function.

The Gauss-Seidel algorithm is discussed in most textbooks on numerical analysis; see for example Golub and Van Loan (1983). For a description of the QR algorithm, see Thisted (1988).

5.6 Further results and exercises 5

5.1 Consider an additive model with symmetric but not necessarily invertible smoother matrices. Show that the minimizers of the penalized least-squares criterion (5.14) are the solutions to the estimating equations (5.6).

[Buja, Hastie and Tibshirani, 1989]

5.2 Complete the proof of convergence of backfitting in the case of orthogonal projections (section 5.3.2) by showing that $\|\mathbf{C}^m\hat{\mathbf{y}}\| \to 0$. (Hint: show that $\|\mathbf{Ca}\| \leq \|\mathbf{a}\|$ for all $\mathbf{a}$, with equality if and only if $\mathbf{a}$ is in the orthogonal complement of V).

5.3 Consider a backfitting procedure with orthogonal projections, and let $\mathbf{D}$ be the overall design matrix whose columns span $V = \mathcal{L}_{col}(\mathbf{S}_1) \oplus \mathcal{L}_{col}(\mathbf{S}_2) \oplus \ldots \oplus \mathcal{L}_{col}(\mathbf{S}_p)$. Show that the estimating equations $\hat{\mathbf{P}}\mathbf{f} = \hat{\mathbf{Q}}\mathbf{y}$ are equivalent to the least-squares normal equations $\mathbf{D}^T\mathbf{D}\boldsymbol{\beta} = \mathbf{D}^T\mathbf{y}$ where $\boldsymbol{\beta}$ is the vector of coefficients.

5.4 In a backfitting procedure with two least-squares projections $\mathbf{S}_1$ and $\mathbf{S}_2$ based on predictors $\mathbf{x}_1$ and $\mathbf{x}_2$, show that if $\theta \neq 0$, then the difference between the ith iterate and the solution converges to zero geometrically at rate $\cos\theta$, where θ is the angle between $\mathbf{x}_1$ and $\mathbf{x}_2$.

[Deutsch, 1983]

5.5 Prove that in the case of the semi parametric model, backfitting converges to the solution (5.21). Give conditions that guarantee the solution is unique.

[Green, 1985]

5.6 Prove item (ii) in section 5.4.2, stating that the modified backfitting procedure provides a solution to the estimating equations (5.6). Derive the explicit solutions (5.28) to the estimating equations from the fixed points of the modified backfitting algorithm.

[Buja, Hastie and Tibshirani, 1989]

5.7 As an extreme case of concurvity, consider an additive model with p identical smoothers $\mathbf{S}$. Assume that $\mathbf{S}$ is centered and symmetric, with eigenvalues in $[0,1]$, and that the dimension of $\mathcal{M}_1(\mathbf{S})$ is q. Show that

(i) $\operatorname{tr}(2\mathbf{S}-\mathbf{S}^2) \le \operatorname{tr}(2\mathbf{R}-\mathbf{R}^2) \le p\operatorname{tr}(2\mathbf{S}-\mathbf{S}^2)-(p-1)q$.

(ii) $\operatorname{tr}(2\mathbf{R}-\mathbf{R}^2)-\operatorname{tr}(2\mathbf{R}_{(j)}-\mathbf{R}_{(j)}^2) \le \operatorname{tr}(2\mathbf{S}-\mathbf{S}^2)-q$.

Hence conclude that

(a) including the same predictor twice or more increases the degrees of freedom (in contrast to projections such as linear regression);

(b) the sum of the degrees of freedom of the individual function estimates provides an upper bound on the degrees of freedom of the fitted model;

(c) $\operatorname{tr}(2\mathbf{S}-\mathbf{S}^2)-q$ provides an upper bound on the degrees of freedom contribution df_j^{err}.

[Buja, Hastie and Tibshirani, 1989]

5.8 Compute the smoother matrix for a running-mean and kernel smoother for a small dataset, and hence find empirically that these smoothers are not symmetric and that their eigenvalues can have moduli larger than one.

5.9 Derive the expressions (i)–(iii) in section 5.4.6 for the Bayes posterior means and variances for a stochastic additive model.

5.10 Starting with equations (5.6) with each of the smoothers a smoothing spline, derive an equivalent form for $\hat{\mathbf{f}}_+$ as in the previous exercise.

5.11 Suppose that the B-spline basis functions are to be used to represent the smoothing splines in an additive model fit.

(i) Derive closed form expressions for the B-spline coefficients of the additive model solution.

(ii) Derive expressions for $\hat{f}_+(\mathbf{x}^0)$ and $\hat{f}_j(x_{0j})$, the fitted functions evaluated at an arbitrary point $\mathbf{x}^0$.

5.12 What is the rank of $\hat{\mathbf{P}}$ in (5.6), if the smoothers are all cubic smoothing splines, and the $\mathbf{x}_j$ span a p dimensional space? Identify the null space.

5.13 Derive the value of $\boldsymbol{\beta}$ that minimizes

$$\|\mathbf{y} - \mathbf{X}\boldsymbol{\beta} - \mathbf{S}(\mathbf{y} - \mathbf{X}\boldsymbol{\beta})\| \tag{5.32}$$

and compare it with the semi-parametric estimator (5.21) in section 5.3.3, using the same smoother $\mathbf{S}$ in both cases.

5.14 Suppose the same smoother $\mathbf{S}$ is used to estimate both the terms in a two-term additive model (that is, both variables are identical). Show that the backfitting residual converges to $(\mathbf{I}+\mathbf{S})^{-1}(\mathbf{I}-\mathbf{S})\mathbf{y}$, and that the residual sum of squares converges upwards. Can the residual sum of squares converge upwards in less structured situations?

5.15 Consider a semi-parametric model with p predictors (including the constant) and an additional smoother $\mathbf{S}$. Derive explicit expressions for $\mathbf{R}$ and $\mathbf{R}_j$, and show that $\text{tr}(\mathbf{R}) \le p + \text{tr}(\mathbf{S}) - 1$.

5.16 Suppose $\mathbf{S}_1$ and $\mathbf{S}_2$ both have the same eigenspaces; this occurs frequently in time-series applications, as in Chapter 8, where smoothing takes place in the Fourier domain. Let λ_{1k} and λ_{2k}, $k = 1, \ldots, n$, be the eigenvalue pair for the kth eigenvector, with $\lambda_{jk} \in [0, 1]\ \forall k, j$. Assume for convenience that the constant term has been removed, and that $\lambda_{1k}\lambda_{2k} < 1$.

(i) Show that the additive-fit operators $\mathbf{R} = \mathbf{R}_1 + \mathbf{R}_2$ have eigenvalues

$$\begin{aligned}
\lambda_k(\mathbf{R}) &= 1 - \frac{(1-\lambda_{1k})(1-\lambda_{2k})}{1-\lambda_{1k}\lambda_{2k}} \\
\lambda_k(\mathbf{R}_1) &= \frac{\lambda_{1k}(1-\lambda_{2k})}{1-\lambda_{1k}\lambda_{2k}} \\
\lambda_k(\mathbf{R}_2) &= \frac{\lambda_{2k}(1-\lambda_{1k})}{1-\lambda_{1k}\lambda_{2k}}
\end{aligned}$$

(ii) Conclude from (i) that components with eigenvalue one for either smoother get totally absorbed into the corresponding term.

(iii) Show that $\text{tr}(\mathbf{R}) - \text{tr}(\mathbf{S}_1) \le \text{tr}(\mathbf{S}_2)$

(iv) Show that $\text{tr}(\mathbf{R}_j\mathbf{R}_j^T) \le \text{tr}(\mathbf{S}_j\mathbf{S}_j^T)$, and interpret this result in terms of pointwise variances.

5.17 Consider two extremely simple smoothers: $\mathbf{S}_1 = \mathbf{u}\lambda\mathbf{u}^T$ and $\mathbf{S}_2 = \mathbf{v}\lambda\mathbf{v}^T$, with $\lambda \in [0, 1]$, and $\|\mathbf{u}\| = \|\mathbf{v}\| = 1$. Let $\mathbf{R}_j$ be the additive operators as above. Show that

$$\operatorname{tr}(\mathbf{R}_j\mathbf{R}_j^T) = \frac{\lambda^2\{1 - c^2\lambda(2 - \lambda)\}}{(1 - c^2\lambda^2)^2},$$

where $c = \langle \mathbf{u}, \mathbf{v} \rangle$. For $c = 0$, $\operatorname{tr}(\mathbf{R}_j\mathbf{R}_j^T) = \lambda^2 = \operatorname{tr}(\mathbf{S}_j\mathbf{S}_j^T)$, while for $c = 1$, $\operatorname{tr}(\mathbf{R}_j\mathbf{R}_j^T) = \{\lambda/(1+\lambda)\}^2$. Investigate for values of $c \in (0, 1)$ and a range of values of λ.

5.18 Consider a balanced additive model as defined in Exercise 4.8. Let $\mathbf{S}_0 = \mathbf{1}\mathbf{1}^T/n$ denote the *mean smoother.* Using the notation of section 5.4.5, show that

(i) $\mathbf{R}_j = \mathbf{S}_j - \mathbf{S}_0$, $\mathbf{R} = \mathbf{S}_0 + \sum_{j=1}^p \mathbf{R}_j$, and $\mathbf{R}_j\mathbf{R}_k = \mathbf{0}$ for $j \neq k$.

(ii) The full and marginal definitions of df^{err} and df^{var} coincide, that is $\operatorname{tr}(2\mathbf{R} - \mathbf{R}\mathbf{R}^T) - \operatorname{tr}(2\mathbf{R}_{(j)} - \mathbf{R}_{(j)}\mathbf{R}_{(j)}^T) = \operatorname{tr}(2\mathbf{S}_j - \mathbf{S}_j\mathbf{S}_j^T) - 1$ and $\operatorname{tr}(\mathbf{R}_j\mathbf{R}_j^T) = \operatorname{tr}(\mathbf{S}_j\mathbf{S}_j^T) - 1$.

5.19 Suppose each of the predictors in an additive model have ties, and smoothing splines are to be used in the fit. Describe the smoother matrices $\mathbf{S}_j$, as well as their eigenstructure. Show how the estimating equations (5.6) can be reduced from the default dimension np to $\sum_{j=1}^p m_j$, where m_j is the number of unique values in $\mathbf{x}_j$.

CHAPTER 6

Generalized additive models

6.1 Introduction

Most of our discussion so far deals with additive models in which the mean of the response is modelled as an additive sum of the predictors. These models extend the standard linear regression model. In this chapter we describe an additive extension of the family of *generalized linear models*, another useful class of linear models.

Generalized linear models are themselves a generalization of linear regression models. Specifically, the predictor effects are assumed to be linear in the parameters, but the distribution of the responses, as well as the *link* between the predictors and this distribution, can be quite general. Many useful models fall into the class, including the linear logistic model for binary data discussed in Chapter 4. Here the response Y is assumed to have a Bernoulli distribution with $\mu = \mathrm{pr}(Y = 1 \mid X_1, \ldots, X_p)$, and μ is linked to the predictors via $\log\{(\mu/(1-\mu)\} = \alpha + \sum_j X_j\beta_j$. Other familiar models in this comprehensive class are log-linear models for categorical data and gamma regression models for responses with constant coefficient of variation. The family of generalized linear models provides an appealing framework for studying the common structure of such models, and as we will see, there is a convenient, unified method for their estimation.

In this chapter we describe an extension of the class of generalized linear models, called *generalized additive models*. They extend generalized linear models in the same manner that the additive model extends the linear regression model, that is, by replacing the linear form $\alpha + \sum_j X_j\beta_j$ with the additive form $\alpha + \sum_j f_j(X_j)$. The logistic additive model, described briefly in Chapter 4, is a prime example. When applied to binary response data, the model takes the form $\log\{\mu/(1-\mu)\} = \alpha + \sum_j f_j(X_j)$.

We begin the chapter by reviewing the iteratively-reweighted least-squares procedure for computing the maximum likelihood estimates in a generalized linear model. Since this procedure is just an iterated (weighted) regression procedure, it becomes clear how to modify it for the estimation of a generalized additive model; in particular, the linear regression step is replaced by a nonparametric additive regression step. The resultant algorithm is called *local scoring* for reasons that will become apparent. After looking at some specific examples of the local-scoring procedure, we give two justifications for it, parallel to those given for backfitting in Chapter 5. Local scoring is derived both as an empirical version of an algorithm for populations, and as a minimizer of a penalized likelihood criterion. Next we discuss smoothing parameter selection and inferential issues; both of these topics are still current areas of research. We then give a simulated example which demonstrates that nonparametric models must be treated with some caution, lest they be overinterpreted. Finally we discuss estimation of the link function and local likelihood estimation, another technique related to local scoring.

6.2 Fisher scoring for generalized linear models

A generalized linear model consists of a *random component*, a *systematic component*, and a *link function*, linking the two components. The response Y is assumed to have exponential family density

$$\rho_Y(y;\theta;\phi) = \exp\left\{\frac{y\theta - b(\theta)}{a(\phi)} + c(y,\phi)\right\} \tag{6.1}$$

where θ is called the natural parameter, and ϕ is the dispersion parameter. This is the random component of the model. It is also assumed that the expectation of Y, denoted by μ, is related to the set of covariates $X_1, \ldots, X_p$ by $g(\mu) = \eta$ where $\eta = \alpha + X_1\beta_1 + \ldots X_p\beta_p$. η is the systematic component, called the linear predictor, and $g(\cdot)$ is the link function. Note that the mean μ is related to the natural parameter θ by $\mu = b'(\theta)$; an obvious link for any given ρ is called the *canonical link*, in which $g(\mu)$ is chosen so that $\eta = \theta$. As an example, for the Bernoulli distribution, $E(Y) = \text{pr}(Y = 1)$ is the mean and commonly used links are the *logit* and *probit*, with the *logit* being the canonical link. It is customary, however, to define the model in terms of μ and $\eta = g(\mu)$ and thus θ does not play a role. Hence, when convenient we write $\rho_Y(y, \theta, \phi)$ as $\rho_Y(y, \mu, \phi)$. Estimation of μ does not involve the dispersion parameter ϕ, so for simplicity this is assumed known.

Given specific choices for the random and systematic components, a link function, a vector of n observations $\mathbf{y}$, and p corresponding predictor vectors $\mathbf{x}_1, \ldots, \mathbf{x}_p$, the maximum likelihood estimate of $\boldsymbol{\beta} = (\beta_0, \beta_1, \ldots, \beta_p)^T$ is defined by the score equations:

$$\sum_{i=1}^{n} x_{ij} \left(\frac{\partial \mu_i}{\partial \eta_i}\right) V_i^{-1}(y_i - \mu_i) = 0, \qquad j = 0, 1, \ldots, p, \tag{6.2}$$

where $V_i = \text{var}(Y_i)$. Note in (6.2) we assume $x_{i0} = 1$.

The Fisher scoring procedure is the standard method for solving these equations; that is, a Newton-Raphson algorithm using the expected rather than observed information matrix. An equivalent procedure that is convenient for this problem is called *adjusted dependent variable regression*, and is a form of iteratively-reweighted least-squares (IRLS). Given a current coefficient vector $\boldsymbol{\beta}^0$, with corresponding linear predictor $\boldsymbol{\eta}^0 = (\eta_1^0, \ldots, \eta_n^0)^T$ and fitted values $\boldsymbol{\mu}^0 = (\mu_1^0, \ldots, \mu_n^0)^T$, construct the *adjusted dependent variable*

$$z_i = \eta_i^0 + (y_i - \mu_i^0)\left(\frac{\partial \eta_i}{\partial \mu_i}\right)_0. \tag{6.3}$$

Define weights w_i by

$$w_i^{-1} = \left(\frac{\partial \eta_i}{\partial \mu_i}\right)_0^2 V_i^0 \tag{6.4}$$

where V_i^0 is the variance of Y at μ_i^0. The algorithm proceeds by regressing z_i on $\mathbf{x}^i$ with weights w_i to obtain a revised estimate $\boldsymbol{\beta}$. Then a new $\boldsymbol{\mu}^0$ and $\boldsymbol{\eta}^0$ are computed, new z_is are computed, and the process is repeated, until the change in the *deviance*

$$D(\mathbf{y}; \boldsymbol{\mu}) = 2\{l(\boldsymbol{\mu}_{\max}; \mathbf{y}) - l(\boldsymbol{\mu}; \mathbf{y})\} \tag{6.5}$$

is sufficiently small, where $\boldsymbol{\mu}_{\max}$ is the parameter value that maximizes $l(\boldsymbol{\mu}; \mathbf{y})$ over all $\boldsymbol{\mu}$ (the *saturated model*). Often $\boldsymbol{\mu}_{\max} = \mathbf{y}$. In the above, $l(\boldsymbol{\mu}; \mathbf{y})$ is the log likelihood $\sum_{i=1}^{n} l(\mu_i; y_i) =$

$\sum_{i=1}^{n} \log \rho_Y(y_i, \mu_i, \phi)$. It can be shown that this procedure is equivalent to the Fisher scoring procedure, that is the sequence of parameter estimates is identical (Exercise 6.1). It is attractive because no special optimization software is required, just a function that computes weighted least-squares estimates. A slight modification of this procedure is obtained if one uses observed, rather than expected, information; this gives a procedure which is essentially equivalent to the Newton-Raphson algorithm. In the case of canonical links, they are equivalent (Exercise 6.1).

In fact, this equivalence between Fisher scoring and adjusted dependent variable regression holds outside of the exponential family. Suppose we have a model with log-likelihood l and n-dimensional parameter $\boldsymbol{\eta}$, with η_i modelled as $\eta_i = \beta_0 + \sum_1^p x_{ij}\beta_j$. Denote the score and expected information for $\boldsymbol{\eta}$ by $\mathcal{S}_\eta = \partial l/\partial \boldsymbol{\eta}$ and $\mathcal{I}_\eta = E(-\partial^2 l/\partial \boldsymbol{\eta}\boldsymbol{\eta}^T)$. Then the general form of the adjusted dependent variable is

$$\boldsymbol{\eta} + \mathcal{I}_\eta^{-1}\mathcal{S}_\eta \tag{6.6}$$

with weight matrix $\mathcal{I}_\eta$. If l is the sum of n independent terms, then $\mathcal{I}_\eta$ is diagonal and the components of the adjusted dependent variable simplify to

$$\eta_i + s_{\eta_i}/i_{\eta_i} \tag{6.7}$$

with weights i_{η_i}. When the model is a (exponential family) generalized linear model, these reduce to expressions (6.3) and (6.4) as they should. Details are given in Exercise 6.1.

Table 6.1 gives the adjusted dependent variable and weights for a number of generalized linear models.

Table 6.1 Adjusted dependent variable and weights for some commonly used models

Distribution	*link*	*Adjusted dependent variable z*	*Weights w*
Normal	*identity*	y	1
Binomial (m,μ)	*logit*	$\eta + (y-\mu)/m\mu(1-\mu)$	$m\mu(1-\mu)$
Gamma	*reciprocal*	$\eta - (y-\mu)/\mu^2$	μ^2
Gamma	*log*	$\eta + (y-\mu)/\mu$	1
Poisson	*log*	$\eta + (y-\mu)/\mu$	μ
Inverse Gaussian	μ^{-2}	$\eta - 2(y-\mu)/\mu^3$	μ^3

6.3 Local scoring for generalized additive models

A generalized additive model differs from a generalized linear model in that an *additive* predictor replaces the *linear* predictor. Specifically, we assume that the response Y has a distribution given by (6.1), with the mean $\mu = E(Y \mid X_1, \ldots, X_p)$ linked to the predictors via

$$g(\mu) = \alpha + \sum_{j=1}^{p} f_j(X_j). \qquad (6.8)$$

Estimation of α and $f_1, \ldots, f_p$ is accomplished by replacing the weighted linear regression in the adjusted dependent variable regression by an appropriate algorithm for fitting a weighted additive model. This procedure is outlined in algorithm 6.1. Although for ease of notation we focus on simple additive models composed of a sum of univariate terms, this once again need not be the case. Any multivariate regression method can be used in the regression step, such as surface smoothers, multi-factorial analysis of variance decompositions, or all of these as components of the additive model.

The name *local scoring* derives from the fact that local averaging is used to generalize the Fisher scoring procedure. If the link function is the identity then $z_i = y_i$ and the procedure is simply an iteratively reweighted additive fit of y_i on $\mathbf{x}^i$, with only the weights changing in the outer loop. If in addition the error distribution is Gaussian, then the weights don't change and the procedure is simply a single additive fit.

The local-scoring procedure described here applies to exponential family models, but it can just as well be applied to other models. All that changes is the definition of the adjusted dependent variable and the weights; general forms for those are given earlier. Difficulties can arise if the weight matrix is nondiagonal: it is not always obvious how to construct a smoother in terms of a general nondiagonal weight matrix (section 3.11.4). Smoothing splines are an exception among the shrinking (nonprojection) smoothers, since they are defined for general nonnegative weight matrices; computationally efficient algorithms, however, are usually no longer available.

These problems arise in both the proportional-hazards model and the matched case-control model, and specialized solutions are described in Chapter 8.

Algorithm 6.1 *The local scoring procedure*

(i) *Initialize:* $\alpha = g(\sum_1^n y_i/n)$; $f_1^0 = \ldots = f_p^0 = 0$.

(ii) *Update:* Construct an adjusted dependent variable

$$z_i = \eta_i^0 + (y_i - \mu_i^0)\left(\frac{\partial \eta_i}{\partial \mu_i}\right)_0$$

with $\eta_i^0 = \alpha^0 + \sum_{j=1}^p f_j^0(x_{ij})$ and $\mu_i^0 = g^{-1}(\eta_i^0)$.

Construct weights

$$w_i = \left(\frac{\partial \mu_i}{\partial \eta_i}\right)_0^2 (V_i^0)^{-1}$$

Fit a weighted additive model to z_i, to obtain estimated functions f_j^1, additive predictor η^1, and fitted values μ_i^1.

Compute the convergence criterion

$$\Delta(\eta^1, \eta^0) = \frac{\sum_{j=1}^p \|f_j^1 - f_j^0\|}{\sum_{j=1}^p \|f_j^0\|}$$

A natural candidate for $\|f\|$ is $\|\mathbf{f}\|$, the length of the vector of evaluations of f at the n sample points.

(iii) Repeat step (ii) replacing η^0 by η^1 until $\Delta(\eta^1, \eta^0)$ is below some small threshold.

6.4 Illustrations

6.4.1 *Clotting times of blood*

Let's look at an example of the additive gamma model. McCullagh and Nelder (1989 p. 300) analysed data on the clotting times of blood, given by Hurn *et al.* (1945). The response is clotting `time` in seconds for normal plasma diluted to nine different percentage `concentrations` with prothrombin-free plasma for different `lots` of thromboplastin.

A plot of the data is shown in Fig. 6.1. McCullagh and Nelder fit a gamma model to the data, presumably because the standard deviation of each measurement was thought to be proportional to its mean. They also used the inverse (canonical) link, and in order

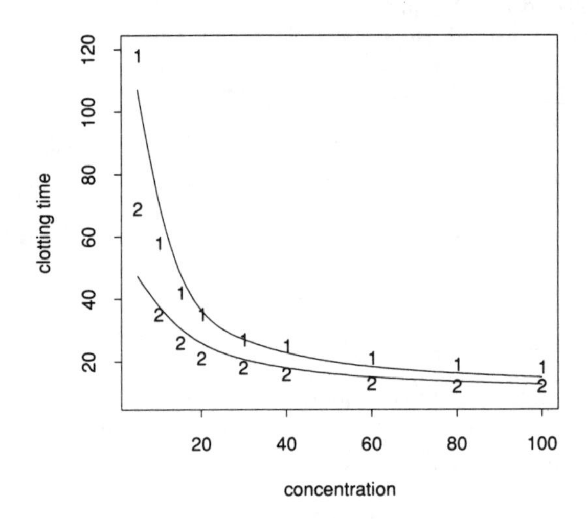

Fig. 6.1. *Plot of clotting* `time` *in seconds versus percent* `concentration`*; the plotting symbol indicates* `lot` *number. The curves are the fitted values based on an additive model with a single smooth term for* `conc` *and a* `lot` *effect.*

to achieve inverse linearity they used `log(concentration)` instead of `concentration`.

Table 6.2 *Analysis of deviance table for clotting time data*

	Model	*Dev*	df^{err}
(i)	1	7.71	17
(ii)	`s(concentration,2)`	0.98	15
(iii)	`s(concentration,2)+lot`	0.29	14
(iv)	`lot/s(concentration,2)`	0.02	10
(v)	`lot/log(concentration)`	0.03	14

In Table 6.2 we use a modelling language similar to that used in McCullagh and Nelder to describe the models fitted (Appendix C). The term `s(concentration,2)+lot` refers to a smooth term in `concentration`, in this case using a smoothing spline, and a level shift for `lot`. The approximate degrees of freedom are integral in this case because we selected the smoothing parameter by prespecifying the value of $df = \mathrm{tr}(\mathbf{S}_\lambda) - 1 = 2$ for the smooth term, and used the approximation $df^{\mathrm{err}} = n - \sum df_j$. Although a value of two for *df* means a lot of smoothing, it is sufficient for these data. The term `lot/s(concentration,2)` means a smooth term for `concentration` nested within `lot`, or in other words a separate smooth for each `lot`. Similarly a linear term in `log(concentration)` nested within `lot` is denoted by `lot/log(concentration)`.

The curves in Fig. 6.1 are the fitted values from the additive model (iii) in the table. The curves are not parallel since they are plotted on the scale of the response. They follow a $\log^{-1}$ shape, corresponding to logarithmic shapes for the transformation of `concentration`. In fact, Table 6.2 suggests separate log fits are best of all those tried for these data, using a rough F-test. The separate smoothing-spline model fits only slightly better, but uses more degrees of freedom. The model using smoothing splines uncovered the log shape successfully, and both models (iv) and (v) essentially interpolate the data. McCullagh and Nelder reached the same model in their analysis.

6.4.2 *Warm cardioplegia*

A second example illustrates the care that must be taken in choosing a generalized additive model. Data on the survival of children, after cardiac surgery for heart defects, was collected by Williams *et al.* (1990) for the period 1983-1988. A pre-operation warm-blood cardioplegia procedure, thought to improve chances for survival, was introduced in February 1988. This was not used on all of the children after February 1988, only on those for which it was thought appropriate and only by surgeons who liked the new procedure. The main question is whether the introduction of the warming procedure improved survival; the importance of risk factors age, weight and diagnostic category is also of interest.

If the warming procedure was given in a randomized manner, we could simply focus on the post-February 1988 data and compare the survival of those who received the new procedure to those who did not. However allocation was not random so we can only try to assess the effectiveness of the warming procedure as it was applied. For this analysis, we use all of the data (1983–1988). To adjust for changes that might have occurred over the five-year period, we

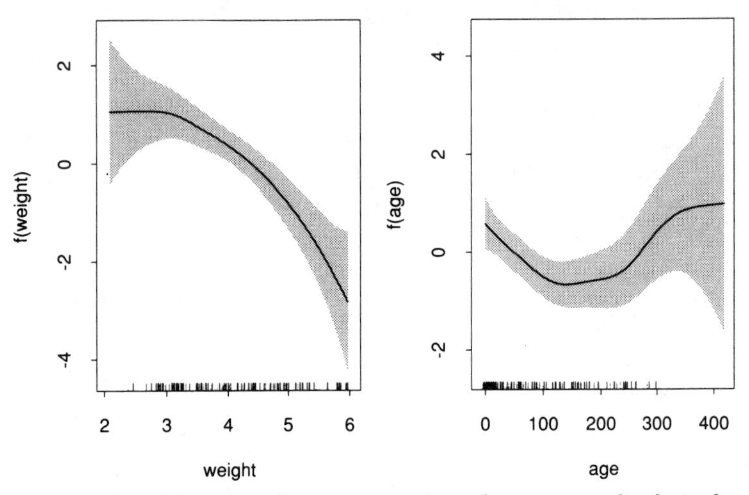

Fig. 6.2. *Fitted functions for* `weight` *and* `age` *for warm cardioplegia data. The shaded region represents twice the pointwise asymptotic standard errors of the estimated curve.*

include the date of the operation as a covariate. However operation date is strongly confounded with the warming operation and thus a general nonparametric fit for date of operation might unduly remove some of the effect attributable to the warming procedure. To avoid this, we allow only a linear effect for operation date. Hence we must assume that any time trend is either a consistently increasing or decreasing trend.

We fit a generalized additive logistic model to the binary response `death`, with smooth terms for `age` and `weight`, a linear term for `operation date`, a categorical variable for `diagnosis`, and a binary variable for the `warming` operation. All the smooth terms are fitted with $df_j = 4$. The resulting curves for `age` and `weight` are shown in Fig. 6.2. As one would expect, the highest risk is for the lighter babies, with a decreasing risk over 3 kg. Somewhat surprisingly, there seems to be a low risk age around 200 days, with higher risk for younger and older children. Interestingly, the function for `weight` is significantly nonlinear (as measured by the corresponding increase in deviance), yet a straight line falls within the pointwise standard-error bands. This is an example of the difficulty in interpreting standard-error bands due to the

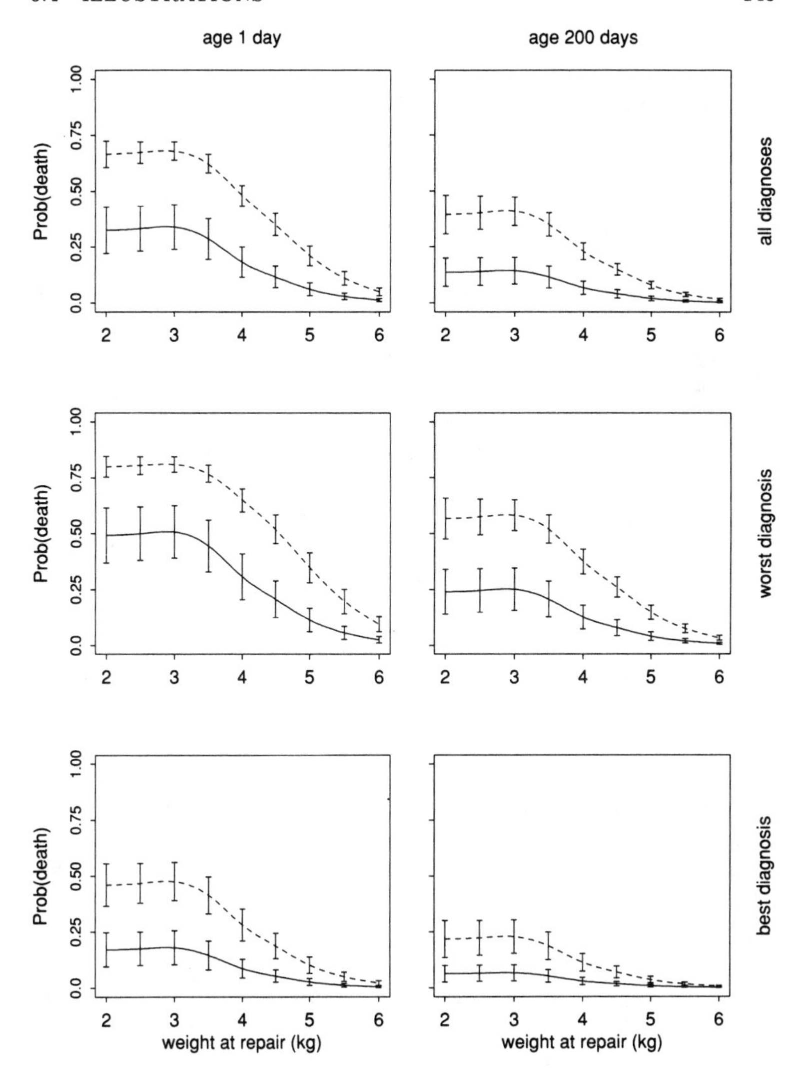

Fig. 6.3. *Fitted probabilities for warm cardioplegia data, conditioned on two ages (columns), and three diagnostic classes (rows). Broken line is standard treatment, solid line is warm cardioplegia. Bars indicate ± one standard error.*

correlation in fitted values along the curve (section 3.8.2).

The `warming` procedure is strongly significant, with an estimated coefficient of 1.43 and a standard-error of 0.45, indicating a survival benefit. There are strong differences in the `diagnosis` categories, while the estimated effect of `operation date` is not large.

As a check of the model-based standard-error for the `warming` coefficient, we estimate the standard-error via the bootstrap. A bootstrap sample is created by drawing 497 observations with replacement from the original observations, then a logistic additive model is fit to this new sample. Repeating this procedure 100 times, the standard deviation of the 100 estimates of the `warming` coefficient is 0.49, just slightly larger than the model-based standard-error reported above.

Figure 6.4 is a diagnostic plot for examining the residuals from the additive logistic fit. The standardized residuals

$$r_i = \frac{y_i - \hat{p}_i}{\sqrt{\hat{p}_i(1 - \hat{p}_i)}}$$

are coded with a dot if they lie within their joint interquartile range, a minus if they lie below the first quartile, and a plus if they lie above the third quartile. These are plotted all together, and then separately for easier identification, as the plot symbols in a scatterplot of `weight` against `age`. A grouping of plusses or minuses would indicate lack of fit and the need for an interaction term in `weight` and `age`. There is no evidence of this here, although the lower left corner seems to have an excess of both plusses and minuses. This may suggest overdispersion in this region, which in turn implies that the mean may not be modelled adequately; we do not pursue the issue further here.

Since a logistic regression is additive on the logit scale but not on the probability scale, a plot of the fitted probabilities is often informative. Figure 6.3 shows the fitted probabilities broken down by `age` and `diagnosis`, and is a concise summary of the findings of this study. The beneficial effect of the treatment at the lower weights is evident. As with all nonrandomized studies, the results here should be interpreted with caution. In particular, one must ensure that the children were not chosen for the warming operation based on their prognosis. To investigate this, we perform a second analysis in which a dummy variable (say `period`), corresponding to

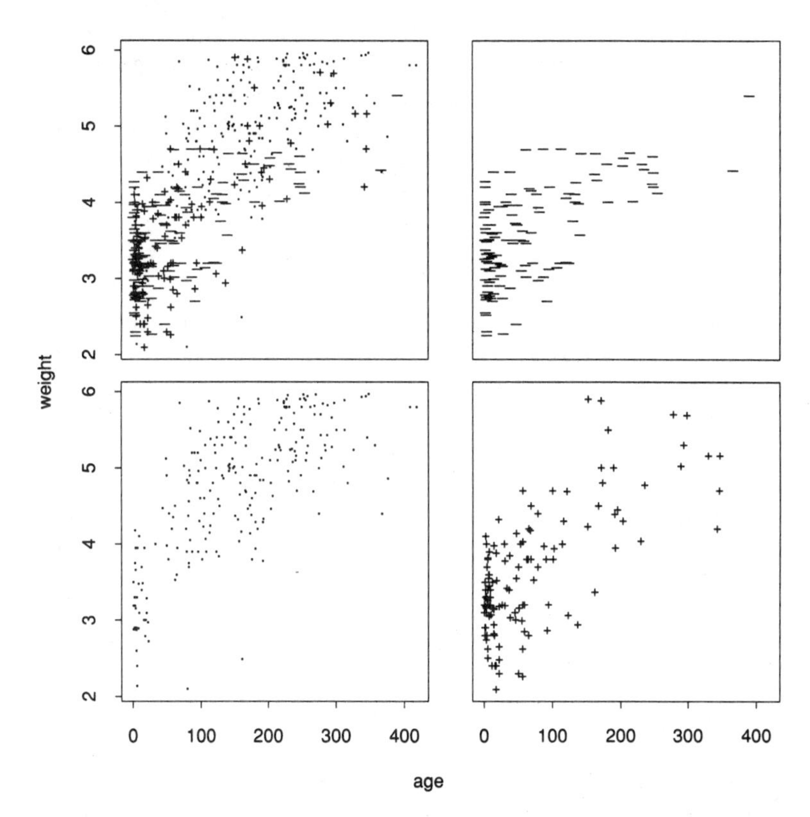

Fig. 6.4. *A diagnostic plot to detect lack of fit or the presence of interactions with respect to the* `weight-age` *pair. The plots show the standardized residuals from the final additive fit. The dots indicate those residuals within their interquartile range, the plus and minus symbols indicate those in the upper and lower quartiles respectively. The top-left plot superimposes all three, and the other three display them separately.*

before versus after February 1988, is inserted in place of the dummy variable for the warming operation. The purpose of this is to investigate whether the overall treatment strategy improved after February 1988. If this turns out not to be the case, it will imply that `warming` was used only for patients with a good prognosis, who would have survived anyway. A linear adjustment for `operation date` is included as before. The results are qualitatively very similar to the first analysis: `age` and `weight` are significant, with effects similar to those in Fig. 6.2; `diagnosis` is significant, while `operation date` (linear effect) is not. `Period` is highly significant, with a coefficient of -1.12 and a standard-error of 0.33. Hence there seems to be a significant overall improvement in survival after February 1988. For more details, see Williams *et al.* (1990).

6.5 Derivation of the local-scoring procedure

As was the case for the backfitting algorithm, one can motivate the local-scoring procedure in a number of ways. Here we justify it as either an empirical version of a technique for populations, or as a finite sample method with smoothness penalties. We discuss these arguments below, for any model in which the likelihood is a sum of n independent terms. Of course the exponential family models are a special but important member of this class. In fact the penalized likelihood arguments can be applied to a wide variety of nonstandard models, and this is illustrated in Chapter 8.

6.5.1 L_2 *function spaces*

Here we assume the same structure as in section 5.2.1. $\mathcal{H}_j$ are Hilbert spaces of measurable functions $\phi(X_j)$, $\mathcal{H}^{add} = \sum_{j=1}^{p} \mathcal{H}_j$, and all are subspaces of $\mathcal{H}$, the space of measurable functions of $X_1, \ldots, X_p$.

Given $\boldsymbol{X} = (X_1, \ldots, X_p)$, the response Y has conditional density $h(y, \zeta)$ where $\zeta = \zeta(\boldsymbol{X})$ is a regression parameter and $\zeta \in \mathcal{H}$. Denote the corresponding log-likelihood for a single observation by $l(\zeta)$, or l for short. Although the true parameter ζ is a function of $\boldsymbol{X}$, we estimate the best additive approximation in the sense to follow. We consider the maximization of the *expected* log-likelihood

$$El\{\eta(\boldsymbol{X}); Y\} \tag{6.9}$$

over $\eta(\boldsymbol{X}) = \sum_{j=1}^{p} f_j(X_j) \in \mathcal{H}^{add}$. This has intuitive appeal since we are choosing the model to maximize the likelihood of all possible *future* observations. In the exponential family the motivation can be strengthened if we consider Kullback-Leibler distance as the generalization of squared error (Exercise 6.3). Stone (1986) gives conditions for the existence and uniqueness of the best additive approximation. The maximizer η of (6.9) is characterized by a score function $\partial l/\partial\eta$ orthogonal to the space of fits, or equivalently,

$$E\left(\frac{\partial l}{\partial \eta} \;\middle|\; X_j\right) = 0 \quad \forall j \tag{6.10}$$

These equations are nonlinear in η and f_j; in order to find a solution we linearize them about a current guess η_0 and obtain

$$E\left\{\left(\frac{\partial l}{\partial \eta_0}\right) + \left(\frac{\partial^2 l}{\partial \eta_0^2}\right)(\eta - \eta_0) \;\middle|\; X_j\right\} = \tag{6.11}$$

$$E\left\{\left(\frac{-\partial^2 l}{\partial \eta_0^2}\right)\left[\eta_0 + \left(-\frac{\partial^2 l}{\partial \eta_0^2}\right)^{-1}\left(\frac{\partial l}{\partial \eta_0}\right) - \eta\right] \;\middle|\; X_j\right\} = 0 \;\forall j$$

Writing $Z_0 = \eta_0 + (-\partial^2 l/\partial\eta_0^2)^{-1}(\partial l/\partial\eta_0)$ and $W_0 = W_0(\boldsymbol{X}) = -\partial^2 l/\partial\eta_0^2$, we see that (6.11) also characterizes the minimum over $\mathcal{H}_+$ of the quadratic criterion $E\{W_0(Z_0 - \eta)^2\}$. Finally, we can rewrite (6.11) as

$$f_j(X_j) = \frac{E\left[W_0(\boldsymbol{X})\{Z_0 - \sum_{k\neq j} f_k(X_k)\} \,|\, X_j\right]}{E\left[W_0(\boldsymbol{X}) \,|\, X_j\right]}. \tag{6.12}$$

The operators $E[W(\boldsymbol{X}) \cdot \,|\, X_j]/E[W(\boldsymbol{X}) \,|\, X_j]$ are weighted conditional expectations, and are naturally estimated by weighted smoothers. The linearized score equations are solved by a weighted backfitting algorithm; at convergence, Z_0 and W_0 are updated and the operation is repeated.

6.5.2 *Penalized likelihood*

One can also generalize the penalized least-squares argument for backfitting to a penalized likelihood argument in this setting, to give a finite sample justification for the local-scoring method. Let $\eta_i = \alpha + \sum_{j=1}^p f_j(x_{ij})$ and consider the log-likelihood l as a function of $\boldsymbol{\eta}$. Now suppose $\mathcal{H}_j$ is the Sobolev space of functions on Ω_j, the domain of the jth predictor, with continuous first and second derivatives, and integrable second derivatives. We consider the following optimization problem. Find $f_1 \in \mathcal{H}_1, \ldots, f_p \in \mathcal{H}_p$ to maximize

$$j(f_1, \ldots, f_p) = l(\boldsymbol{\eta}; \mathbf{y}) - \tfrac{1}{2}\sum_{j=1}^p \lambda_j \int \{f_j''(x)\}^2\, dx \tag{6.13}$$

where $\lambda_j \geq 0,\ \ j = 1, \ldots, p$, are smoothing parameters.

The theory outlined in sections 5.2.2–3 applies here as well. The solution is an additive cubic spline, that is, each coordinate function is a cubic spline. The solution is unique as long as the corresponding linear problem has a unique solution (Cox and O'Sullivan, 1985). As before we can parametrize by the evaluations of the cubic splines $f_j(x)$ at the observed points $x_{1j}, \ldots, x_{nj}$. Using the same notation as in equation (5.12), (6.13) reduces to

$$\jmath(\mathbf{f}_1, \ldots, \mathbf{f}_p) = l(\boldsymbol{\eta}; \mathbf{y}) - \tfrac{1}{2} \sum_{j=1}^{p} \lambda_j\, \mathbf{f}_j^T\, \mathbf{K}_j\, \mathbf{f}_j. \tag{6.14}$$

Letting $\mathbf{u} = \partial l / \partial \boldsymbol{\eta}$ and $\mathbf{A} = -\partial^2 l / \partial \boldsymbol{\eta} \boldsymbol{\eta}^T$, we now show that the local-scoring procedure corresponds to a Newton-Raphson algorithm for maximizing $\jmath(\mathbf{f}_1, \ldots, \mathbf{f}_p)$ over $\mathbf{f}_1, \ldots, \mathbf{f}_p$. Straightforward calculations show that the Newton-Raphson step to go from $\mathbf{f}_1^0, \ldots, \mathbf{f}_p^0$ to $\mathbf{f}_1^1, \ldots, \mathbf{f}_p^1$ is

$$\begin{pmatrix} \mathbf{A} + \lambda_1 \mathbf{K}_1 & \mathbf{A} & \ldots & \mathbf{A} \\ \mathbf{A} & \mathbf{A} + \lambda_2 \mathbf{K}_2 & \ldots & \mathbf{A} \\ \vdots & \vdots & \ddots & \vdots \\ \mathbf{A} & \mathbf{A} & \ldots & \mathbf{A} + \lambda_p \mathbf{K}_p \end{pmatrix} \begin{pmatrix} \mathbf{f}_1^1 - \mathbf{f}_1^0 \\ \mathbf{f}_2^1 - \mathbf{f}_2^0 \\ \vdots \\ \mathbf{f}_p^1 - \mathbf{f}_p^0 \end{pmatrix}$$

$$= \begin{pmatrix} \mathbf{u} - \lambda_1 \mathbf{K}_1 \mathbf{f}_1^0 \\ \mathbf{u} - \lambda_2 \mathbf{K}_2 \mathbf{f}_2^0 \\ \vdots \\ \mathbf{u} - \lambda_p \mathbf{K}_p \mathbf{f}_p^0 \end{pmatrix}, \tag{6.15}$$

where both $\mathbf{A}$ and $\mathbf{u}$ are evaluated at $\boldsymbol{\eta}^0$. In the exponential family, the score $\mathbf{u} - \lambda_j \mathbf{K}_j \mathbf{f}_j$ has the particularly simple form $(\mathbf{y} - \boldsymbol{\mu}) - \lambda_j \mathbf{K}_j \mathbf{f}_j$, and the matrix $\mathbf{A}$ is diagonal with diagonal elements $a_{ii} = (\partial \mu_i / \partial \eta_i)^2 V_i^{-1}$.

To simplify (6.15) further, we let $\mathbf{z} = \boldsymbol{\eta}^0 + \mathbf{A}^{-1}\mathbf{u}$, and $\mathbf{S}_j = (\mathbf{A} + \lambda_j \mathbf{K}_j)^{-1} \mathbf{A}$, a weighted cubic smoothing-spline operator. Then (6.15) can then be written as

$$\begin{pmatrix} \mathbf{I} & \mathbf{S}_1 & \mathbf{S}_1 & \cdots & \mathbf{S}_1 \\ \mathbf{S}_2 & \mathbf{I} & \mathbf{S}_2 & \cdots & \mathbf{S}_2 \\ \vdots & \vdots & \vdots & \ddots & \vdots \\ \mathbf{S}_p & \mathbf{S}_p & \mathbf{S}_p & \cdots & \mathbf{I} \end{pmatrix} \begin{pmatrix} \mathbf{f}_1^1 \\ \mathbf{f}_2^1 \\ \vdots \\ \mathbf{f}_p^1 \end{pmatrix} = \begin{pmatrix} \mathbf{S}_1 \mathbf{z} \\ \mathbf{S}_2 \mathbf{z} \\ \vdots \\ \mathbf{S}_p \mathbf{z} \end{pmatrix}. \tag{6.16}$$

Finally we may rewrite (6.16) as

$$\begin{pmatrix} \mathbf{f}_1^1 \\ \mathbf{f}_2^1 \\ \vdots \\ \mathbf{f}_p^1 \end{pmatrix} = \begin{pmatrix} \mathbf{S}_1(\mathbf{z} - \sum_{j\neq 1} \mathbf{f}_j^1) \\ \mathbf{S}_2(\mathbf{z} - \sum_{j\neq 2} \mathbf{f}_j^1) \\ \vdots \\ \mathbf{S}_p(\mathbf{z} - \sum_{j\neq p} \mathbf{f}_j^1) \end{pmatrix} \tag{6.17}$$

Thus the Newton-Raphson updates are an additive model fit; in fact they solve a weighted and penalized quadratic criterion which is the local approximation to the penalized log-likelihood. The backfitting algorithm can be used to solve them, yielding updated versions of $\boldsymbol{\eta}$, and hence $\mathbf{z}$ and $\mathbf{A}$, and the loop is repeated. In the exponential family $\mathbf{A}$ is diagonal, which means the weighted cubic smoothing splines can be computed in $O(n)$ operations. Thus the overall procedure is a local-scoring algorithm that uses weighted cubic smoothing splines.

6.6 Convergence of the local-scoring algorithm

As we have seen in Chapter 5, the inner (backfitting) loop can be shown to converge when cubic smoothing splines are used, or any smoothers that satisfy the conditions set out there. The outer loop is simply a Newton-Raphson step. Thus if step size optimization is performed, the outer loop will converge as well. Specifically, consider a trial value of the form

$$\boldsymbol{\eta}^{(\gamma)} = \gamma\boldsymbol{\eta}^1 + (1-\gamma)\boldsymbol{\eta}^0, \tag{6.18}$$

with the $\mathbf{f}_j^{(\gamma)}$ defined similarly. Then (6.18) corresponds to a Newton-Raphson step of size γ and we can maximize $j(\boldsymbol{\eta}^{(\gamma)})$ over γ. Standard results on the Newton-Raphson procedure ensure convergence (for example, Ortega and Rheinboldt, 1970). In practice it appears that step size optimization is rarely be necessary in this context.

Note that we monitor convergence by the change in the fitted functions rather than the deviance (Exercise 6.5). Since it is a penalized deviance that is being minimized, the deviance itself may increase during the iterations; this occurs if the starting functions are too rough.

6.7 Semi-parametric generalized linear models

The semi-parametric model described briefly in section 5.3.3 has also been studied in the context of nonlinear models. This is a simple extension of a generalized linear model in that the usual parametric terms are augmented by a single nonparametric component. The model is

$$\eta(\mathbf{X}, t) = \alpha + \sum_{j=1}^{p} X_j\beta_j + \gamma(t) \tag{6.19}$$

where t is a scalar vector covariate and γ is an arbitrary function. Any or all of the X_js may be among the predictors in the vector t, but the emphasis has been on models for which the X_js are the predictors of interest, while t is an extraneous or *nuisance* variable. In this context (6.19) may be viewed as a semi-parametric analysis of covariance model. Indeed, some of the earliest work on these models concerned the analysis of agricultural field trials. When smoothing splines are used to estimate γ, (6.19) is often referred to as a partially-splined model.

We present an often quoted example here in which the analysis of `treatment` effects of three different insecticides on grain `yield` is confounded by a trend in the physical plot order, possibly a fertility trend. The three treatments and the control are shown in Fig. 6.5; it is fairly clear from the figure that if we ignore the effect of `plot number`, the four `treatment` means are not significantly different. Actually, the pooled `treatment` effect is different from the control ($p < 0.01$), but no other differences are significant at the 5% level. The model we fit is

$$\texttt{yield} = \sum_{j\in(0,1,2,R)} \beta_j I(\texttt{treatment} = j) + f(\texttt{plot number}) + \varepsilon \tag{6.20}$$

where $I(\texttt{treatment} = j)$ is an indicator function that is one for the jth treatment and zero otherwise. We describe the estimation for models such as this below, where we focus on smoothing splines. Of course, (6.20) represents a rather simple Gaussian additive model, and the backfitting algorithm could be used to fit it. The two smoothers would be a bin type smoother that computes means in the four `treatment` categories, and any scatterplot smoother for

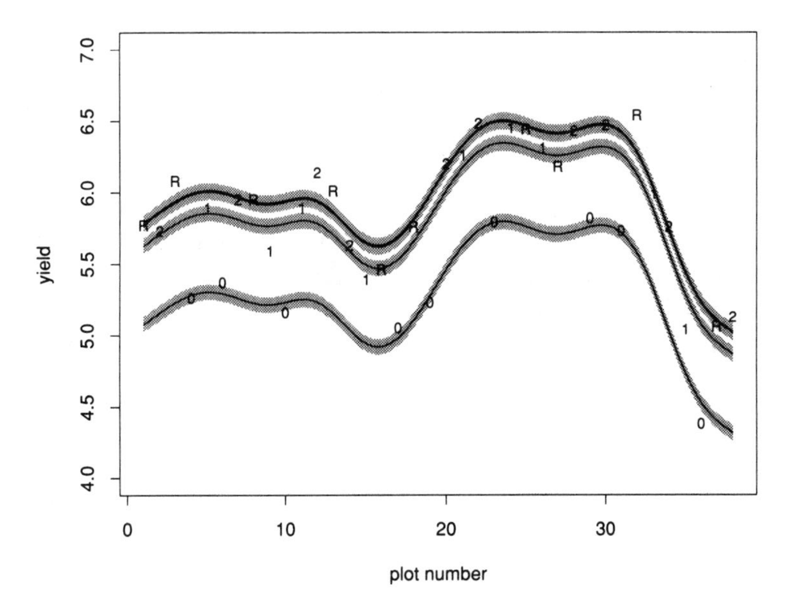

Fig. 6.5. *The four* `treatments` *are labelled 0 (none) and 1,2, R. The four curves represent the fitted treatment effects together with the additive* `plot` *effect. The shaded bands around the curves represent ±1 standard-error of the* `treatment` *differences. This means that two treatments will be significantly different if their bands do not overlap. Note that treatments 2 and R nearly coincide, and cannot be distinguished in the plot.*

`plot number`. The simple structure of the model enables us to compute the fit and standard errors more efficiently. Here we use a smoothing spline with $df = \text{tr}(\mathbf{S}) - 1 = 12$ for the `plot number` effect. We chose 12 *df* by inspecting the residuals from the fitted model.

The four parallel curves represent the estimated `plot number` function, plotted at the level of the estimated `treatment` effects: $\hat{\beta}_j + \hat{f}$, $j \in (0, 1, 2, R)$. The vertical shifts in the curves represent the estimated `treatment` effects, after adjustment for the estimated `plot number` effect. The shaded regions about each curve represent ±1 standard-error of the treatment differences. This allows us to make rough comparisons between the `treatment` effects at the 5% level by simply observing whether the bands overlap. Treatments

0, 1 and $(2, R)$ are all different, while 2 and R are indistinguishable.

Now to the efficient algorithms, which we present for the generalized case. The penalized-likelihood criterion (6.13) for the model (6.19) and scalar covariate t is

$$j(\boldsymbol{\beta}, \gamma) = l(\boldsymbol{\eta}; \mathbf{y}) - \tfrac{1}{2}\lambda \int \{\gamma''(t)\}^2 \, dt. \tag{6.21}$$

Once again the solution to (6.21) can be computed using a local-scoring algorithm using a weighted cubic smoothing spline for t_i and a weighted least-squares regression for the X_is. However, as in 5.3.3, iteration can be avoided in the inner (backfitting) loop because $\hat{\boldsymbol{\beta}}$ and $\hat{\gamma}$ can be solved for explicitly. Using a derivation similar to that leading up to (6.16), an explicit expression for the Newton-Raphson update is

$$\begin{aligned} \hat{\boldsymbol{\beta}} &= \{\mathbf{X}^T\mathbf{A}(\mathbf{I} - \mathbf{S}_2)\mathbf{X}\}^{-1}\mathbf{X}^T\mathbf{A}(\mathbf{I} - \mathbf{S}_2)\mathbf{z} \\ \hat{\boldsymbol{\gamma}} &= \mathbf{S}_2(\mathbf{z} - \mathbf{X}\hat{\boldsymbol{\beta}}) \end{aligned} \tag{6.22}$$

where $\mathbf{X}$ represents the regression matrix for the $\mathbf{x}_j$s, $\mathbf{S}_2$ computes a weighted cubic smoothing spline on the variable t_i with weights given by $\mathbf{A} = -\partial^2 l/\partial\boldsymbol{\eta}\boldsymbol{\eta}^T$, $\mathbf{z}$ is the adjusted dependent variable $\boldsymbol{\eta}^0 + \mathbf{A}^{-1}\mathbf{u}$, and $\mathbf{u} = \partial l/\partial\boldsymbol{\eta}$. The outer loop must still be iterated, updating $\mathbf{u}$, $\boldsymbol{\eta}$, $\mathbf{A}$ and $\mathbf{z}$ each time (Green and Yandell, 1985).

When t_i is a vector, penalties appropriate for smoothing splines of dimension larger than one can be used, and will lead to the same equations with $\mathbf{S}_2$ denoting the higher dimensional spline smoother. Alternatively, any one or higher dimensional smoother can be used to replace $\mathbf{S}_2$ in (6.22).

One consideration that makes the semi-parametric generalized linear model attractive (relative to the full additive model) is that simplifications occur in the computation of degrees of freedom and standard errors (next section and Exercise 6.7).

The asymptotic properties of these models have been studied in some detail. Of interest is whether $\hat{\boldsymbol{\beta}}$ has the usual $\sqrt{n}$ rate of convergence, despite the nonparametric adjustment. This is not the case for the estimator (6.22), while a close relative does achieve the optimal rate. The modified estimate (Denby, 1986; Speckman, 1988) is

$$\begin{aligned} \hat{\boldsymbol{\beta}} &= \{\mathbf{X}^T\mathbf{A}(\mathbf{I} - \mathbf{S}_2)^2\mathbf{X}\}^{-1}\mathbf{X}^T\mathbf{A}(\mathbf{I} - \mathbf{S}_2)^2\mathbf{z} \\ \hat{\boldsymbol{\gamma}} &= \mathbf{S}_2(\mathbf{z} - \mathbf{X}\hat{\boldsymbol{\beta}}). \end{aligned} \tag{6.23}$$

Notice that $\hat{\boldsymbol{\beta}}$ is the coefficient of the regression of $(\mathbf{I} - \mathbf{S}_2)\mathbf{z}$ on $(\mathbf{I} - \mathbf{S}_2)\mathbf{X}$, using weight matrix $\mathbf{A}$, and hence has a "partial-residual" flavour. One way of interpreting the difference between (6.23) and (6.22) is that in order to achieve the optimal rate for $\hat{\boldsymbol{\beta}}$, we need to *undersmooth* when adjusting $\mathbf{X}$ for the covariate t, while the estimate of γ is not undersmoothed.

6.8 Inference

6.8.1 *Analysis of deviance*

The deviance, or likelihood-ratio statistic, for a fitted model $\hat{\boldsymbol{\mu}}$ is defined by

$$D(\mathbf{y}; \hat{\boldsymbol{\mu}}) = 2\{l(\boldsymbol{\mu}_{\max}; \mathbf{y}) - l(\hat{\boldsymbol{\mu}}; \mathbf{y})\} \tag{6.24}$$

where $\boldsymbol{\mu}_{\max}$ is the parameter value that maximizes $l(\boldsymbol{\mu}; \mathbf{y})$ over all $\boldsymbol{\mu}$ (the *saturated model*). We sometimes unambiguously use $\hat{\boldsymbol{\eta}}$ as the argument of the deviance rather than $\hat{\boldsymbol{\mu}}$. The deviance plays the role of the residual sum of squares for generalized models, and can be used for assessing goodness-of-fit and for comparing models.

The asymptotic distribution theory for generalized linear models is well known. Suppose η_1 and η_2 are two linear models, with η_1 nested within η_2. Then under appropriate regularity conditions, and assuming η_1 is correct, $D(\hat{\boldsymbol{\eta}}_2; \hat{\boldsymbol{\eta}}_1) = D(\mathbf{y}; \hat{\boldsymbol{\eta}}_1) - D(\mathbf{y}; \hat{\boldsymbol{\eta}}_2)$ has an asymptotic χ^2 distribution with degrees of freedom equal to the difference in the dimensions of the two models. This result is used extensively for comparing models, and often presented in the form of an analysis of deviance table such as Table 6.2. If the dispersion parameter is unknown, an approximate F-test can be derived in a similar fashion.

For nonparametric and additive models, the deviance still makes sense as a means for assessing models and their differences. The distribution theory, however, is undeveloped. We nevertheless perform informal deviance tests with some heuristic justification.

In section 6.8.3 below we define df^{err} for generalized additive models as the expected value of a quadratic approximation to the deviance. If the asymptotic distribution of the deviance were χ^2, then df^{err} would be its degrees of freedom. Although the deviance is not χ^2 distributed, not even asymptotically, simulations have shown that the χ^2 distribution is still a useful approximation for screening models. At the end of section 10.2 we summarize one such simulation. We therefore use the χ^2 as the reference distribution in an informal way for comparing models.

There are other ways to make inferences about additive fits. The nonparametric additive model may be viewed as a diagnostic for identifying functional forms. The fitted functions can be used to inspire parsimonious reparametrizations of some of the variables, for example, using logs, inverse, polynomial terms, etc. Further selection and deviance tests can then be based on the reparametrized model.

Another approach is to use the bootstrap to estimate the distribution of the deviance for an effect, under the null hypothesis that it is absent. The simulation in section 10.2 is an example. Although useful for justifying the simpler χ^2 approximation, such simulations are unlikely to be used routinely in model building because they are computationally expensive.

6.8.2 *Standard error bands*

The methodology discussed in Chapter 5 for degrees of freedom and standard-error bands can be extended in a heuristic fashion to the generalized additive model, by expanding the solution $\hat{\boldsymbol{\eta}}$ of the local-scoring algorithm about the true value $\boldsymbol{\eta}_0$.

Each step of local-scoring consists of a backfitting loop applied to the adjusted dependent variable $\mathbf{z}$, with weights $\mathbf{A}$ given by the estimated information matrix. If $\mathbf{R}$ is the weighted additive-fit operator, then at convergence

$$\begin{aligned}\hat{\boldsymbol{\eta}} &= \mathbf{R}(\hat{\boldsymbol{\eta}} + \mathbf{A}^{-1}\hat{\mathbf{u}}) \\ &= \mathbf{Rz},\end{aligned} \tag{6.25}$$

where $\hat{\mathbf{u}} = \partial l/\partial\hat{\boldsymbol{\eta}}$. The idea is to approximate $\mathbf{z}$ by an asymptotically equivalent quantity $\mathbf{z}_0$, assuming the model is consistent. For the calculations here, this amounts to approximating the covariance of $\mathbf{z}$ by that of $\mathbf{z}_0$. In what follows, $\approx$ means "asymptotically equal to" (under unspecified assumptions). Expanding $\hat{\mathbf{u}}$ to first order about the true $\boldsymbol{\eta}_0$, we get $\mathbf{z} \approx \mathbf{z}_0 = \boldsymbol{\eta}_0 + \mathbf{A}_0{}^{-1}\mathbf{u}_0$, which has mean $\boldsymbol{\eta}_0$ and variance $\mathbf{A}_0{}^{-1}\phi \approx \mathbf{A}^{-1}\phi$. The situation is almost the same as in section 5.4.4. There the fitted additive predictor is $\hat{\boldsymbol{\eta}} = \mathbf{Ry}$, where $\mathbf{y}$ has covariance $\sigma^2\mathbf{I}$. Here $\hat{\boldsymbol{\eta}} = \mathbf{Rz}$, and $\mathbf{z}$ has asymptotic covariance $\mathbf{A}_0^{-1}$. $\mathbf{R}$ is not a linear operator due to its dependence on $\hat{\boldsymbol{\mu}}$ and thus $\mathbf{y}$ through the weights, so we need to use its asymptotic version $\mathbf{R}_0$ as well. We therefore have

$$\begin{aligned}\operatorname{cov}(\hat{\boldsymbol{\eta}}) &\approx \mathbf{R}_0\mathbf{A}_0^{-1}\mathbf{R}_0^T\phi \\ &\approx \mathbf{R}\mathbf{A}^{-1}\mathbf{R}^T\phi.\end{aligned} \tag{6.26}$$

Similarly,

$$\operatorname{cov}(\hat{\mathbf{f}}_j) \approx \mathbf{R}_j\mathbf{A}^{-1}\mathbf{R}_j^T\phi,$$

where $\mathbf{R}_j$ is the matrix that produces $\hat{\mathbf{f}}_j$ from $\mathbf{z}$. In some models, such as the logistic model, the dispersion parameter ϕ is known and taken equal to one; in other models, it must be estimated, for example, by the mean error deviance.

The development so far mimics the asymptotics for generalized linear models. The usual regularity conditions are required, including consistency. Consistency implicitly requires that the amount of smoothing decreases at an appropriate rate (section 3.10). These arguments can be extended to show that $\hat{\boldsymbol{\eta}}$ is asymptotically distributed as $N(\boldsymbol{\eta}_0, \mathbf{R}_0\mathbf{A}_0^{-1}\mathbf{R}_0^T\phi)$. Recently Gu (1989) developed a Bayesian argument to approximate the posterior distribution of $\boldsymbol{\eta}$; for the situation he considered, which used smoothing splines, $\mathcal{L}(\boldsymbol{\eta}\,|\,\mathbf{y}) \approx N(\hat{\boldsymbol{\eta}}, \mathbf{R}\mathbf{A}^{-1}\phi)$.

6.8.3 *Degrees of freedom*

In Chapter 5 we define df^{err} in terms of the the expected value of the residual sum of squares. The analogous quantity here is the deviance. In order to generalize the results for the Gaussian additive model, it is convenient to use the asymptotic approximation to the deviance:

$$\begin{aligned}D(\mathbf{y};\boldsymbol{\mu}) &\approx (\mathbf{y}-\hat{\boldsymbol{\mu}})^T\mathbf{A}^{-1}(\mathbf{y}-\hat{\boldsymbol{\mu}}) \\ &\approx (\mathbf{z}-\hat{\boldsymbol{\eta}})^T\mathbf{A}(\mathbf{z}-\hat{\boldsymbol{\eta}}).\end{aligned} \tag{6.27}$$

Applying the same definition of df^{err} to the last expression in (6.27), we obtain

$$df^{\text{err}} = n - \operatorname{tr}(2\mathbf{R} - \mathbf{R}^T\mathbf{A}\mathbf{R}\mathbf{A}^{-1}) \tag{6.28}$$

(Exercise 6.2). If the model is unbiased, then $E(D) \approx df^{\text{err}}\phi$. Of more interest is the difference in deviance between two nested models. Suppose $\hat{\boldsymbol{\eta}}_1$ and $\hat{\boldsymbol{\eta}}_2$ differ by a single term, for example a nonparametric term in X_j. If the smaller model $\boldsymbol{\eta}_1$ is correct, then

$$\begin{aligned} ED(\hat{\boldsymbol{\eta}}_2;\hat{\boldsymbol{\eta}}_1)/\phi &= E\left\{D(\mathbf{y};\hat{\boldsymbol{\eta}}_1) - D(\mathbf{y};\hat{\boldsymbol{\eta}}_2)\right\}/\phi \\ &\approx \operatorname{tr}(2\mathbf{R}_1 - \mathbf{R}_1^T\mathbf{A}_1\mathbf{R}_1\mathbf{A}_1^{-1}) \\ &\qquad - \operatorname{tr}(2\mathbf{R}_2 - \mathbf{R}_2^T\mathbf{A}_2\mathbf{R}_2\mathbf{A}_2^{-1}) \\ &= df^{\,\mathrm{err}}(\hat{\boldsymbol{\eta}}_1) - df^{\,\mathrm{err}}(\hat{\boldsymbol{\eta}}_2) \\ &= df_j^{\,\mathrm{err}}, \end{aligned}$$

since the bias terms cancel. If the dispersion parameter ϕ is known, as is the case for the binomial and Poisson models, then we can approximate the asymptotic distribution of $D(\hat{\boldsymbol{\eta}}_2;\hat{\boldsymbol{\eta}}_1)$ by a $\chi^2_{df_j^{\,\mathrm{err}}}$ distribution. Two-moment corrections may be computed as in section 3.9 to improve the approximations. When the dispersion parameter ϕ is unknown, an approximate F-test is more appropriate.

Recall the motivation for $df = \operatorname{tr}(\mathbf{R})$ in terms of the C_p statistic; adding $(2df/n)\sigma^2$ to the average squared residual ASR makes it unbiased for the average squared prediction error. Here one can start with the weighted version of ASR, and proceed in a similar fashion. Alternatively, we can use the Kullback-Leibler distance to measure prediction errors:

$$PE = E\left\{\frac{1}{n}\sum_{i=1}^{n} D(Y_i^0;\hat{\mu}_i)\right\} \tag{6.29}$$

where Y_i^0 has the same distribution as the realization y_i, and $\hat{\mu}_i = \hat{\mu}(\mathbf{x}^i)$ is the additive fit based on $\mathbf{y}$. The AIC statistic is then defined to be

$$AIC = D(\mathbf{y};\hat{\boldsymbol{\mu}})/n + 2df\phi/n \tag{6.30}$$

where $df = \operatorname{tr}(\mathbf{R})$ makes AIC asymptotically unbiased for prediction error in (6.29) (Exercise 6.6). This quantity has the form of an Akaike-information criterion, and hence the name AIC.

As was the case for the additive model, routine computation of any of these quantities is time consuming in multiple predictor models. Some efficiency can be achieved for the semi-parametric model (Exercise 6.7). In the general case we often use as crude approximations $df_j = \operatorname{tr}(\mathbf{S}_j) - 1$ to measure the individual effect contributions $df_j^{\,\mathrm{err}}$, and approximate $df^{\,\mathrm{err}}$ by $n-1-\sum_{j=1}^{p}\{\operatorname{tr}(\mathbf{S}_j)-1\}$.

6.9 Smoothing parameter selection

There are a number of possible methods for automatic selection of the smoothing parameters in a generalized additive model. We discuss several approaches here, and postpone a more general treatment of model selection strategies until Chapter 9.

Suppose our model uses linear smoothers $\mathbf{S}_1, \ldots, \mathbf{S}_p$ with corresponding smoothing parameters $\lambda_1, \ldots \lambda_p$. In Chapter 3 we find that one can not simply minimize the residual sum of squares in order to choose a smoothing parameter in a single predictor setting, as this results in an overfit solution. The same applies here: if we simply minimize the deviance $D(\mathbf{y}; \boldsymbol{\mu}) = \sum_{i=1}^{n} D(y_i; \mu_i)$ over $\lambda_1, \ldots, \lambda_p$ we will drive the deviance to zero with an interpolating solution $\hat{\mu}_i = y_i$ for every i. We can avoid this by using cross-validation as in Chapter 3. Let $\hat{\mu}_i^{-i}$ be the fitted value at the ith design point, obtained by leaving the ith point out of the sample. Then the cross-validated deviance is defined to be

$$CV = \frac{1}{n} \sum_{i=1}^{n} D(y_i; \hat{\mu}_i^{-i}).$$

In principle one could minimize this quantity over $\lambda_1, \lambda_2, \ldots \lambda_p$, but this would be computationally expensive, requiring n complete applications of the local-scoring procedure for each trial value of $\lambda_1, \lambda_2, \ldots \lambda_p$. The simple computing formula (3.18) in section 3.4.3 is no longer applicable except for the Gaussian case.

As in the single smoother case there are a number of approximations to cross-validation that are more feasible computationally. As before denote by $\mathbf{R}$ the weighted additive-fit operator corresponding to the last iteration of the local-scoring procedure. By analogy to the generalized cross-validation score defined in section 3.4.5, we define the generalized cross-validated deviance:

$$GCV = \frac{\frac{1}{n} \sum_{i=1}^{n} D(y_i; \hat{\mu}_i)}{\{1 - \operatorname{tr}(\mathbf{R})/n\}^2}. \tag{6.31}$$

To derive this quantity, the argument of section 3.4.5, leading to generalized cross-validation as an approximation to cross-validation, can be adapted to this setting by considering the final step of the local-scoring procedure as a weighted additive fit.

Alternatively, we can use the AIC statistic defined in the previous section:

$$AIC = \frac{1}{n}\sum_{i=1}^{n} D(y_i; \hat{\mu}_i) + 2\text{tr}(\mathbf{R})\phi/n. \qquad (6.32)$$

As is true for generalized cross-validation and C_p (section 3.4.5), the AIC and generalized cross-validated deviance are equal to first order, since $1/(1-x)^2 \approx 1 + 2x$.

Both GCV and AIC require only a single application of the local scoring procedure for each value of $\boldsymbol{\lambda} = (\lambda_1, \lambda_2, \ldots \lambda_p)$, and hence are much more feasible to compute than is the cross-validated deviance. Gu and Wahba (1988) describe a Newton algorithm for minimizing GCV for the additive spline model; their algorithm requires $O(n^3)$ operations to compute.

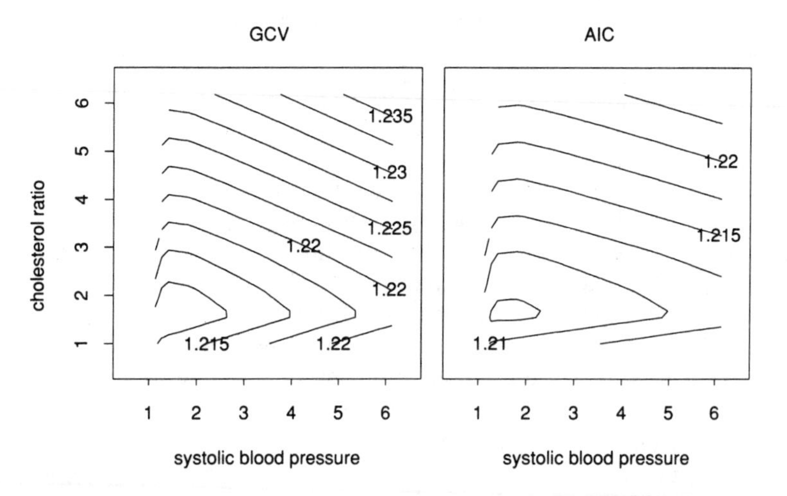

Fig. 6.6. *Contour plots of GCV and AIC as a function of the individual component degrees of freedom* $\text{tr}(\mathbf{S}_1) - 1$ *and* $\text{tr}(\mathbf{S}_2) - 1$ *for the heart attack data.*

Figure 6.6 shows contour plots of the generalized cross-validated deviance and AIC criteria as a function of the individual degrees of freedom $\text{tr}(\mathbf{S}_1) - 1$ and $\text{tr}(\mathbf{S}_2) - 1$ for the heart-attack data.

GCV has a minimum around (1.3,1.5), while AIC is around (1.8,1.7); a linear fit would have one df, so these numbers represent a lot of smoothing. Both criteria are quite flat in the region around their minima, especially in the northwest–southeast direction. This is most likely a consequence of the correlation of 0.16 between `systolic blood pressure` and `cholesterol ratio`. A rougher fit for either variable can pick up some of the effect of the other.

While there has been some mathematical study of these criteria, there has not, to our knowledge, been enough practical study of their use in additive models. In a sense, optimization of GCV is analogous to all-subsets regression. A different approach to this problem recognizes the interdependence of smoothing parameter selection and model selection. Techniques that tackle both problems together are discussed in Chapter 9.

6.10 Overinterpreting additive fits

With the added flexibility of nonparametric and additive regression models, there is always the risk of overfitting the data and interpreting spurious features in the fitted curves. Typically the width of standard-error bands, approximate deviance tests and residual plots give support to the important features of the fit, and give the analyst warnings about those features that are likely to be spurious.

Logistic regression models for 0-1 data are probably the most frequently used generalized linear or additive models. Unfortunately they are more prone to exhibit the problems outlined above, as we demonstrate in this section.

From one viewpoint the amount of information in a sample of binary observations is typically less than in a sample of quantitative observations. For example, the sample mean based on n realizations from a $N(\mu, \sigma^2)$ distribution has variance σ^2/n. In logistic regression, the parameters of interest are usually on the logit scale; the logit of the sample mean based on n observations from a $B(1, p)$ distribution has approximate variance $1/np(1-p) \geq 4/n$. So with binary data it is as if we are always limited to data with measurement error variance of at least four units, which in turn limits the signal to noise ratio.

The simulated example in this section demonstrates a more serious bias problem inherent in fitting logistic regression models, which is made worse as the model becomes more flexible. Figure 6.7

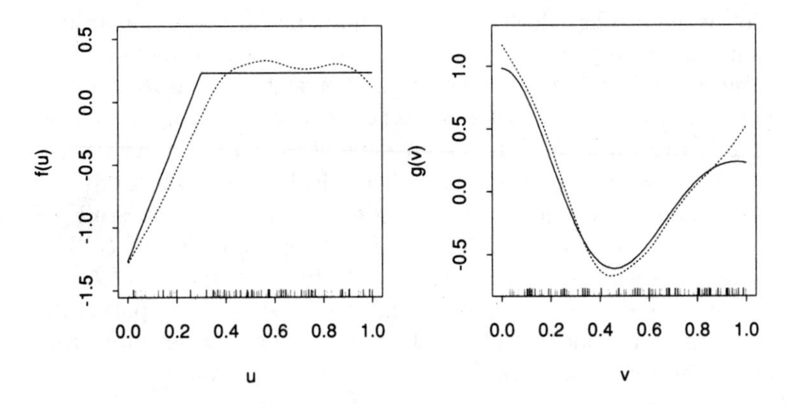

Fig. 6.7. *The solid curves define an additive model* $\eta(u,v) = -1 + f(u) + g(v)$. *The dotted curves represent the additive logistic smoothing spline fit to 250 binary responses generated with probability* $\mu(u,v) = \exp\{\eta(u,v)\}/[1+\exp\{\eta(u,v)\}]$, *where* u *and* v *were both generated from the uniform distribution.*

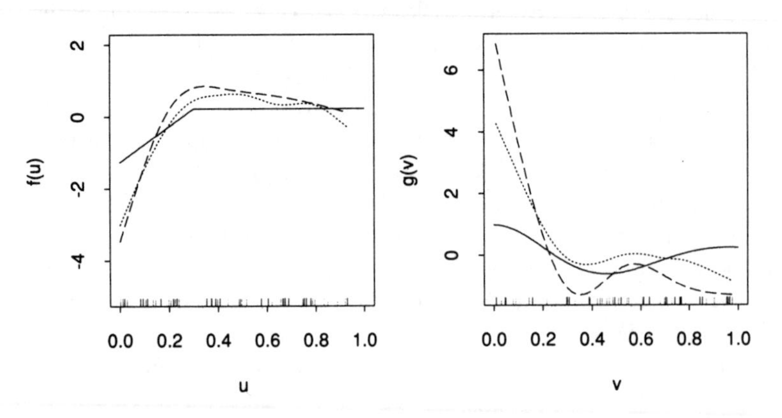

Fig. 6.8. *The solid curves are the model curves. The dotted curves represent the additive logistic smoothing spline fitted to 50 binary responses generated from the model. The dashed curves are similar, except natural regression splines were used in the fitting. Each estimate uses about 4 df.*

shows the two functions that define the additive logit model

$$\mu(u,v) = \frac{\exp\{-1 + f(u) + g(v)\}}{1 + \exp\{-1 + f(u) + g(v)\}}. \tag{6.33}$$

The dotted curves in Fig. 6.7 represent the additive fit to 250 binary responses generated from (6.33), where u and v were sampled independently and randomly from a uniform distribution. An additive logistic model with smoothing splines was used, with approximately 4 *df* for each term.

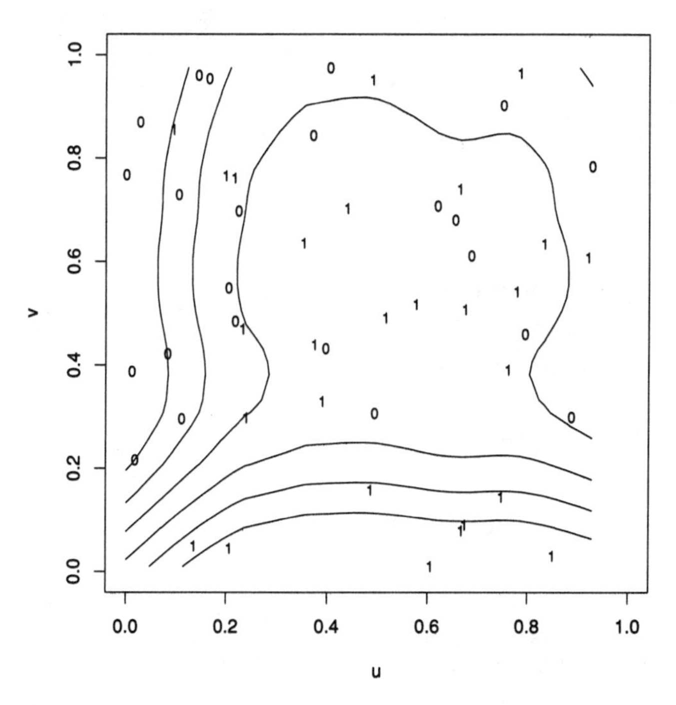

Fig. 6.9. *The 50 0-1 observations, together with contours of constant fitted probability, based on the additive smoothing spline model.*

Figure 6.8 is similar to Fig. 6.7, except here only 50 observations were generated from the model. Notice that the fitted functions seem to be blowing up, with the estimates for g about three times the size of g itself. The dashed curves in Fig. 6.8 represent a similar fit, except there natural splines with three interior knots were used for each term (also resulting in 4 *df*). The same phenomenon occurs.

In order to understand this behaviour, consider a simple linear logistic regression model with only one term, say x. It is easy to see

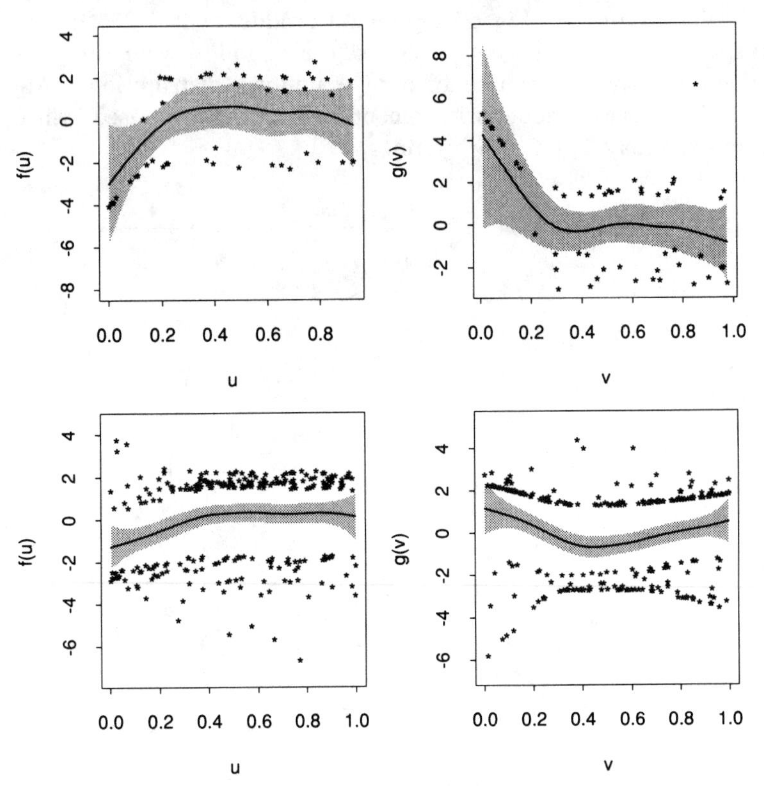

Fig. 6.10. *The top panel shows the fitted additive spline model for the size 50 sample, together with pointwise 2× standard-error bands and partial residuals. The lower panel is the same plot for the size 250 sample.*

that if all the zeros and ones in the sample are perfectly separated when plotted against x, then the maximum likelihood estimate of the slope parameter is infinite. Since there is a nonzero probability of such configurations occurring, the bias of $\hat{\beta}$ is infinite. Usually the bias is studied asymptotically; McCullagh and Nelder (1989, Exercise 4.29) give an expression for this asymptotic bias, and an approximation $E(\hat{\boldsymbol{\beta}}) \approx \boldsymbol{\beta} \times (1 + p/n)$ where p is the number of predictors in the linear model. This is rather modest compared to the small sample bias, which is infinite.

Typically there is some overlap in the configuration of zeros and ones, but if it is small, the estimated slope may still be inflated.

Clearly the less data there is, the more likely it is that these situations can occur.

The regression spline fit has eight basis functions, or variables, to model the bivariate response function. It seems clear that if *locally* there are pure regions of zeros and ones, then by inflating some of the coefficients, the model might be capable of producing extreme fitted probabilities in those regions. Although the analogy between regression splines and the simple single-variable example is crisper, the same phenomenon can occur with smoothing splines or other shrinking smoothers.

Figure 6.9 shows that this is indeed the case. There we see the 50 zeros and ones, and contours of constant fitted probability from the additive spline model. The responses are all one below $v = 0.2$, and the contours climb steeply there, as does the fitted function for g in Fig. 6.8. Similarly, left of $u = 0.2$, the responses are almost all zeros, and this accounts for the plunge of the fitted function for f in this region of the data. An even richer regression spline model (not shown) attempted to model the local clusters of zeros and ones in the interior.

There are some warnings to guide us in situations like this. The top panel of Fig. 6.10 shows the additive spline fit to the 50 observations, together with added partial residuals and approximate pointwise standard-error bands. The partial residuals are simply $(y_i - \hat{\mu}_i)/\{\hat{\mu}_i(1 - \hat{\mu}_i)\}$; what is plotted is $\hat{f}_j(x_{ij}) + (y_i - \hat{\mu}_i)/\{\hat{\mu}_i(1 - \hat{\mu}_i)\}$, whose weighted smooth is $\hat{f}_j$. The bands are wide in the offending areas, which is partly due to the boundary influence, but also due to the inflationary effect of small or large probabilities on the asymptotic standard errors. What is most noticeable is how the residuals track the fitted functions in the pure regions, and serve as a warning. The lower panel is the same plot for the fit to the 250 observations, plotted on the same scale. The bands are much narrower, but most noticeably there are both positive and negative residuals everywhere.

6.11 Missing predictor values

Missing values in the predictors are a thorn in the side of every data analyst. If only a few cases have missing observations on one or more predictors, they are usually omitted from an analysis, unless special circumstances require their presence. A messier situation is when many variables have missing components, but for different cases. Deleting entire cases that have missing observations on *any* variable can dramatically reduce the number of cases. Often the final model depends only on a small subset of the predictors, which makes this latter strategy even more wasteful.

A large variety of procedures have been proposed to handle missing data, and many of them operate by filling in, or *imputing* the missing values. The simplest case to handle is that in which the data are assumed to be *missing at random*, which in the present problem means roughly that the chance of a predictor value being missing is not a function of the response for that observation. We make that assumption here. Some procedures use all the available data, including the response, while others use only the predictors. The technique we describe here for additive models falls into the latter category, and is particularly convenient when the backfitting algorithm is used to fit the additive model.

All observations missing for a given predictor are omitted when the partial residuals are smoothed against that predictor in the backfitting algorithm, and their fitted values are set to zero. Since the fitted curves have mean zero (when the model includes an intercept), this amounts to assigning the average fitted value to the missing observations. This approach is analogous to the strategy in linear-models of replacing the missing observations for a given predictor by the mean of the available observations for that predictor.

6.12 Estimation of the link function

The models in this chapter assume a known link function between the mean of the response and the additive predictor. In some instances it is of interest to estimate the link function from the data, to increase the flexibility of the generalized additive model. For example, the logit function exhibits a certain symmetry: $g(\mu) =$

$\log\{\mu/(1-\mu)\}$, $g(1-\mu) = -g(\mu)$. It is not clear a priori that such symmetry should be assumed, and asymmetric link functions, such as the *complimentary log-log*, might be a useful alternative.

There are a number of approaches that have been suggested. Parametric link function estimation has been developed for generalized linear models, and could be easily incorporated into the estimation of a generalized additive model. Nonparametric estimation of the link function is possible, through a Gauss-Newton procedure, leading to a generalized version of the local-scoring procedure. We have experimented with this approach and had limited success (Hastie and Tibshirani, 1984); the algorithm exhibited numerical instabilities when the estimated link function became very flat. Monotone regression splines might provide a more stable basis for estimation.

There are two other approaches, quite different from those above, that have been suggested for estimation of the link function in a linear model (extensions to generalized linear or generalized additive models seem plausible but to our knowledge have not been attempted). The *Average Derivative Estimation* technique uses a multivariate density estimate of the predictors to derive an estimate of the link function. *Slicing regression* uses inverse regression of the predictors on the response to obtain a link-free estimate of the regression parameters; an estimate of the link function can then be obtained through a simple smoothing operation.

This is a relatively new area, and much work is needed in the development and comparison of these methods. A number of references are given in the bibliographic notes.

6.13 Local-likelihood estimation

We end this chapter with brief description of a somewhat different method for estimating a generalized additive model, called *local-likelihood* estimation. Suppose we have a model with a log-likelihood of the form $l(\boldsymbol{\eta}; \mathbf{y}) = \sum_{i=1}^{n} l(\eta_i; y_i)$. For simplicity, assume first that we have a single predictor X so that $\eta_i = f(x_i)$ for some function f, and let's focus on the estimation of $f(x_0)$ for some point x_0. Local-likelihood estimation assumes a linear approximation of the form

$$f(x) \approx \alpha_0 + \beta_0 x \quad \text{for} \quad x \in N(x_0), \tag{6.34}$$

where $N(x_0)$ is some neighbourhood of x_0. The parameters α_0 and β_0 are estimated by maximizing the local log-likelihood

$$\sum_{i=1}^{n} w_{0i}\, l(\alpha_0 + \beta_0 x_i; y_i). \tag{6.35}$$

The w_{0i} are weights that decrease as x_i increases in distance from x_0, so that (6.35) is somewhat *local* to x_0. The idea is that for x_i near x_0, approximation (6.34) should be a reasonable one. This forms a natural extension to the locally-weighted running-line smoothers of section 10.2. A particularly simple choice of weights is to set $w_{0j} = 1$ if $j \in N(x_0)$ and zero otherwise. In that case (6.35) is just a standard log-likelihood for the reduced data set consisting of points in the neighbourhood $N(x_0)$ of x_0. Alternatively, one may use weights that decrease in a smooth fashion away from x_0, defined by a kernel as in Chapter 2.

Denoting the maximizers by $\hat{\alpha}_0$ and $\hat{\beta}_0$, the estimated function is given by $\hat{f}(x_0) = \hat{\alpha}_0 + \hat{\beta}_0 x_0$. The estimation procedure is repeated for each x_0 for which an estimate at f is required, typically $x_0 \in \{x_1, \ldots, x_n\}$. The simpler approximation $f(x_0) \approx \alpha_0$ may be used in place of the linear approximation above; the slopes are included to reduce bias at the endpoints, as is the case for the running-line smoothers in Chapter 2.

With multiple predictors, one can use multi-dimensional neighbourhoods, and estimate surfaces in the same way. It is not necessary that the log-likelihood be of the form (6.35) that arises from a parametric model for independent observations. In particular extensions to partial-likelihood models, and models with complicated dependence structures fall easily into this framework. Whenever a likelihood or criterion is available, it can be converted to a local-likelihood. Local-likelihood models, on the other hand, are not naturally suitable for additive modelling, although additive versions have been proposed.

Not surprisingly, local-likelihood estimation bears a strong similarity to the local-scoring procedure. In the Gaussian case, local-likelihood estimation is exactly the same as a locally-weighted running-line smoothing. In other nonlinear models, the first iteration of the local-scoring procedure coincides with local-likelihood estimation, but the two methods differ in subsequent iterations. In our view, local-likelihood estimation suffers from a number of drawbacks, due largely to the fact that it is an implicit, rather than explicit, smoothing method. Specifically,

(i) it does not have a clear justification through penalized likelihood considerations,
(ii) it cannot easily incorporate different smoothing methods, and
(iii) its estimation procedure is costly.

Despite these points, local-likelihood estimation deserves further study.

6.14 Bibliographic notes

Generalized linear models were proposed as a comprehensive class of models by Nelder and Wedderburn (1972); McCullagh and Nelder (1989) gave a thorough and up-to-date account. Generalized additive models were introduced by Hastie and Tibshirani (1984, 1986). Further developments of the models and some detailed applications were given in Hastie and Tibshirani (1985, 1987a and 1987b). Penalized-likelihood estimation was proposed by Good and Gaskins (1971). Green and Yandell (1985) and Green (1987) proposed the semi-parametric generalized linear model. Green gave a more general framework than that given in section 6.7, with arbitrary likelihood and a general penalty functional. Green discussed many aspects of the problem including methods for inference and smoothing parameter selection. Yandell and Green (1986) developed diagnostic methods for semi-parametric models. O'Sullivan (1983) studied the theoretical properties of models consisting of a single nonparametric term. He established conditions for the existence and uniqueness of the generalized penalized likelihood problem. O'Sullivan, Yandell and Raynor (1986) described an algorithm for the estimation of this model, essentially a local-scoring procedure with a single *thin plate* spline smoother. A simple extension of O'Sullivan's results for additive spline models was given by Buja, Hastie and Tibshirani (1989).

For the semi-parametric additive model, Heckman (1986) proved consistency of the estimate of $\boldsymbol{\beta}$ and gave a Bayesian interpretation. Rice (1986) gave a further analysis of this model, giving rates of convergence for the bias and variance. Denby (1986) and Speckman (1988) proposed the alternative estimator (6.23), and Speckman showed its superior convergence properties.

Stone (1986) extended his results (Stone, 1985) on the rates of convergence for additive models to generalized additive models, using regression smoothers. Burman (1985) also studied the estimation of generalized additive models. Cox and O'Sullivan (1989) studied convergence rates for penalized log-likelihood estimates. Gu (1989) derived the approximate posterior distribution in section 6.8.2. Cross-validation and related topics for generalized additive models (or similar settings) were discussed by O'Sullivan, Yandell and Raynor (1986), Green (1987), Gu and Wahba (1988), Gu, Bates, Chen and Wahba (1988), and Gu (1989). Akaike (1973) proposed the *AIC* criterion for model checking.

Bacchetti (1989) studied an additive model for binary data in which the dependence on each predictor was assumed to be monotone. He suggested a cyclical algorithm for its estimation. Ramsay (1988) used monotone splines in a variety of settings, including likelihood-based regression models.

Little and Rubin (1986) discuss a variety of methods for dealing with missing data. Brant and Tibshirani (1990) explore approaches more sophisticated than the method discussed here, which involve modelling of the joint distribution of the predictors. This is an area of current research.

Estimation of the link function and score tests for assessing the need for a link function modification were discussed in Pregibon (1979, 1980). Hastie and Tibshirani (1984) described nonparametric estimation of the link function in generalized additive models. The average derivative estimate (ADE) was proposed by Härdle (1989); slicing regression is due to Duan and Li (1988). Li and Duan (1989) studied the effects of incorrect link function specification.

Some recent applications of smoothing and generalized additive models include Fowlkes (1987), Azzalini, Bowman and Härdle (1989) and Lambert (1989). Hastie and Herman (1990) gave an an analysis in which markedly different results are obtained using a generalized additive model compared to simply examining the univariate smooths for covariate effects.

Local-likelihood estimation was studied by Brillinger (1977), Tibshirani (1984), Tibshirani and Hastie (1987) and Staniswalis (1989). Gentleman (1988) presented a modification of the local-likelihood method for the proportional hazards model.

6.15 Further results and exercises 6

6.1 Consider a model with log-likelihood l and n-dimensional parameter $\boldsymbol{\eta}$, with $\eta_i = \eta(x_i)$ modelled as $\eta_i = \beta_0 + \sum_{j=1}^{p} x_{ij}\beta_j$, $i = 1, \ldots, n$. Let $\mathbf{X}$ be the $n \times p + 1$ design matrix. Denote the score and expected Fisher-information for $\boldsymbol{\beta}$ by $\mathcal{S}_\beta = \partial l / \partial \boldsymbol{\beta}$ and $\mathcal{I}_\beta = E(-\partial^2 l / \partial \boldsymbol{\beta} \boldsymbol{\beta}^T)$, and that for $\boldsymbol{\eta}$ by $\mathcal{S}_\eta = \partial l / \partial \boldsymbol{\eta}$ and $\mathcal{I}_\eta = E(-\partial^2 l / \partial \boldsymbol{\eta} \boldsymbol{\eta}^T)$. Given a starting estimate $\boldsymbol{\beta}^0$, the Fisher scoring step for computing the maximum likelihood estimate of $\boldsymbol{\beta}$ is given by $\boldsymbol{\beta}^1 = \boldsymbol{\beta}^0 + \mathcal{I}_{\beta^0}^{-1} \mathcal{S}_{\beta^0}$.

(i) Show that

$$\boldsymbol{\beta}^1 = (\mathbf{X}^\mathbf{T} \mathcal{I}_{\eta^0} \mathbf{X})^{-1} \mathbf{X}^\mathbf{T} \mathcal{I}_{\eta^0} (\boldsymbol{\eta}^\mathbf{0} + \mathcal{I}_{\eta^0}^{-1} \mathcal{S}_{\eta^0})$$

where $\boldsymbol{\eta}^0 = \mathbf{X}\boldsymbol{\beta}^\mathbf{0}$. Interpret this as a weighted regression of the adjusted dependent variable $\boldsymbol{\eta}^0 + \mathcal{I}_{\eta^0}^{-1} \mathcal{S}_{\eta^0}$ on $\mathbf{X}$ with weight matrix $\mathcal{I}_{\eta^0}$. Note that we haven't assumed that the model is an (exponential family) generalized linear model.

(ii) Show that $\mathcal{I}_\beta$ can be replaced by the observed Fisher information $I_\beta = -\partial^2 l / \partial \boldsymbol{\beta} \boldsymbol{\beta}^T$ and the resulting algorithm is equivalent to the Newton-Raphson procedure. [Green, 1984, Jørgensen, 1984].

(iii) Conversely, show that $\mathcal{I}_\eta$ can be replaced by any positive definite matrix and this will produce some new algorithm for finding $\hat{\boldsymbol{\beta}}$. This general procedure is called the *delta algorithm* [Jørgensen, 1984].

(iv) When the model is an exponential family generalized linear model, show that $\mathcal{S}_\eta$ has components $(y_i - \mu_i) V_i^{-1} (\partial \mu_i / \partial \eta_i)$ and that $\mathcal{I}_\eta$ is a diagonal matrix with diagonal elements $(\partial \mu_i / \partial \eta_i)^2 V_i^{-1}$, where μ_i and V_i are the mean and variance of Y. Hence show that the adjusted dependent variable and weights are simply

$$z_i = \eta_i^0 + (y_i - \mu_i^0) \left(\frac{\partial \eta_i}{\partial \mu_i} \right)_0$$

and

$$w_i^{-1} = \left(\frac{\partial \eta_i}{\partial \mu_i} \right)_0^2 V_i^0$$

as stated in section 6.1. (Wedderburn, 1974; McCullagh and Nelder, 1989)

6.2 Verify expression (6.28) for the degrees of freedom in a generalized additive model. Show that if $\mathbf{R}$ defines a weighted projection operator, that $df^{\text{err}} = n - p$, where p is the dimension of the projection space. If $\mathbf{R}$ defines an additive smoothing-spline fit, then $\text{tr}(\mathbf{R}^T\mathbf{A}\mathbf{R}\mathbf{A}^{-1}) = \text{tr}(\mathbf{R}^2)$.

6.3 The Kullback-Leibler distance between a model with true parameter η^* and one with parameter η is defined as $K(\eta^*;\eta) = E_{\eta^*}\log\{h(Y;\eta^*)/h(Y;\eta)\}$ where h is the density of Y. We can regard this equivalently as a measure of distance between the two parameters η^* and η, or even the associated means μ^* and μ. Derive the following decompositions, one for squared error, the other for Kullback-Leibler distance in the exponential family:

$$E_{XY}\{Y-\mu(X)\}^2 = E_{XY}\{Y-\mu^*(X)\}^2 + E_X\{\mu^*(X)-\mu(X)\}^2$$
$$E_{XY}K\{Y;\mu(X)\} = E_{XY}K\{Y;\mu^*(X)\} + E_XK\{\mu^*(X);\mu(X)\}$$

where $\mu^*(X)$ is the true conditional mean. Hence show that if we minimize the expected Kullback-Leibler distance from future observations $EK\{Y,\mu(X)\}$, then we obtain the model $\mu(X)$ closest to $\mu^*(X)$. If μ is unrestricted, the minimum is achieved at $\mu = \mu^*$. If the distribution is Gaussian, the Kullback-Leibler distance becomes 1/2 times squared error. Since $EK\{Y,\mu(X)\} = E\log h(Y,Y) - E\log h(Y,\mu)$, we see that this is equivalent to maximizing the expected log-likelihood.
[Hastie and Tibshirani, 1986]

6.4 Derive formula (6.12) leading to the general form of the local scoring procedure. In the exponential family case show that this simplifies to the local scoring procedure of section 6.3.

6.5 Consider a local-scoring procedure for an additive model using smoothing splines. In order to track the estimation criterion for convergence, we need to compute the penalty terms at each iteration. Suppose the iterate $\boldsymbol{\eta}$ is obtained by fitting a weighted additive spline fit (using backfitting) to the adjusted dependent vector $\mathbf{z}$ with diagonal weight matrix $\mathbf{W}$. Show that the additive penalty is given by $\sum_{j=1}^{p}\lambda_j\mathbf{f}_j^T\mathbf{K}_j\mathbf{f}_j = \boldsymbol{\eta}^T\mathbf{W}(\mathbf{z}-\boldsymbol{\eta})$.

6.6 Show that $AIC = D(\mathbf{y}; \hat{\boldsymbol{\mu}})/n + 2df\,\phi/n$ in section 6.8.3 where $df = \mathrm{tr}(\mathbf{R})$ is approximately unbiased for the prediction error

$$PE = E\Big\{\frac{1}{n}\sum_{i=1}^{n} D(Y_i^0; \hat{\mu}_i)\Big\}.$$

6.7 Show that the additive operator $\mathbf{R}$ in $\hat{\boldsymbol{\eta}} = \mathbf{Rz}$ for the semi-parametric model (6.22) is given by

$$\mathbf{R} = (\mathbf{I} - \mathbf{S}_2)\mathbf{X}^T\{\mathbf{X}^T\mathbf{A}(\mathbf{I} - \mathbf{S}_2)\mathbf{X}\}^{-1}\mathbf{X}^T\mathbf{A}(\mathbf{I} - \mathbf{S}_2) + \mathbf{S}_2.$$

What is the order of computations required to compute $\mathrm{tr}(\mathbf{R})$? Give an asymptotic expression for $\mathrm{cov}(\hat{\boldsymbol{\beta}})$; what does it cost to compute?

6.8 Fit a Poisson generalized additive model to the ozone data, and compare the fit to a Gaussian additive model applied to the logarithm of the response `Daggert pressure gradient`.

6.9 Show that if the n-vector $\mathbf{z}$ has a diagonal covariance matrix $\mathbf{V}$ and $\mathbf{H}$ is the weighted least-squares operator matrix for regressing $\mathbf{z}$ onto the columns of $\mathbf{X}$ with weights $\mathbf{W} = \mathbf{V}^{-1}$, then $\sum_{i=1}^{n} w_i \mathrm{var}(\hat{z}_i) = \mathrm{tr}(\mathbf{H}) = \mathrm{rank}(\mathbf{X})$ where $\hat{z}_i$ is the ith fitted value. If instead $\mathbf{R}$ is a weighted additive spline operator, show that $\sum_{i=1}^{n} w_i \mathrm{var}(\hat{z}_i) = \mathrm{tr}(\mathbf{R}^2)$, and thus $df^{\,\mathrm{err}} = n - \mathrm{tr}(2\mathbf{R} - \mathbf{R}^2)$.

CHAPTER 7

Response transformation models

7.1 Introduction

In the previous chapter we describe models that generalize the additive model by allowing the response Y to have any distribution in the exponential family, and further by incorporating a monotone relationship or link function between the conditional expectation of Y given the predictors $X_1, X_2, \ldots X_p$ and the additive predictor $\alpha + \sum_{j=1}^{p} f_j(X_j)$. The transformation models described in this chapter generalize the additive model in a different way: they include a transformation of the response Y. The working model here is

$$\theta(Y) = \alpha + \sum_{j=1}^{p} f_j(X_j) + \varepsilon \tag{7.1}$$

where ε has mean zero and is independent of the X_js. The transformation $\theta(Y)$, like the $f_j(X_j)$s, is an arbitrary smooth function.

Before discussing this class of models, it is important to consider why we would want to transform the response, and the pros and cons of this approach. The main reason for allowing a transformation $\theta(Y)$ is that in some cases a simple model may not be appropriate for $E\{Y \mid X_1, \ldots, X_p\}$, but may be quite appropriate for $E\{\theta(Y) \mid X_1, \ldots, X_p\}$. For example if $Y = \exp(X_1 + X_2^2)\varepsilon$, then a simple additive model describes $\log(Y)$ but not Y. Depending on our application, we may be quite interested in this simple relationship between $\log(Y)$ and X_1 and X_2, because it tells us something about the underlying process that generated the data. On the other hand, we might be interested exclusively in describing the mean of Y (and not the mean of some transformation of Y) and therefore transforming the response would not make sense. What if our purpose is prediction? It depends on what we want to predict. Model (7.1) could be effective for predicting $\theta(Y)$, but the obvious predictor for Y, $\hat{\theta}^{-1}\{\alpha+\sum_1^p \hat{f}_j(X_j)\}$ (assuming $\hat{\theta}(Y)$ is invertible) may or may not work well. The reason is that the estimation procedures that we discuss for $\theta(Y)$ and $f_1(X_1),\ldots,f_p(X_p)$ are not driven by prediction error for Y. We return to the prediction issue later in the chapter.

The first part of the chapter is devoted to the "ACE" (Alternating Conditional Expectation) algorithm, a nonparametric generalization of the additive model that fits an additive model as part of an alternating estimation procedure. Some very interesting theory exists for this method and we describe this theory and explore the close connection of ACE to (standard) canonical-correlation analysis.

The ACE procedure doesn't require the assumption of a model such as (7.1), but works only on the joint distribution of Y and $X_1,\ldots,X_p$. This makes the method very flexible but may cause some of the anomalies of ACE in regression settings. In fact, we argue that ACE is more of a tool for correlation than regression. We then describe alternative methods that are designed for regression models, including the "AVAS" technique, and generalizations of the familiar Box-Cox procedure for parametric-regression analysis. Finally, prediction and inference are discussed for response-transformation models.

7.2 The ACE algorithm

7.2.1 *Introduction*

Suppose first we have two random variables Y and X, and we seek transformations $\theta(Y)$ and $f(X)$ so that $E\{\theta(Y)\,|\,X\}\approx f(X)$. The ACE algorithm approaches this problem by minimizing the squared-error loss

$$E\{\theta(Y)-f(X)\}^2.$$

Note that for fixed θ, the minimizing f is $f(X)=E\{\theta(Y)\,|\,X\}$, and conversely, for fixed f the minimizing θ is $\theta(Y)=E\{f(X)\,|\,Y\}$. This is the key idea in the ACE algorithm: it begins with some starting functions and alternates between these two steps until convergence. Note however that the zero functions trivially minimize squared error, so we must not allow the iterates in the procedure to shrink down to zero. ACE does this by standardizing $\theta(Y)$ so that it has unit variance at each step. We summarize ACE in algorithm 7.1.

Algorithm 7.1 *The ACE algorithm for two random variables*

(i) *Initialize:* set $\theta(Y) = \{Y - E(Y)\}/\{\text{var}(Y)\}^{1/2}$
(ii) *Compute:* $f(X) = E\{\theta(Y) \mid X\}$ to obtain a new f
(iii) *Compute:* $\tilde{\theta}(Y) = E\{f(X) \mid Y\}$ and standardize $\theta(Y) = \tilde{\theta}(Y)/\text{var}\{\tilde{\theta}(Y)\}^{1/2}$ to obtain a new θ.
(iv) *Alternate:* steps (ii) and (iii) until $E\{\theta(Y) - f(X)\}^2$ doesn't change.

Since $E[\{\theta(Y)+c\}-\{f(X)+c\}]^2 = E[\theta(Y)-f(X)]^2$, there is a free location parameter in the mean squared-error criterion. Without loss of generality we can therefore fix each function to have mean 0; in the algorithm this is ensured by starting with $\theta(Y)$ having mean zero. If we have data instead of random variables, the conditional expectation operators are replaced by smoothers, as in other algorithms in this book. The ACE algorithm alternates between smoothing $\theta(Y)$ on X to get a new $f(X)$, and $f(X)$ on Y to get a new $\theta(Y)$, until the mean-squared error doesn't change. This is illustrated in Fig. 7.1.

With multiple predictors $X_1, \ldots, X_p$, the ACE procedure seeks to minimize

$$E\Big\{\theta(Y) - \sum_{j=1}^{p} f_j(X_j)\Big\}^2.$$

The only change needed to the above algorithm is in step (ii), where we now fit an additive model to $\theta(Y)$. Details are given in algorithm 7.2.

In the next section we study the theoretical properties of the ACE procedure; here we first give an example of ACE applied to the ozone concentration data, which is analysed in more detail in Chapter 10. The response is `ozone concentration` with nine predictors in all. The ACE algorithm produced the transformations shown in Fig. 7.2 (for brevity we show only a few of the predictor

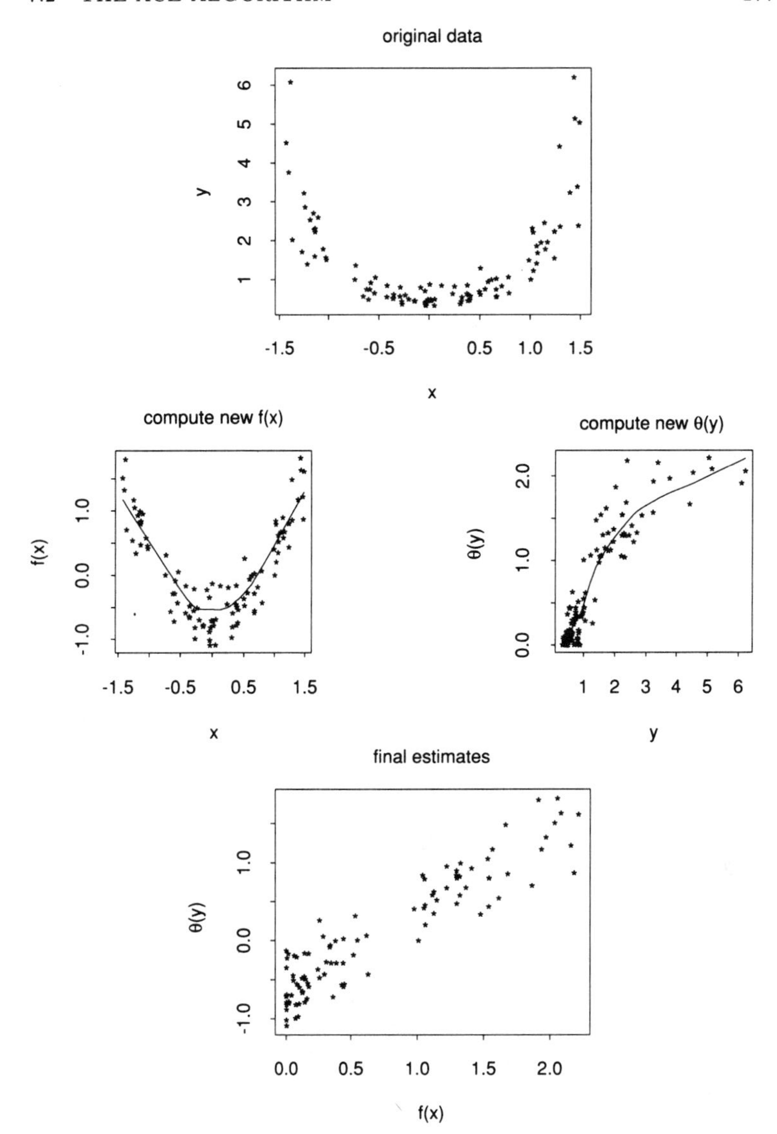

Fig. 7.1. *Illustration of the ACE algorithm. The top panel shows the untransformed data. The two middle panels show the smoothing steps that are alternated until convergence. The bottom panel shows a plot of the final estimates, $\theta(Y)$ plotted against $f(X)$. Note that $\theta(Y) \approx f(X)$.*

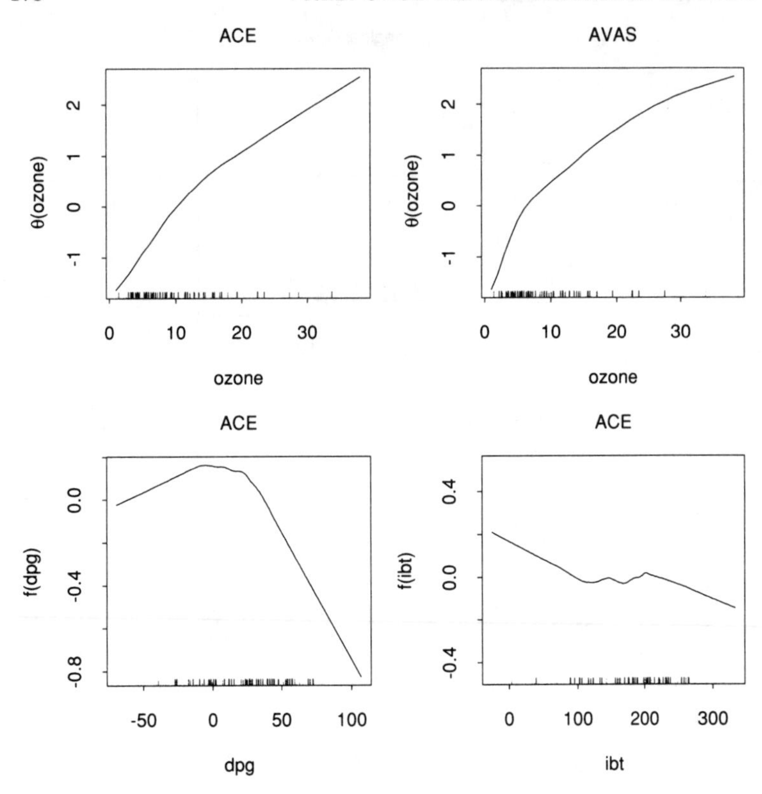

Fig. 7.2. *Transformations for the ozone data. Notice that the transformation of the response* **ozone** *is logarithmic in shape. The top right panel is the* **ozone** *transformation from AVAS, also logarithmic in shape. The predictor transformations from AVAS are similar to those from ACE and are not shown.*

Algorithm 7.2 *The ACE algorithm for multiple predictors*

(i) *Initialize:* set $\theta(Y) = \{Y - E(Y)\}/\{\text{var}(Y)\}^{1/2}$

(ii) Fit an additive model to $\theta(Y)$ to obtain new functions $f_1(X_1), \ldots, f_p(X_p)$.

(iii) *Compute:* $\tilde{\theta}(Y) = E\{\sum_j f_j(X_j) \mid Y\}$ and standardize $\theta(Y) = \tilde{\theta}(Y)/[\text{var}\{\tilde{\theta}(Y)\}]^{1/2}$ to give a new $\theta(Y)$.

(iv) *Alternate:* steps (ii) and (iii) until $E\{\theta(Y) - \sum_j f_j(X_j)\}^2$ doesn't change.

---

transformations). The transformation of `ozone concentration` looks logarithmic in shape, `Daggert pressure gradient` has a quadratic effect, while `inversion base temperature` is also non-linear.

7.2.2 *ACE in L_2 function spaces*

In Chapter 5 we provide two different kinds of justification for the backfitting algorithm: an empirical version of a population or Hilbert space procedure, and a penalized least-squares method. We have already motivated ACE as a method for minimizing mean squared error over a population. Before discussing the properties of ACE in this setting, we note a striking feature of the procedure. Unlike the traditional approaches reviewed earlier, there is no assumption of an underlying model relating $E(Y)$ to $X_1, \ldots, X_p$. Although one can think of ACE as a method for estimating the transformation model $\theta(Y) = \sum_j f_j(X_j) + \varepsilon$, it can also be thought of, somewhat more generally, as a procedure that operates on the joint distribution of Y and $X_1, \ldots, X_p$. This may be considered an advantage or disadvantage. On the one hand it makes ACE mathematically appealing and facilitates some elegant, uncluttered theory; on the other hand, it may be responsible for some of the problems with ACE that we will discuss later in the chapter.

Let us now summarize the properties of ACE in the Hilbert space setting. We refer the reader to Breiman and Friedman (1985) for further details and proofs. As in Chapter 5, we let $\mathcal{H}_j$, $j = 1, \ldots, p$ denote the Hilbert spaces of measurable functions $f_j(X_j)$ with $E\{f_j(X_j)\} = 0$, $E\{f_j^2(X_j)\} < \infty$, and inner product $\langle f_j(X_j), f_j'(X_j)\rangle = E\{f_j(X_j)f_j'(X_j)\}$. In addition, denote by $\mathcal{H}_Y$ the corresponding Hilbert space of functions of Y. Denote by $\mathcal{H}^{add} \subset \mathcal{H}$ the closed linear subspace of additive functions, and by P_j the conditional expectation operator $E(\cdot \mid X_j)$; P_j is an orthogonal projection onto $\mathcal{H}_j$. Finally, let P_{add} orthogonally project onto $\mathcal{H}^{add}$. The following statements hold under suitable regularity conditions:

(i) There exists a unique set of square-integrable functions θ^* and $f_1^*, \ldots, f_p^*$ that minimize $E\{\theta(Y) - \sum_j f_j(X_j)\}^2$ subject to $\text{var}\{\theta(Y)\} = 1$. These transformations are said to be the *optimal transformations for regression.* Equivalently, consider the transformations θ^{**} and $f_1^{**}, \ldots, f_p^{**}$ that maximize $p^{**} =$

corr$\{\theta^{**}(Y), \sum_1^p f_j^{**}(X_j)\}$ over all square-integrable functions satisfying var$\{\theta^{**}(Y)\}$ = var$\{\sum_1^p f_j^{**}(X_j)\}$. Then the solutions, called the *optimal transformations for correlation*, are just multiples of the optimal regression transformations, that is $\theta^{**} = \theta^*, f_j^{**} = f_j^*/p^{**}$ (Exercise 7.2). The quantity p^{**} is called the *maximal correlation*. One can show that $p^{**} = 0$ if and only if Y is independent of $X_1, \ldots, X_p$. For the remainder of this chapter, *optimal transformations* refer to the optimal transformations for regression, unless otherwise stated.

(ii) Let $U = P_Y P_{add}$, $V = P_{add} P_Y$. Then U and V are self-adjoint, nonnegative definite, and have the same positive eigenvalues. Letting $\bar{\lambda}$ be the largest eigenvalue, we have

$$U\theta^*(Y) = \bar{\lambda}\theta^*(Y)$$

$$V\sum_{j=1}^{p} f_j^*(X_j) = \bar{\lambda}\sum_{j=1}^{p} f_j^*(X_j)$$

$$\text{corr}^2\Big\{\theta^*(Y), \sum_{j=1}^{p} f_j^*(X_j)\Big\} = \bar{\lambda}. \tag{7.2}$$

(iii) Let the initial function for Y be $\theta_0(Y)$ and let E be the eigenspace of U corresponding to the largest eigenvalue $\bar{\lambda}$. Then if $P_E\theta_0(Y) \neq 0$, the ACE algorithm converges to $\theta^*(Y) = P_E\theta_0(Y)/\|P_E\theta_0(Y)\|$ in mean square and $\sum_j f_j^*(X_j) = P_{add}\theta^*(Y)$.

In summary, the transformations minimizing $E\{\theta(Y) - \sum_1^p f_j(X_j)\}^2$, subject to var$\{\theta(Y)\} = 1$, are the eigenfunctions corresponding to the largest eigenvalue of both $P_Y P_{add}$ and $P_{add} P_Y$, and the corresponding eigenvalue is the square of the maximal correlation corr$\{\theta^*(Y), \sum_1^p f_j^*(X_j)\}$. The ACE algorithm converges to these optimal transformations as long as it doesn't start orthogonal to them. Rigorous proofs of these results are difficult but the main ideas can be grasped by looking at the discrete case (Exercise 7.3).

In fact, ACE works by *powering up* the operators $P_Y P_{add}$ and $P_{add} P_Y$; this will become clear when we analyse the data version of ACE using linear smoothers.

7.2.3 *ACE and penalized least-squares*

We can augment the penalized least-squares justification derived for the backfitting procedure to create a similar criterion for the ACE algorithm.

Consider minimization of

$$\sum_{i=1}^{n}\left\{\theta(y_i)-\textstyle\sum_j f_j(x_{ij})\right\}^2+\lambda_0\int\{\theta''(t)\}^2\,dt+\sum_{j=1}^{p}\lambda_j\int\{f_j''(t)\}^2\,dt \tag{7.3}$$

over all twice continuously differentiable functions f_j and θ with $\text{var}\{\theta(y_i)\}=1$. Any solution can be shown to consist of cubic splines for all the functions (Exercise 7.11); rewriting (7.3) in terms of cubic-spline bases, we get

$$\left(\boldsymbol{\theta}-\sum_{j=1}^{p}\mathbf{f}_j\right)^T\left(\boldsymbol{\theta}-\sum_{j=1}^{p}\mathbf{f}_j\right)+\lambda_0\boldsymbol{\theta}^T\mathbf{K}_0\boldsymbol{\theta}+\sum_{j=1}^{p}\lambda_j\mathbf{f}_j^T\mathbf{K}_j\mathbf{f}_j \tag{7.4}$$

where $\boldsymbol{\theta}=\{\theta(y_1),\ldots,\theta(y_n)\}^T$, and the $\mathbf{f}_j$ and $\mathbf{K}_j$ are defined in sections 2.10 and 5.2.2. Now it is clear from (7.4) that we may restrict attention to functions with mean zero. Then if we add a Lagrangian term $\gamma(\boldsymbol{\theta}^T\boldsymbol{\theta}/n-1)$ to ensure $\text{var}\{\theta(y_i)\}=1$, and differentiate with respect to each of the functions, we find (Exercise 7.4)

$$\hat{\mathbf{f}}_j=(\mathbf{I}+\lambda_j\mathbf{K}_j)^{-1}\left(\hat{\boldsymbol{\theta}}-\sum_{k\neq j}\hat{\mathbf{f}}_k\right);\qquad j=1,\ldots,p$$

$$\hat{\boldsymbol{\theta}}=(1+\gamma/n)^{-1}\left\{\mathbf{I}+\frac{\lambda_0}{1+\gamma/n}\mathbf{K}_0\right\}^{-1}\sum_{j=1}^{p}\hat{\mathbf{f}}_j. \tag{7.5}$$

Thus the recipe for $\hat{\mathbf{f}}_j$ is a backfitting step, using a cubic spline smoother on the response $\hat{\boldsymbol{\theta}}$; the solution $\hat{\boldsymbol{\theta}}$ is proportional to a cubic smoothing spline, with smoothing parameter not λ_0 but $\lambda_0/(1+\gamma/n)$. The value of γ is determined by the requirement $\text{var}\{\theta(y_i)\}=1$. We may modify the above derivation for any set of linear, symmetric, invertible smoother matrices, using an argument similar to that used for backfitting in section 5.2.2.

7.2.4 *Convergence of ACE with linear smoothers*

In this section we demonstrate the convergence of ACE for a certain class of linear smoothers.

Let the smoother matrices for the predictors be $\mathbf{S}_1, \ldots, \mathbf{S}_p$ and that for the response be $\mathbf{S}_y$. Suppose $\mathbf{S}_1, \ldots, \mathbf{S}_p$ are such that the backfitting procedure always converges. In Chapter 5 we show that the overall backfitting operator is itself linear, and we denote the operator here by $\mathbf{R}_{add}$. Denoting the transformation of y at the kth full cycle of ACE by $\boldsymbol{\theta}^{(k)}$, it is easy to see that

$$\boldsymbol{\theta}^{(k+1)} = \mathbf{S}_y \mathbf{R}_{add} \boldsymbol{\theta}^{(k)} \Big/ \left\| \mathbf{S}_y \mathbf{R}_{add} \theta^{(k)} \right\|. \tag{7.6}$$

The matrix $\mathbf{U} = \mathbf{S}_y \mathbf{R}_{add}$ is the analogue of the operator U that appears in the Hilbert space version of ACE, and (7.6) says that ACE *powers up* $\mathbf{U}$. Specifically, if the eigenvalue $\bar{l}$ of $\mathbf{U}$ having largest absolute value is real and positive, then $\boldsymbol{\theta}^{(k)}$ converges to the projection of the starting function

$$\boldsymbol{\theta}^{(0)} = \frac{\mathbf{y} - \bar{y}\mathbf{1}}{\sqrt{\sum_i (y_i - \bar{y})^2 / n}}$$

onto the eigenspace corresponding to $\bar{l}$. That is, the ACE solution is a multiple of the eigenvector corresponding to the largest eigenvalue $\bar{l}$ of $\mathbf{U}$, analogous to the Hilbert space result.

For convergence we require that $\mathbf{S}_1, \ldots, \mathbf{S}_p$ be such that the backfitting algorithm converges, and also that the largest eigenvalue of $\mathbf{U} = \mathbf{S}_y \mathbf{R}_{add}$ be real and positive. The largest class of smoothers for which we can establish this fact are symmetric smoothers with eigenvalues in $[0, 1]$ for the predictors and a symmetric, nonnegative definitive smoother $\mathbf{S}_y$ for the response. When $\mathbf{S}_1, \ldots, \mathbf{S}_p$ are symmetric with eigenvalues in $[0, 1]$, we show in Chapter 5 that backfitting converges. In fact, the resultant operator $\mathbf{R}_{add}$ is itself symmetric and nonnegative definite (Exercise 7.5). Thus $\mathbf{U} = \mathbf{S}_y \mathbf{R}_{add}$ is the product of two symmetric, nonnegative definite matrices and hence has real, nonnegative eigenvalues (Exercise 7.6).

Probably the most common application of this convergence result is the case in which the smoothers are either orthogonal projections, cubic smoothing splines, or a mixture of the two (for example, cubic smoothing splines for some variables, orthogonal projections for the others).

7.2.5 *A close ancestor to ACE: canonical correlation*

Suppose we have two sets of random variables $A_1, \ldots, A_{m_1}$ and $B_1, \ldots, B_{m_2}$. The method of *canonical correlation* finds the unit vectors $\mathbf{a} = (a_1, \ldots, a_{m_1})^T$ and $\mathbf{b} = (b_1, \ldots, b_{m_2})^T$ such that

$$\mathrm{corr}\Big(\sum_{j=1}^{m_1} a_j A_j, \sum_{k=1}^{m_2} b_k B_k\Big)$$

is maximal. How does this technique relate to the ACE algorithm? ACE is applied to regression data, for which we have only one B_k ($B_1 = Y$), and p A_js ($A_1 = X_1, \ldots, A_p = X_p$). In that case, canonical correlation simply produces the least-squares regression of Y on $X_1, \ldots, X_p$: that is, $\hat{\mathbf{a}}$ is proportional to the least-squares coefficient vector. Notice that canonical correlation finds the linear combination of the X_js that is maximally correlated with Y, while ACE generalizes this by finding the linear combination of *transformed* X_js that is maximally correlated with a *transformation* of Y. The transformations are arbitrary, the only restriction being that the transformed variable must be square integrable.

One can construct a parametric version of ACE, based on the method of canonical correlation, that further exposes the similarities between the two methods. The idea is to apply canonical correlation to a few functions of each variable. That is, we take for example, $B_1 = Y, B_2 = Y^2, B_3 = Y^3$, $A_1 = X_1, A_2 = X_1^2, A_3 = X_1^3$, $A_4 = X_2, A_5 = X_2^2, A_6 = X_2^3$, etc. The actual choice and number of functions is unimportant in the discussion below.

One can then show that the ACE algorithm, using smoothers corresponding to least-squares fitting on the B_ks and A_js, solves the canonical correlation problem. In particular, the ACE alternation corresponds to *powering up* the matrix that is the product of the projection matrix for the B_ks and the projection matrix for the A_js. This power method converges to the eigenvector corresponding to the largest eigenvalue of the product matrix, as described in the previous section for general linear smoothers. In the special case of projection smoothers considered here, this eigenvector is also the solution to the canonical correlation problem. Details are given in Exercise 7.8.

Not only does this derivation expose the similarity between ACE and canonical correlation, it tells us that for some choice of linear smoothers, it may be unnecessary to use the ACE algorithm because it may be easier to solve the eigenvalue problem explicitly. This is the case if the smoothers are low-dimensional orthogonal projections; for example, regression splines with a small number of knots. In fact it is sufficient that the smoother for Y have low dimension m_2, for then one can solve the m_2-dimensional eigenvalue problem to obtain $\hat{\theta}$, then regress $\hat{\theta}(Y)$ on $X_1, \ldots, X_p$ to obtain the $\hat{f}_j$s.

7.2.6 *Some anomalies of ACE*

There are a number of properties of the ACE procedure that are quite curious if one views ACE as a regression (rather than a correlation) tool. Specifically, let's consider what happens if we view ACE as a method for estimating either the single predictor model

$$\theta(Y) = f(X) + \varepsilon$$

or the transformed additive model

$$\theta(Y) = \sum_{j=1}^{p} f_j(X_j) + \varepsilon.$$

We list some facts about ACE below.

(i) *For a single predictor, ACE is symmetric in X and Y.* That is, if we interchange X and Y, the ACE solutions are the same up to a constant. This is surprising because one expects a regression procedure to treat Y differently from X; for example, it is well known that the linear regression of Y on X is different from the regression of X on Y.

(ii) *ACE does not reproduce model transformations.* Suppose $\theta(Y) = f(X) + \varepsilon$, where X and ε are independent. Then if θ is invertible, specification of the distributions of X and ε, for fixed θ and f, defines a joint distribution of X and Y. The optimal transformations for this joint distribution are not in general f and θ. In fact f and θ are optimal when $f(X)$ and ε are independent and identically distributed or both have a Gaussian distribution, possibly with different variances, but in no other cases that we know of. As an important example, if $Y = X + \varepsilon$ where X is uniform on $[0, 1]$ and ε is $N(0, 1)$, then the identity transformations are not fixed points of ACE and are therefore not the optimal ones. To see this, note that $E(X \mid Y) \neq Y$ but *bends* for Y near zero or one (Exercise 7.10). This phenomenon occurs when X has any distribution with bounded support and the conditional distribution of Y given X has unbounded support.

(iii) *ACE is not equivariant under monotone transformations of the predictors.* Consider the single predictor case. We may hope that under a monotone mapping $h(X)$, the transformation of Y will be unchanged and the transformation of X will simply reflect the new *labelling* of X, that is, if f is the optimal transformation for X, the new optimal transformation of X will simply be $f\{h^{-1}(X)\}$. Why might we want this equivariance? Of course it doesn't make sense to talk about equivariance in the linear regression setting, because the model transformations are restricted to be linear. However, saying a procedure is equivariant under monotone transformations of the predictors is another way of saying that it doesn't depend on the marginal distribution of the predictors. And we may want this latter property to hold, since typically regression analysis is conditional on the observed predictors. Unfortunately, this doesn't hold in general for ACE. As an example, if $Y = X + \varepsilon$ where both X and ε are $N(0,1)$, the optimal transformations are the identity transformations, because $X = E(Y \mid X)$ and $Y \propto E(X \mid Y)$, the latter following by the basic properties of the bivariate Gaussian distribution. However, we can apply a transformation to X so that the transformed X is uniform on $[0,1]$, and from point (ii) the identity transformation are no longer optimal for Y.

(iv) *Disjoint clusters are collapsed by ACE.* Suppose Y and X have all their support on two diagonally-disjoint regions, for example $\{X < 0, Y < 0\}$ and $\{X \geq 0, Y \geq 0\}$. Then the random variable version of ACE produces transformations of X and Y that are constant in each of the two regions, with correlation one (Exercise 7.7 is an extreme example). In essence, ACE is finding a function of Y, namely a piecewise constant function, that is perfectly predicted by a function of X. This phenomenon occurs for any number (≥ 2) of disjoint clusters. In the data case, the actual transformations are not exactly step functions but some approximations to them, depending on the smoothers used. This collapsing of disjoint clusters may or may not be desirable. It may be useful in identifying clumps of data points but we don't want it to unknowingly determine the shape of the transformations.

(v) *ACE exhibits strange behaviour in low-correlation settings.* Consider X and Y uniform in the unit disk $X^2 + Y^2 < 1$. Note that X and Y are uncorrelated. Surprisingly, the optimal transformations are parabolic in shape, with correlation about 1/3. We might think of this as a null situation but when we measure the dependence by maximal correlation, it is nonnull. Some researchers warn that nonmonotone response transformations, coupled with moderately-low maximal correlations, are signs of anomalous behaviour.

(vi) *The crossing of eigenvalues can cause discontinuous behaviour.* In low-correlation settings there is another problem that can complicate the interpretation of the results. In our analysis of ACE, we show that the maximal correlation is the square root of the largest eigenvalue of a certain product operator. Just as one does in canonical correlation analysis, we can modify ACE so that it finds the second, third and remaining sets of optimal transformations. These are the eigenfunctions of the product operator that correspond to the next largest eigenvalues. In the ACE algorithm this is accomplished by forcing the response transformation to be uncorrelated with the lower-order optimal transformations at each stage. Now in low-correlation settings, the largest eigenvalues are often close together, with quite different eigenfunctions. This can lead to discontinuous behaviour of ACE because one can change a joint distribution of X and Y only slightly causing the largest two eigenvalues to change and hence change the optimal transformations substantially. This phenomenon is not mysterious, however, if one routinely checks a few of the sub-optimal eigenfunctions.

Note that all of the above anomalies can be thought of as properties not of the ACE procedure but of its objective, minimizing squared error $E\{\theta(Y) - \sum_j f_j(X_j)\}^2$ subject to $\text{var}\{\theta(Y)\} = 1$.

Examples can be constructed that illustrate problems with maximal correlation even as an overall measure of dependence between two random variables. In particular, the maximal correlation can be one when the variables are *almost* independent (Exercise 7.7).

7.3 Response transformations for regression

7.3.1 *Introduction*

In this section we consider two approaches for estimating transformations that are designed specifically for regression problems. The first approach generalizes the well known (parametric) Box-Cox procedure, while the second aims at variance stabilization.

7.3.2 *Generalizations of the Box-Cox procedure*

Here we consider a model of the form

$$\theta_\lambda(Y) = \alpha + \sum_{j=1}^{p} X_j\beta_j + \sigma Z \tag{7.7}$$

where Z is a standard Gaussian random variable. The response transformation $\theta_\lambda(Y)$ is parametrized by a k-dimensional parameter $\boldsymbol{\lambda}$. In addition, we only consider response transformations $\theta_\lambda(Y)$ that are monotone in Y. We include a constant in the model for convenience; it could just as well be incorporated in the transformation $\theta_\lambda(Y)$.

The reason for this restriction is that for nonmonotone θ, model (7.7) is of limited usefulness; in particular a unique prediction of Y may not exist for a given set of predictor values. Without loss of generality we assume that $\theta_\lambda(Y)$ is strictly increasing. Probably the most popular choice for $\theta_\lambda(Y)$ is the power transformation $(Y^\lambda - 1)/\lambda$, written in this form to give $\log(Y)$ as $\lambda \to 0$.

Now suppose we have regression data $(y_1, \mathbf{x}^1), \ldots (y_n, \mathbf{x}^n)$ and we want to fit model (7.7). Then we can estimate $\{\boldsymbol{\lambda}, \boldsymbol{\beta}\}$ by maximizing the log-likelihood

$$l(\boldsymbol{\lambda}, \boldsymbol{\beta}, \sigma) = -\frac{n}{2}\log\sigma - \frac{1}{2\sigma^2}\sum_{i=1}^{n}\Big\{\theta_\lambda(y_i) - \textstyle\sum_j x_{ij}\beta_j\Big\}^2 + \displaystyle\sum_{i=1}^{n}\log\left|\frac{\partial\theta_\lambda}{\partial y_i}\right| \tag{7.8}$$

where we absorb the constant α into the linear predictor. The term $\sum_{i=1}^{n}\log|\partial\theta_\lambda/\partial y_i|$ is the log of the Jacobian of the mapping from $y_1, \ldots, y_n$ to $\theta_\lambda(y_1), \ldots, \theta_\lambda(y_n)$. This Jacobian is necessary because the log-likelihood is in units of $y_1, \ldots, y_n$ and not

$\theta_\lambda(y_1), \ldots, \theta_\lambda(y_n)$. It makes an approximate adjustment for scale changes inherent in the transformation θ_λ.

For fixed $\boldsymbol{\lambda}$ and $\boldsymbol{\beta}$, the maximizing value of σ^2 is

$$\hat{\sigma}^2 = \frac{1}{n}\sum_{i=1}^{n}\{\theta_\lambda(y_i) - \sum_j x_{ij}\beta_j\}^2.$$

We can therefore replace σ^2 by $\hat{\sigma}^2$, to obtain the *profile log-likelihood*

$$l(\boldsymbol{\lambda}, \boldsymbol{\beta}) = -n \log \left[\sum_{i=1}^{n}\left\{\theta_\lambda(y_i) - \sum_j x_{ij}\beta_j\right\}^2\right] + \sum_{i=1}^{n} \log \left|\frac{\partial \theta_\lambda}{\partial y_i}\right| \tag{7.9}$$

For the actual computation of the maximum likelihood estimates, we note that for fixed $\boldsymbol{\lambda}$, the maximizers over $\boldsymbol{\beta}$ are simply the least-squares estimates that result when $\theta_\lambda(y_i)$ is regressed on $x_{i1}, \ldots, x_{ip}$. The standard approach is to construct a profile log-likelihood for $\boldsymbol{\lambda}$ alone and maximize it (often graphically). Instead we could find the maximum-likelihood estimates in an alternating fashion, in a similar manner to the ACE algorithm. To update θ we would have to maximize (7.9) with $\boldsymbol{\beta}$ held fixed.

What are the pros and cons of the Box-Cox transformation method? Like the linear-regression model, it is useful when the assumed model is appropriate, but may be ineffective when it is inappropriate. The assumptions inherent in the method are

(i) linearity, normality and homogeneous variance on the transformed scale, and

(ii) a parametric form for the transformation of Y.

The nonparametric methods described in this chapter make less restrictive assumptions and instead work with a general model of the form (7.1).

How can we modify the Box-Cox method to allow general smooth transformations of a variable? A straightforward approach is to use a regression spline for such a variable. One can use a regression spline for the response and linear terms for the predictors, or regression splines for all of the variables, in the spirit of ACE.

One additional difficulty is that the transformation for the response must be constrained to be monotone. If B-splines are used, one must constrain the coefficients of the B-splines to be monotone. Details of this can be found in Kelly and Rice (1988). A different approach is to use a basis consisting of monotone functions, for example I-splines which are essentially integrated B-splines. The coefficients of the I-splines need only be positive to ensure monotonicity of the estimate. Details are in Ramsay (1988).

The spline coefficients are estimated by maximizing the log-likelihood subject to the constraints. This can be difficult numerically, requiring special algorithms for constrained optimization. Ramsay (1988) has written software for this problem that uses I-splines for all of the predictors.

7.3.3 *Comparison of generalized Box-Cox and ACE*

By considering the population versions of the generalized Box-Cox procedure and ACE, some interesting facts arise. In particular, we show that the scaling method used by ACE, that is, forcing $\text{var}\{\theta(Y)\} = 1$, is what causes its undesirable behaviour for regression problems.

For simplicity we assume a single predictor problem (the results carry over easily to the multiple predictor case). The population version of the generalized Box-Cox procedure is obtained by considering maximization of the expected log-likelihood (section 6.5.1)

$$\frac{-E\{\theta(Y) - f(X)\}^2}{2\sigma^2} - \tfrac{1}{2}\log\sigma^2 + E\log\left|\frac{\partial\theta(Y)}{\partial Y}\right|$$

where the expectation is taken with respect to the joint distribution of X and Y. Substituting the maximizing value $E\{\theta(Y) - f(X)\}^2$ for σ^2 and multiplying by -2, we obtain the quantity

$$\log E\{\theta(Y) - f(X)\}^2 - 2E\log\left|\frac{\partial\theta(Y)}{\partial Y}\right| \qquad (7.10)$$

to be minimized. ACE, on the other hand, minimizes $E\{\theta(Y) - f(X)\}^2$ subject to $\text{var}\{\theta(Y)\} = 1$, or equivalently, minimizes

$$\log E\{\theta(Y) - f(X)\}^2 - \log[\text{var}\{\theta(Y)\}] \qquad (7.11)$$

(Exercise 7.9).

Both criteria (7.10) and (7.11) have the same form: a goodness-of-fit measure minus a factor that adjusts for the scale of $\theta(Y)$.

Specifically, both criteria are unchanged if we transform $\theta(Y)$ to $c\theta(Y)$ and $f(X)$ to $cf(X)$, and each criterion is the logarithm of the ratio of two scale functionals. Note that by a first order Taylor series argument $\log[\text{var}\{\theta(Y)\}] \approx 2E \log \mid \partial\theta(Y)/\partial Y \mid + \textit{constant}$, so the criteria seem similar. However, the choice of scaling term is important:

The generalized Box-Cox procedure minimizes the expected log-likelihood under the model $\theta^0(Y) = f^0(X) + \sigma^0 Z$ *where* Z *is distributed as* $N(0,1)$. *By standard likelihood arguments, the true functions* $\theta^0(Y)$, $f^0(X)$, *and* σ^0, *minimize the criterion* (7.10).

As we have seen earlier, this is not true for ACE; that is, ACE does not in general reproduce model transformations. Hence we see that ACE is an approximate generalized maximum likelihood method for the Gaussian error model. The approximation used, namely $2E \log \mid \partial\theta(Y)/\partial Y \mid \approx \log[\text{var}\{\theta(Y)\}]$ turns the criterion into a quadratic one and thus facilitates the use of an alternating projection algorithm for carrying out the minimization. It does not appear possible to derive an "ACE-like" algorithm for minimizing the generalized Box-Cox criterion (7.10). For example, if we simply change ACE so that the standardization of $\theta(Y)$ is done by $2E \log \mid \partial\theta(Y)/\partial Y \mid$ rather than $\text{var}\{\theta(Y)\}$, we obtain a multiple of the usual ACE solutions, *not* the maximizer of the expected log-likelihood. Furthermore, it is easy to check that the ACE criterion (7.11) does not correspond to the expected log-likelihood for any model.

Finally, we note that under a monotone transformation $X \to h(X)$, the generalized Box-Cox criterion (7.10) is equivariant in the sense that it only causes a relabelling of the X variable and the minimizing function $\theta(Y)$ is unchanged. This equivariance is appropriate for a regression procedure and does not hold for the ACE criterion (7.11).

7.4 Additivity and variance stabilization

Another method for estimating regression transformations uses the notion of variance stabilization. It is a modification of ACE designed directly for regression problems. The algorithm implementing this procedure is called AVAS (Additivity and Variance

Stabilization). It differs from ACE in that instead of using the estimate $\hat{\theta}(Y) = E\{\sum_{j=1}^{p} f_j(X_j) \mid Y\}$, it uses an *asymptotic variance stabilizing transformation.* We first describe the AVAS algorithm for random variables, then give details of a data implementation.

Given random variables Y and $X_1, \ldots, X_p$, the goal is to find real-valued, measurable transformations $f_1(X_1), \ldots, f_p(X_p)$ and $\theta(Y)$ such that

$$E\{\theta(Y) \mid X_1, \ldots, X_p\} = \sum_{j=1}^{p} f_j(X_j)$$

$$\mathrm{var}\left\{\theta(Y) \,\middle|\, \sum_{j=1}^{p} f_j(X_j)\right\} = \textit{constant} \tag{7.12}$$

The transformation θ is assumed to be strictly monotone, and without loss of generality we assume it to be strictly increasing. One seeks these transformations because a model is postulated of the form

$$\theta(Y) = \sum_{j=1}^{p} f_j(X_j) + \varepsilon \tag{7.13}$$

where $\theta(Y)$ is strictly increasing, and ε has mean zero and is independent of $X_1, \ldots, X_p$. The AVAS algorithm for finding $f_1, \ldots, f_p$ and $\theta(Y)$ works in an alternating fashion, as does ACE. If we know θ, we can obtain $f_1, \ldots, f_p$ by fitting an additive model to $\theta(Y)$ based on $X_1, \ldots, X_p$. Now recall that if the family of distributions of a random variable W has mean u and variance $V(u)$, then the asymptotic variance-stabilizing transformation for W is given by

$$h(t) = \int_0^t \frac{1}{\sqrt{V(u)}}\, du. \tag{7.14}$$

This can be derived from a Taylor series ("delta method") argument. One can use this fact to find θ given $\sum_{j=1}^{p} f_j(X_j)$. The idea is to apply the asymptotic variance-stabilizing transformation h to $\theta(Y)$. If θ is already a solution to our problem, then $\mathrm{var}\{\theta(Y) \mid \sum_1^p f_j(X_j) = u\}$ is constant, h is the identity function and hence θ is unchanged. Otherwise the new transformation of Y, namely $h\{\theta(Y)\}$, should have a more constant variance as a function of u than $\theta(Y)$. The AVAS procedure alternates between these two steps: one to find the mean function given the response transformation, the other to find the approximate variance stabilizing transformation given the mean. Now it is clear that if $f_1, , \ldots, f_p$ and θ satisfy (7.12), then so do $a + bf_1, \ldots, a + bf_p$ and $a + b\theta$ for any a and b. Thus we assume $E\{f_j(X_j)\} = E\{\theta(Y)\} = 0$ and $\text{var}\{\theta(Y)\} = 1$ to ensure uniqueness of our solution.

Algorithm 7.3 *The AVAS algorithm for random variables*

(i) *Initialize:* set $\theta(Y) = (Y - EY)/\{\text{var}(Y)\}^{1/2}$ and fit an additive model to $\theta(Y)$ based on $X_1, \ldots, X_p$ to give $f_1, \ldots f_p$.

(ii) *Find new transformation of Y:* let $m(X_1, \ldots, X_p) = \sum_1^p f_j(X_j)$ and compute the variance function

$$V(u) = \text{var}\{\theta(Y) \,|\, m(X_1, \ldots, X_p) = u\}.$$

Compute the variance stabilizing transformation

$$h(t) = \int_0^t \frac{1}{\sqrt{V(u)}}\, du$$

and set $\theta(t) \leftarrow h\{\theta(t)\}$. Renormalize θ:

$$\theta(t) \leftarrow \frac{\theta(t) - E\theta(Y)}{\sqrt{\text{var}\,\theta(Y)}}$$

(iii) *Find new transformations of the X_js:* fit an additive model to $\theta(Y)$ based on $X_1, \ldots, X_p$ to give new functions $f_1, \ldots, f_p$.

(iv) Alternate steps (ii) and (iii) until $E\{\theta(Y) - \sum_1^p f_j(X_j)\}^2$ doesn't change.

We now briefly describe the existing data implementation of AVAS. As in ACE, the conditional expectations $E(\cdot \,|\, X_j)$ are replaced by scatterplot smoothers, and quantities like $E(Y)$ and $\text{var}(Y)$ are replaced by their sample versions. To compute the variance function $V(u)$, the log of (the sample version of) the squared residual $r^2 = E\{\theta(Y) - \sum_1^p f_j(X_j)\}^2$ is smoothed against $u = \sum_1^p f_j(X_j)$ and the result is exponentiated. Finally, the integral in the AVAS procedure is replaced by a numerical quadrature rule, for example the trapezoid or Simpson's rule.

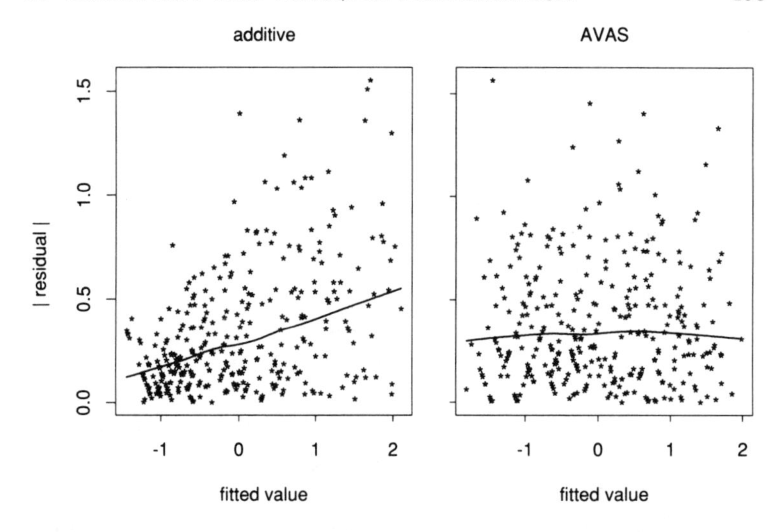

Fig. 7.3. *Left figure shows the absolute residual versus fitted value, from an additive fit to the ozone concentration data (no response transformation). A scatterplot smooth has been added to the plot and shows that the variance increases with the fitted value. On the right is the absolute residual versus fitted value, from an AVAS fit to the ozone concentration data. The smooth shows that the variance is fairly constant as a function of the fitted value.*

We applied the AVAS procedure to the ozone concentration data, producing the response transformation shown in the top right panel of Fig. 7.2. Again, a logarithmic shaped transformation results. Figure 7.3 shows the absolute value of the residual plotted against the fitted value, from a simple additive fit (no response transformation) and the AVAS fit. The response transformation has stabilized the variance. Of course, ACE and AVAS do not always produce similar transformations; see Tibshirani (1988) for examples in which they differ.

7.4.1 *Some properties of AVAS*

The AVAS procedure seeks transformations that achieve additivity and a homogeneous variance, and is more directed towards regression problems than is ACE. A key feature is that the operations in AVAS are all *conditional* on the predictors. In addition, the response transformation is always monotone, being the integral of a positive function.

Not surprisingly AVAS does not share many of ACE's regression anomalies. Specifically:

(i) it is not symmetric in X and Y, the roles of the response and predictor(s) being quite distinct.

(ii) it reproduces model transformations; that is, when it is applied to model (7.13), θ and f_j, $j = 1, \ldots, p$, are fixed points.

(iii) it is equivariant under monotone transformations of the predictors, in the sense described in section 7.2.6.

(iv) it does not collapse disjoint clusters, and there is some empirical evidence that it does not behave poorly in low correlation situations, as does ACE.

While some theoretical investigation into AVAS has been made, there is not nearly as much theoretical support for the technique as there is for ACE. In particular, global convergence of AVAS has not been established, and the behaviour of the algorithm when model (7.12) fails needs to be carefully studied. In addition, it would be useful to compare AVAS to the generalized Box-Cox procedure described earlier. From our discussion here, it seems likely that the two methods will have similar qualitative behaviour.

7.5 Further topics

7.5.1 *Prediction from a transformation model*

Given an estimated transformation model $\hat{\theta}(Y) = \sum_{j=1}^{p} \hat{f}_j(X_j)$, how do we predict Y for a given set of predictors $X_1, \ldots, X_p$? There are at least three methods that have been suggested:

(i) $\hat{Y} = \hat{\theta}^{-1}\{\sum_{j=1}^{p} \hat{f}_j(X_j)\}$.

(ii) $\hat{Y} = E\{Y \mid \sum_{j=1}^{p} \hat{f}_j(X_j)\}$, estimated by smoothing the observed y_i against $\sum_{j=1}^{p} \hat{f}_j(x_{ij})$.

(iii) The *smearing estimate:*

$$\begin{aligned}\hat{Y} &= E(Y \mid X_1, \ldots, X_p)\\ &= E\Big\{\hat{\theta}^{-1}\big(\textstyle\sum_j \hat{f}_j(X_j) + \varepsilon_i\big) \mid X_1, \ldots, X_p\Big\}\end{aligned}$$

estimated by $(1/n)\sum_1^n \hat{\theta}^{-1}(\sum_{j=1}^{p} \hat{f}_j(X_j) + r_i)$, where r_i are the observed residuals.

The first method is the naive approach, and simply back-transforms the prediction on the $\theta(Y)$ scale; the second method improves on the first because if one is to base a prediction of Y on $Z = \sum_{j=1}^{p} \hat{f}_j(X_j)$, then the best predictor in terms of mean squared error is $E(Y \mid Z)$. Suggestion (iii) goes directly at the best predictor of Y based on $X_1, \ldots, X_p$, namely $E(Y \mid X_1, \ldots, X_p)$. While some theoretical study of the smearing estimate has been carried out (Duan, 1983), a comparison of methods (i)–(iii) has not, to our knowledge, been undertaken.

A more fundamental question is whether or not it is worthwhile transforming the response at all, if the ultimate goal is to minimize prediction errors on the original scale. The bibliographic notes list a few papers that study this important issue.

7.5.2 *Methods for inference*

The estimation of a response transformation makes it more difficult to make inferences from the fitted model about the predictors. In particular, the variance and degrees of freedom of each of the fitted functions are more difficult to compute. Indeed, even in a *simple* setting like the Box-Cox transformation model, there is considerable controversy over what variance we should be computing! An attractive approach is to ignore the estimation of θ and pretend it is known. In the Box-Cox transformation model, this would mean that linear least-squares theory could be used, considering $\hat{\theta}(Y)$ to be the response. In ACE or AVAS, this would allow us to apply the methods for inference for the additive model described in Chapters 5 and 6. Some authors argue that this *conditioning* on the estimated value of $\theta(Y)$ is not only convenient but appropriate. Others argue that the unconditional inferences are the correct ones, that is we should take into account the variability in estimating $\theta(Y)$ in our inferences. To achieve this, a bootstrap approach could be used, resampling from the original data. However relatively little work has been done on this problem.

7.6 Bibliographical notes

The Box-Cox transformation method was proposed by Box and Cox (1964). The papers by Bickel and Doksum (1981), Box and Cox (1982), and Hinkley and Runger (1984) contain a lively series of arguments about inference for the model. Carroll and Ruppert (1981) studied the cost of data transformation in terms of prediction error on the original scale. See also Cox and Reid (1987) for a treatment of *orthogonal parameters*, and their role in this discussion. Kruskal (1965) proposed a method that used isotonic regression for the response transformation and linear regression for the predictor. His estimation procedure had the same form as the ACE algorithm. Details on isotonic regression may be found in Barlow *et al.* (1972).

Other empirical methods for finding transformations are given by Anscombe and Tukey (1963), Fraser (1967), Andrews (1971), Linsey (1972), Box and Hill (1974), Mosteller and Tukey (1977), Tukey (1982), and Hoaglin *et al.* (1983). Chapter 5 of Mosteller and Tukey (1977) contains an excellent discussion of strategies for selecting transformations. The monographs by Cook and Weisberg (1982), Atkinson (1985), and Carroll and Ruppert (1988) contain (among other things) a comprehensive study of transformations in parametric linear regression models. Carroll and Ruppert (1984) looked at the problem of transforming both sides of a regression model simultaneously and in the same way.

The ACE algorithm was introduced by Breiman and Friedman (1985). Similar ideas were proposed earlier by Young, de Leeuw, and Takane (1976) (also de Leeuw, Young and Takane, 1976) in their MORALS procedure. Many interesting properties of ACE were pointed out by the discussants of the Breiman and Friedman paper, namely Pregibon and Vardi (1985), Buja and Kass (1985), and Fowlkes and Kettenring (1985). Buja (1989) gave a penetrating study of ACE and the corresponding eigenvalue problem. Buja pointed out the connection of ACE with canonical correlation, and an account of the latter may be found in most books on multivariate analysis, for example Anderson (1958), Srivastava and Khatri (1979), or Mardia, Kent, and Bibby (1979). Brillinger and Preisler (1984) extended the ACE idea to mixed effects models.

The comparison of ACE and the Box-Cox procedure in section 7.3.3 was taken from unpublished work by the authors and

Daryl Pregibon. The AVAS procedure was proposed by Tibshirani (1988). The methods for prediction from a transformation model are discussed in Breiman and Friedman (1985) (attributing method (ii) of section 7.5.1 to A. Owen), while the smearing estimate (iii) is due to Duan (1983).

7.7 Further results and exercises 7

7.1 Apply the ACE and AVAS algorithms to the diabetes data, with response `C-peptide`, and predictors `age` and `base deficit`. What response transformation does each method suggest?

7.2 Show that transformations $\theta^{**} = \theta^*$ and $f_j^{**} = f_j^*/p^{**}$ are optimal for correlation if and only if θ^* and f_j^*, $j = 1, \ldots, p$ are optimal for regression, p^{**} being the maximal correlation (section 7.2.2).

[Breiman and Friedman, 1985]

7.3 *ACE and correspondence analysis.* Consider the ACE algorithm applied to a discrete joint distribution of X and Y. Let $p_{ij} = P(Y = i, X = j)$ for $i = 1, \ldots, m$ and $j = 1, \ldots, n$, and $\theta_i = \theta(i)$, $f_j = f(j)$.

(i) Using a Lagrangian-multiplier λ to force $E(\theta_i^2) = 1$, show that the Lagrangian function for minimizing the expected sum of squares is equivalent to

$$1 - 2\sum_{i,j} \theta_i f_j p_{ij} + \sum_{i,j} f_j^2 p_{ij} - \lambda\Big(1 - \sum_{i,j} \theta_i^2 p_{ij}\Big)$$

(ii) Differentiate the Lagrangian and hence show the solutions satisfy

$$\theta_k = \frac{1}{\lambda}\sum_j f_j \frac{p_{kj}}{p_{k+}} = \frac{1}{\lambda} E\{f(X) \,|\, Y = k\}$$

$$f_k = \sum_i \theta_i \frac{p_{ik}}{p_{+k}} = E\{\theta(Y) \,|\, X = k\}$$

where the subscript $+$ denotes summation over the corresponding index.

(iii) From this show that $\lambda = E\{\theta(Y)f(X)\} = \|E\{f(X) \,|\, Y\}\| = Ef(X)^2$.

(iv) Let $\mathbf{P}$ be an $m \times n$ matrix with elements p_{ij}, $\mathbf{D}_{+n}$ be a diagonal matrix with elements p_{+j} and $\mathbf{D}_{m+}$ be a diagonal matrix with elements p_{i+}. Then by substituting the expression for θ_k into that for f_k and vice versa, show

$$\lambda\boldsymbol{\theta} = \mathbf{D}_{m+}^{-1}\mathbf{P}\mathbf{D}_{+n}^{-1}\mathbf{P}^T\boldsymbol{\theta}$$

$$\lambda\mathbf{f} = \mathbf{D}_{+n}^{-1}\mathbf{P}^T\mathbf{D}_{m+}^{-1}\mathbf{P}\mathbf{f}.$$

These equations define the *correspondence analysis factorization* of the table $\mathbf{P}$ (Greenacre, 1984).

(v) Show that the correlation between the solutions θ_i and f_j is $\lambda^{1/2}$.

[Pregibon and Vardi, 1985].

7.4 Verify the penalized least-squares derivation of ACE, equations (7.5).

7.5 Construct the backfitting operator $\mathbf{R}_{add}$ from equation (5.28) in section 5.4.3 and show that $\mathbf{R}_{add}$ is symmetric and nonnegative definite.

7.6 Show that the product of two symmetric, nonnegative definite matrices has real, nonnegative eigenvalues. This fact and the result of the previous exercise are used in section 7.2.4 to establish the convergence of ACE.

7.7 Let (X, Y) equal (X', Y') with probability $1-\alpha$ and (X'', Y'') with probability α $(0 < \alpha < 1)$, where (X', Y') are two independent random variables with absolutely continuous distributions and (X'', Y'') are two discrete random variables. Show that the optimal correlation is one regardless of the value of α and regardless of whether X'' and Y'' are independent. [Hint: think about the *collapsing clusters* example in section 7.2.6; Pregibon and Vardi, 1985]

7.8 *ACE and canonical correlation.* Consider the setup of section 7.2.5.

(i) Let the covariance matrices of the A_js and B_ks be $\boldsymbol{\Sigma}_{11}$ and $\boldsymbol{\Sigma}_{22}$, respectively, and let $\boldsymbol{\Sigma}_{12}$ be the cross covariance between the two sets of variates. Let $\mathbf{L} = \boldsymbol{\Sigma}_{11}^{-1/2}\boldsymbol{\Sigma}_{12}\boldsymbol{\Sigma}_{22}$ and $\mathbf{N}_1 = \mathbf{L}\mathbf{L}^T$, $\mathbf{N}_2 = \mathbf{L}^T\mathbf{L}$. Denote by γ_i and $\boldsymbol{\tau}_i$ the eigenvectors of $\mathbf{N}_1$ and $\mathbf{N}_2$ respectively, ordered according to decreasing value of their common eigenvalues λ_i. Without loss of generality assume $\text{var}(\sum_{j=1}^{m_1} a_j A_j) = \text{var}(\sum_{k=1}^{m_2} b_k B_k) = 1$. Show that the vectors $\hat{\mathbf{a}}$ and $\hat{\mathbf{b}}$ maximizing $\text{corr}(\sum_{j=1}^{m_1} a_j A_j, \sum_{k=1}^{m_2} b_k B_k)$ are given by

$$\begin{aligned}\hat{\mathbf{a}} &= \boldsymbol{\Sigma}_{11}^{-1/2}\boldsymbol{\gamma}_1\\ \hat{\mathbf{b}} &= \boldsymbol{\Sigma}_{22}^{-1/2}\boldsymbol{\tau}_1.\end{aligned} \tag{7.15}$$

Show that corresponding canonical variates, or maximally correlated linear combinations of the variates, are $\sum_{j=1}^{m_1} \hat{a}_j A_j$ and $\sum_{k=1}^{m_2} \hat{b}_k B_k$, and have correlation $\lambda_1^{1/2}$.

(ii) Now consider the data case. Suppose that the variables are centered, and let the data matrices be $\mathbf{A}$ and $\mathbf{B}$, of dimension $n \times m_1$ and $n \times m_2$ respectively. To relate more closely to ACE, $\mathbf{A}$ could be the regression matrix consisting of evaluated basis matrices for representing each of the functions $f_1, \ldots, f_p$ as cubic splines; similarly $\mathbf{B}$ would be a basis matrix for representing $\theta(Y)$ as a cubic spline. Then the canonical correlation results are obtained simply by replacing $\boldsymbol{\Sigma}_{11}$, $\boldsymbol{\Sigma}_{22}$ and $\boldsymbol{\Sigma}_{12}$ by their sample estimates.

Hence show that the optimal unit vectors are $\hat{\mathbf{a}} = (\mathbf{A}^T\mathbf{A})^{-1/2}\boldsymbol{\gamma}_1$ and $\hat{\mathbf{b}} = (\mathbf{B}^T\mathbf{B})^{-1/2}\boldsymbol{\tau}_1$ where $\boldsymbol{\gamma}_1$ and $\boldsymbol{\tau}_1$ are the leading eigenvectors of

$$\mathbf{N}_1 = (\mathbf{A}^T\mathbf{A})^{-1/2}\mathbf{A}^T\mathbf{B}(\mathbf{B}^T\mathbf{B})^{-1}\mathbf{B}^T\mathbf{A}(\mathbf{A}^T\mathbf{A})^{-1/2}$$

and

$$\mathbf{N}_2 = (\mathbf{B}^T\mathbf{B})^{-1/2}\mathbf{B}^T\mathbf{A}(\mathbf{A}^T\mathbf{A})^{-1}\mathbf{A}^T\mathbf{B}(\mathbf{B}^T\mathbf{B})^{-1/2}.$$

These matrices are of dimension $m_1 \times m_1$ and $m_2 \times m_2$ respectively. The canonical variates are $\mathbf{A}\hat{\mathbf{a}}$ and $\mathbf{B}\hat{\mathbf{b}}$.

(iii) Relate this to the ACE algorithm by showing the canonical variate $\mathbf{f}_1 = \mathbf{A}\hat{\mathbf{a}}$ is the leading eigenvector of

$$\mathbf{A}(\mathbf{A}^T\mathbf{A})^{-1}\mathbf{A}^T\mathbf{B}(\mathbf{B}^T\mathbf{B})^{-1}\mathbf{B}^T, \tag{7.16}$$

and similarly for $\boldsymbol{\theta}_1 = \mathbf{B}\hat{\mathbf{b}}$. Expression (7.16) is the product of the projection matrices that project onto the A_i and B_i regression spaces. This says that if we carry out a parametric version of ACE, that is, alternately regressing onto the A_is and B_is, we solve the canonical correlation problem. In the example considered above, these regressions would be alternating-spline regressions on the X_is and Y.

7.9 Show that the ACE problem, that is, minimizing $E\{\theta(Y) - f(X)\}^2$ subject to $\text{var}\{\theta(Y)\} = 1$, is equivalent to minimizing $\log E\{\theta(Y) - f(X)\}^2 - \log[\text{var}\{\theta(Y)\}]$ (section 7.3.3).

7.10 Suppose $Y = X + \varepsilon$ where X is uniform on $[0,1]$ independent of $\varepsilon \sim N(0,1)$. Show that

$$E(X \mid Y) = Y - \{\phi(Y) - \phi(Y-1)\}/\{\Phi(Y) - \Phi(Y-1)\}$$

where ϕ and Φ are the density and distribution function of the standard Gaussian distribution, and hence the identity transformations are not optimal.

7.11 Show that if a solution to (7.3) exists, then it has to consist of natural cubic splines for θ and each of the f_j [Hint: see exercise 2.12].

CHAPTER 8

Extensions to other settings

8.1 Introduction

The procedures developed so far work out particularly nicely for exponential family models. In that case an efficient local-scoring algorithm emerges, in which each iteration requires a weighted additive fit. If there is only one nonparametric term in the additive predictor, this fit can be obtained in closed form; for the general nonparametric additive model the modular backfitting algorithm is used. Apart from being simple, the backfitting modules add insight into the method of estimation; for example, wherever a smoother is applied to partial residuals as part of the iterative fitting process, the residuals themselves have diagnostic information, and can be used for identifying outliers and influential points.

Of course, these nonparametric regression techniques can be used in any generalized regression situation. Specifically, whenever one has a likelihood-based regression model, a nonparametric version can be derived by augmenting the likelihood with appropriate penalty terms. However, it is not always the case that a simple and modular algorithm emerges. Fortunately, in many cases we can develop an efficient algorithm with a little work, and as a byproduct gain insight into the problem at hand. In this chapter we look at four models of this kind: additive models for survival data, matched case-control data, ordinal response data and time series.

The first three models are special because they do not fit into the univariate exponential family framework. From a technical point of view, they each seem to require a local-scoring algorithm that involves a nondiagonal weight matrix. Although in theory this is not a problem, the algorithms would be expensive and require specialized software. We see that in each case this can be overcome by using a special modification of the local-scoring and backfitting algorithms. Besides being interesting in themselves, these examples illustrate some of the techniques needed to construct effective modular algorithms. They also give a different perspective on the usual linear version of these problems.

The matched case-control and proportional-hazards models for censored survival data are similar in flavour since they both use conditioning arguments for estimation. The proportional-hazards model and the proportional-odds model have a similar model structure, and the latter can be viewed as a discrete version of the transformation models discussed earlier.

The fourth model is really a technique for decomposing a time series into a sum of seasonal, trend and remainder components. Unlike other additive models, the predictors are all the same variable — time. What distinguishes the different terms in the model are the smoothers used for each; indeed, they define the seasonal and trend components.

8.2 Matched case-control data

8.2.1 *Background*

The linear logistic regression model is often used by epidemiologists and biostatisticians for the analysis of matched case-control data. Breslow and Day (1980) give a clear and detailed description. The notation below is very close to that used in the discussion of logistic regression in section 4.5.

Often the linear logistic regression model

$$\log \frac{\mathrm{pr}(Y=1\,|\,\boldsymbol{X})}{1-\mathrm{pr}(Y=1\,|\,\boldsymbol{X})} = \alpha + X_1\beta_1 + \cdots + X_p\beta_p$$

is expressed in an equivalent way

$$\log\{\psi(\boldsymbol{X})\} = X_1\beta_1 + \cdots + X_p\beta_p \tag{8.1}$$

where the odds-ratio $\psi(\boldsymbol{X})$ is defined as the odds of an individual with exposure $\boldsymbol{X}$ developing the disease relative to an individual with baseline exposure $\mathbf{0}$:

$$\psi(\boldsymbol{X}) = \frac{\mathrm{pr}(Y=1\,|\,\boldsymbol{X})}{1-\mathrm{pr}(Y=1\,|\,\boldsymbol{X})}\bigg/\frac{\mathrm{pr}(Y=1\,|\,\mathbf{0})}{1-\mathrm{pr}(Y=1\,|\,\mathbf{0})}. \tag{8.2}$$

Sometimes certain exposure variables are known a priori to affect prevalence but are of limited interest otherwise. The idea is to control for the effect of these variables in order to focus on the variables of interest. Let M_k denote a partition of the space of these controlled variables into *matched sets*. For example, M_1 might be all women aged 30–35, M_2 might be women aged 36–40 and so on. In this case the logistic regression model can be written as

$$\log \frac{\{\text{pr}(Y=1 \mid \boldsymbol{X}, M_k)\}}{\{1-\text{pr}(Y=1 \mid \boldsymbol{X}, M_k)\}} = \alpha_k + X_1\beta_1 + \cdots + X_p\beta_p$$

for the log-odds of success in group k, where α_k is an intercept term specific to M_k. The parameters of interest, β_j, are assumed to be constant across matched sets. In terms of odds-ratios

$$\log\{\psi(\boldsymbol{X}, M_k)\} = X_1\beta_1 + \cdots + X_p\beta_p, \tag{8.3}$$

which has the same form as (8.1).

The natural additive extension of (8.3) is

$$\log\{\psi(\boldsymbol{X}, M_k)\} = f_1(X_1) + \cdots + f_p(X_p). \tag{8.4}$$

8.2.2 *Estimation*

The sampling schemes available for models of this kind are interesting and deserve a short discussion.

If we were free to design the experiment (with no cost constraints), we would construct cohorts of subjects sampled randomly from $\boldsymbol{X}$ given M_k. These cohorts would then be followed up and their disease status Y recorded. With sufficient observations in each set, the nuisance parameters α_k, and the parameters of interest β_j could be estimated by maximum likelihood.

Often such prospective studies are impracticable (for low-incidence diseases, for example), expensive, and perhaps even unethical. An alternative is to sample retrospectively from the cases and controls. For the kth of K cases with predictor vector $\mathbf{x}^{0k}$, a pool of controls is formed having the same values of the matching variables, and R_k of these matched controls are randomly selected with predictor vectors $\mathbf{x}^{rk}$ for $r = 1, \ldots, R_k$. For notational simplicity, assume further that $R_k = R$ for all k.

Using (8.4), the *conditional* probability that within a matched set, the assignment of the $R+1$ values $\mathbf{x}^{rk}$ to case and controls is as observed is given by

$$\mu_{0k} = \frac{\exp\{\eta(\mathbf{x}^{0k})\}}{\sum_{r=0}^{R} \exp\{\eta(\mathbf{x}^{rk})\}} \tag{8.5}$$

where $\eta(\boldsymbol{X}) = \sum_{j=1}^{p} f_j(X_j)$ is the additive predictor. The full conditional likelihood is simply the product over matched sets $L(f_1, \ldots, f_p) = \prod_{k=1}^{K} \mu_{0k}$; we typically work with its logarithm:

$$l(f_1, \ldots, f_p) = \sum_{k=1}^{K} \Bigl(\eta(\mathbf{x}^{0k}) - \log\Bigl[\sum_{r=0}^{R} \exp\{\eta(\mathbf{x}^{rk})\}\Bigr]\Bigr). \tag{8.6}$$

This conditional likelihood forms the basis for inference, both for the linear and additive case-control models.

Before describing an algorithm for estimating this model, it is useful to outline a general strategy in situations like these. So far our algorithms have been modular, consisting of repeated smoothing within a backfitting algorithm, perhaps within a local-scoring algorithm. Since the scatterplot smoother is at the core, the choice of smoother is open and this leads to a very general methodology.

A useful approach, which we follow here, is to penalize the conditional likelihood (8.6) and from it derive an estimation procedure. Once the score equations and Newton-Raphson iterations have been obtained, we try to rewrite the equations in the form of a repeated (usually weighted) cubic spline smoothing step. If this is possible, one may feel justified in using any smoother in place of a cubic smoothing spline.

8.2.3 *Maximizing the conditional likelihood for the linear model*

Let's first see what the iterations look like for the linear model. Differentiating (8.6) with $\eta(\mathbf{x}^{rk}) = \boldsymbol{\beta}^T \mathbf{x}^{rk}$, the score for $\boldsymbol{\beta}$ can be written as

$$\begin{aligned} \frac{\partial l}{\partial \boldsymbol{\beta}} &= \sum_{k=1}^{K} \sum_{r=0}^{R} \left(y_{rk} \mathbf{x}^{rk} - \mu_{rk} \mathbf{x}^{rk} \right) \\ &= \mathbf{X}^T (\mathbf{y} - \boldsymbol{\mu}) \end{aligned} \tag{8.7}$$

where $\boldsymbol{\mu} = \{\mu_{rk}\}$, and $\mathbf{y} = \{y_{rk}\}$ indicates cases (1) or controls (0). It is also convenient to define $\boldsymbol{\mu}_k$, the sub-vector of these probabilities (which sum to 1) for each matched set, and $\mathbf{U} = \text{diag}(\boldsymbol{\mu})$.

Similarly the information matrix can be written

$$\begin{aligned}\frac{-\partial^2 l}{\partial\boldsymbol{\beta}\boldsymbol{\beta}^T} &= \sum_{k=1}^{K}\Big[\sum_{r=0}^{R}\mu_{rk}\mathbf{x}^{rk}\mathbf{x}^{rk^T} - \Big(\sum_{r=0}^{R}\mu_{rk}\mathbf{x}^{rk}\Big)\Big(\sum_{r=0}^{R}\mu_{rk}\mathbf{x}^{rk}\Big)^T\Big] \\ &= \mathbf{X}^T\mathbf{A}\mathbf{X} \qquad (8.8)\end{aligned}$$

where $\mathbf{A}$ is an $n \times n$ block-diagonal matrix with kth block $\mathbf{A}_k = \mathbf{U}_k - \boldsymbol{\mu}_k\boldsymbol{\mu}_k^T$, $n = (R+1)K$ being the total sample size.

The Newton-Raphson update can be expressed in iterative reweighted least-squares form as

$$\boldsymbol{\beta}^1 = \left(\mathbf{X}^T\mathbf{A}^0\mathbf{X}\right)^{-1}\mathbf{X}^T\mathbf{A}^0\left\{\mathbf{X}\boldsymbol{\beta}^0 + (\mathbf{A}^0)^-(\mathbf{y}-\boldsymbol{\mu}^0)\right\} \qquad (8.9)$$

where $\boldsymbol{\mu}^0$ and $\mathbf{A}^0$ denote $\boldsymbol{\mu}$ and $\mathbf{A}$ is evaluated at $\boldsymbol{\beta}^0$, and $\mathbf{A}^-$ denotes a generalized inverse of $\mathbf{A}$ and is equivalent to $\text{diag}(\boldsymbol{\mu}^{-1})$.

Even the algorithm for the linear case is not simple! The iterations are repeated regressions, but with nondiagonal weight matrices. A simple transformation reduces the linear algorithm to a suitable diagonal iterative reweighted least-squares problem (Exercise 8.3). A similar problem arises in the additive algorithm, but this trick no longer applies.

8.2.4 *Spline estimation for a single function*

Consider a single exposure variable X. Without loss of generality assume there are no ties in the x_{rk}. The logarithm of the penalized conditional likelihood for the model $\log\{\psi(X)\} = f(X)$ is

$$j(f) = \sum_{k=1}^{K}\Big(f(x_{0k}) - \log\Big[\sum_{r=0}^{R}\exp\{f(x_{rk})\}\Big]\Big) - \tfrac{1}{2}\lambda\int\{f''(s)\}^2\,ds. \qquad (8.10)$$

As before, the criterion has two components: the likelihood component measures fidelity of the function to the data, and the integrated squared second derivative component measures its smoothness. The results of O'Sullivan (1983) show that the solution exists and is a natural cubic spline.

The details are the same as in section 6.5.2, and once again we use the evaluated functions to represent the solution. Differentiating the penalized log-likelihood (8.10) with respect to the n components $f(x_{rk})$ of $\mathbf{f}$ leads to the score equation

$$\frac{\partial j}{\partial \mathbf{f}} = \mathbf{y} - \boldsymbol{\mu} - \lambda \mathbf{K}\mathbf{f}$$

where $\mu_{rk} = \exp\{f(x_{rk})\}/\sum_{r=0}^{R}\exp\{f(x_{rk})\}$ as in (8.7). The information matrix is $\mathbf{A} + \lambda\mathbf{K}$, where $\mathbf{A}$ is the information matrix of the conditional log-likelihood component, and is the same as in (8.8).

The Newton-Raphson update for $\mathbf{f}$ is

$$(\mathbf{A}^0 + \lambda\mathbf{K})(\mathbf{f}^1 - \mathbf{f}^0) = \mathbf{y} - \boldsymbol{\mu} - \lambda\mathbf{K}\mathbf{f} \tag{8.11}$$

which can be written as

$$\begin{aligned} \mathbf{f}^1 &= (\mathbf{A}^0 + \lambda\mathbf{K})^{-1}\mathbf{A}^0\left[\mathbf{f}^0 + (\mathbf{A}^0)^{-}(\mathbf{y} - \boldsymbol{\mu}^0)\right] \\ &= \mathbf{S}_{A^0}\mathbf{z}, \end{aligned} \tag{8.12}$$

where $\mathbf{z}$ is the adjusted dependent variable in the square brackets. The matrix $\mathbf{S}_A = (\mathbf{A} + \lambda\mathbf{K})^{-1}\mathbf{A}$ is a cubic smoothing-spline operator matrix (equation (3.38) in section 3.11.4). In all of the applications we have encountered thus far, $\mathbf{A}$ is diagonal; this together with the special banded structure of $\mathbf{K}$ allows one to apply the smoother in $O(n)$ operations. This is not the case here, even though $\mathbf{A}$ is block diagonal and hence is banded. Now $\mathbf{K}$ is banded if the rows are ordered with $\mathbf{x}$, but this ordering destroys the block diagonal structure of $\mathbf{A}$. Thus $\mathbf{A}+\lambda\mathbf{K}$ is a full matrix and expensive to invert ($O(n^3)$ operations). So we need a trick to simplify the calculations.

Our solution is to replace $\mathbf{A}$ by its diagonal

$$\mathbf{D} = \text{diag}(\boldsymbol{\mu})\{\mathbf{I} - \text{diag}(\boldsymbol{\mu})\}.$$

The update formula for $\mathbf{f}$ is

$$\begin{aligned} \mathbf{f}^1 &= (\mathbf{D}^0 + \lambda\mathbf{K})^{-1}\mathbf{D}^0\left\{\mathbf{f}^0 + {\mathbf{D}^0}^{-1}(\mathbf{y} - \boldsymbol{\mu}^0)\right\} \\ &= \mathbf{S}_{\mathbf{D}^0}\mathbf{z}, \end{aligned} \tag{8.13}$$

where $\mathbf{S}_{\mathbf{D}^0}$ computes a diagonally-weighted cubic smoothing spline. If this modified algorithm converges, it is clear from (8.11) that the solutions it produces are solutions to the original problem. This is because the left-most matrix of (8.11) is nonsingular with $\mathbf{A}$ replaced by $\mathbf{D}$ and thus convergence implies that the score $\mathbf{y} - \boldsymbol{\mu} - \lambda\mathbf{K}\mathbf{f} = \mathbf{0}$. Jørgenson (1984) gave the name "delta" to modified Newton algorithms of this kind, where the weight matrix is replaced by an approximation. From the above, it is clear that delta algorithms such as this are ascent methods so they will also converge with step size optimization (section 6.6).

8.2.5 *Algorithm for the additive model*

The conditional penalized log-likelihood for the additive model has a penalty for each of the functions:

$$j(f_1, \ldots, f_p) = \sum_{k=1}^{K} \log \mu_{0k} - \tfrac{1}{2} \sum_{j=1}^{p} \lambda_j \int \{f_j''(s)\}^2 \, ds.$$

The derivation of the local-scoring step is an extension of the univariate case in the previous section, and follows the steps outlined in section 6.5.2. We end up with the system of estimating equations

$$\begin{pmatrix} \mathbf{I} & \mathbf{S}_1 & \mathbf{S}_1 & \cdots & \mathbf{S}_1 \\ \mathbf{S}_2 & \mathbf{I} & \mathbf{S}_2 & \cdots & \mathbf{S}_2 \\ \vdots & \vdots & \vdots & \ddots & \vdots \\ \mathbf{S}_p & \mathbf{S}_p & \mathbf{S}_p & \cdots & \mathbf{I} \end{pmatrix} \begin{pmatrix} \mathbf{f}_1^1 \\ \mathbf{f}_2^1 \\ \vdots \\ \mathbf{f}_p^1 \end{pmatrix} = \begin{pmatrix} \mathbf{S}_1\mathbf{z} \\ \mathbf{S}_2\mathbf{z} \\ \vdots \\ \mathbf{S}_p\mathbf{z} \end{pmatrix}, \tag{8.14}$$

for the Newton-Raphson updates $f_1, \ldots, f_p$, where each of the $\mathbf{S}_j$ are $n \times n$ cubic smoothing-spline matrices of the form $\mathbf{S}_A$ for smoothing against variable X_j, and $\mathbf{z}$ is once again the adjusted dependent variable $\mathbf{z} = \boldsymbol{\eta}^0 + (\mathbf{A}^0)^-(\mathbf{y} - \boldsymbol{\mu}^0)$.

If we again use the diagonal approximation $\mathbf{D}$ for $\mathbf{A}$, we can solve the system using a weighted spline backfitting algorithm. However a more efficient procedure can be derived by exploiting the fact that the linear step can be done exactly. Notationally, each function

f_j is split into linear and nonlinear components, $f_j(X_j) = X_j\beta_j + g_j(X_j)$, resulting in the additive predictor $\eta_f(\boldsymbol{X}) = \eta_L(\boldsymbol{X}) + \eta_g(\boldsymbol{X})$. The linear and additive components are estimated separately, much as in the modified backfitting algorithm given in Chapter 5; details are given in algorithm 8.1.

Algorithm 8.1 *A specialized local-scoring algorithm for additive matched case-control models*

(i) *Initialization.* Fit the linear model $\eta^L(\mathbf{x}^{rk}) = \boldsymbol{\beta}^T\mathbf{x}^{rk}$, using either the Newton-Raphson or one of the delta algorithms described in section 3. Set the nonlinear component $\eta_g = 0$. Compute

$$\mu_{rk} = \exp\{\eta(\mathbf{x}^{rk})\}/\textstyle\sum_k \exp\{\eta(\mathbf{x}^{rk})\},$$

and the deviance $D(\mathbf{y};\boldsymbol{\mu})$.

(ii) *Additive step.* Compute weights $d_{rk} = \mu_{rk}(1-\mu_{rk})$ and adjusted dependent variates

$$z_{rk} = \eta_g(\mathbf{x}^{rk}) + \frac{y_{rk} - \mu_{rk}}{\mu_{rk}(1-\mu_{rk})}.$$

Fit a weighted additive model $\eta_g(\mathbf{x}^{rk}) = \sum_{j=1}^{p} g_j(\mathbf{x}^{rkj})$ to the z_{rk} using, for example, a weighted backfitting algorithm. Compute μ_{rk}.

(iii) *Linear step.* Compute weights $w_{rk} = \mu_{rk}$ and adjusted dependent variates

$$z_{rk} = \eta_L(\mathbf{x}^{rk}) + \frac{y_{rk} - \mu_{rk}}{\mu_{rk}} \quad \text{(Exercise 8.5)}.$$

Centre the predictors in each matched set, $\tilde{\mathbf{X}}_k = (\mathbf{I} - \mathbf{1}\boldsymbol{\mu}_k^T)\mathbf{X}_k$, and fit the linear model $\eta_L(\mathbf{x}^{rk}) = \boldsymbol{\beta}^T\tilde{\mathbf{x}}^{rk}$ to z_{rk} by weighted least-squares. Compute μ_{rk} and the deviance.

(iv) Repeat steps (ii) and (iii) until the fitted functions and coefficients do not change. The fitted functions are given by $f_j(X_j) = \beta_j X_j + g_j(X_j)$, $j = 1, \ldots, p$.

8.2.6 *A simulated example*

Matched case-control data are somewhat elusive, since it's not easy to produce plots that reveal the important relationships. Rather than give the results of fitting a model to some real data (where we don't know what the truth should be), we instead demonstrate the algorithm on some simulated data. The method of data generation itself is of interest.

In the first experiment we use two predictors and generate data for an additive model with two terms:

$$f_1(x) = \begin{cases} x, & \text{for } x < 1; \\ 1 & \text{otherwise;} \end{cases}$$

$$f_2(z) = z^2.$$

Both these terms represent functions often encountered in practice (Fig. 8.1). The former represents an effect that increases up to a certain dose and then levels off. The latter represents the normal dose effect, with either extreme representing higher risk.

To produce a sample, we generate 200 independent $N(0,1)$ vectors and construct the additive predictor for each observation. These are grouped at random into 100 *matched* pairs, (η_{1k}, η_{2k}), $k = 1, \ldots, 100$. We construct the conditional probabilities

$$\mu_{1k} = \frac{\exp(\eta_{1k})}{\exp(\eta_{1k}) + \exp(\eta_{2k})}; \qquad k = 1, \ldots, 100.$$

For each pair a $U(0,1)$ random number U_k is generated, and if $U_k \leq \mu_{1k}$, we assign the *case* label to the first element of the pair, otherwise we assign it to the second.

Figure 8.1 (top two panels) shows the results of fitting the matched case-control additive model to one hundred such samples. A running-line smoother was used in the local-scoring algorithm 8.1, with a span of 50% of the data in any given neighbourhood. The light shaded regions show the pointwise 95th percentiles of all the fitted functions. The dark shaded regions show the pointwise quartiles, and the solid curve in the interior is the pointwise median. The broken curve is the generating function in each case.

The figures exhibit a fair amount of variability, especially in the tails of the predictors. Although we expect (median) bias, it is not too serious, the largest bias being in the tails of the quadratic. The

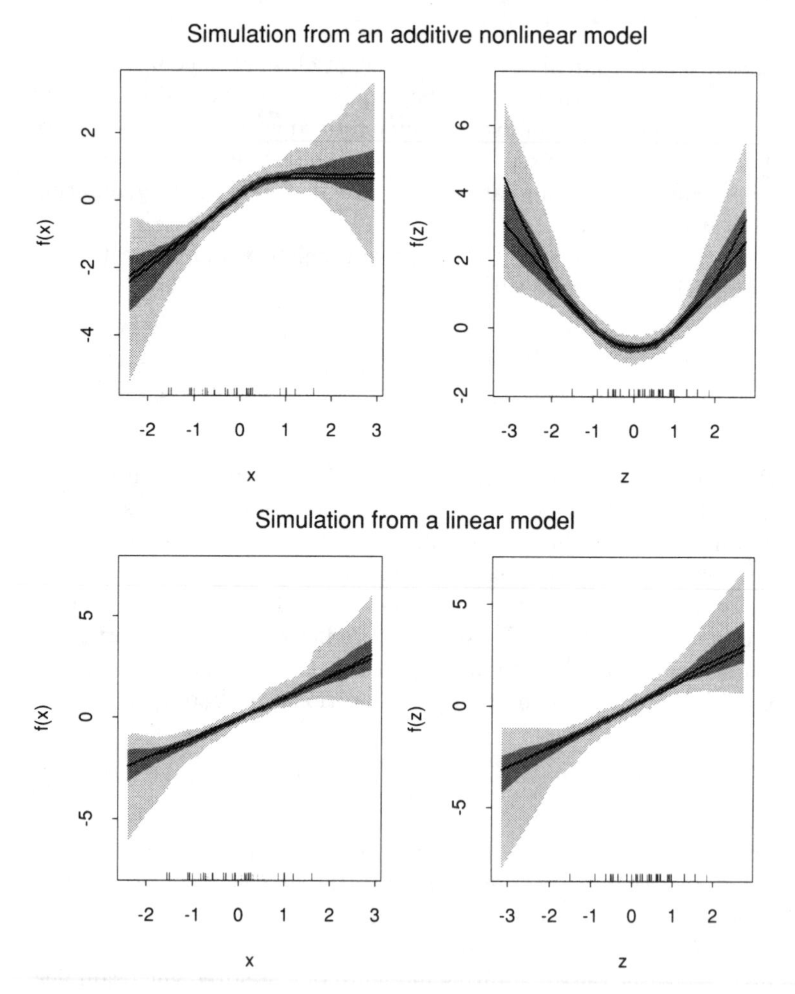

Fig. 8.1. *Results of a simulation study to demonstrate the additive model algorithm for matched case-control data. The top two panels summarize 100 fits each based on 100 matched pairs with a nonlinear underlying model; the lower two are based on a linear underlying model. The light shaded regions show the pointwise 95th percentiles of all the fitted functions. The dark shaded regions show the pointwise quartiles, and the solid curve in the middle of the darker band is the pointwise median. The other curve drawn in each figure is the generating function.*

lower two panels in Fig. 8.1 show the same experiment, except that the generating functions are linear. It shows that the procedure does not find structure when there is none.

8.3 The proportional-hazards model

8.3.1 *Background*

In this section we consider the usual survival data setup. The data available are of the form $(y_1, \mathbf{x}^1, \delta_1,), \ldots, (y_n, \mathbf{x}^n, \delta_n)$, the survival time y_i being complete if $\delta_i = 1$ and censored if $\delta_i = 0$, and $\mathbf{x}^i$ denoting the usual vector of predictors for the ith individual. The distinct failure times are denoted by $t_1 < \cdots < t_k$, there being d_i failures at time t_i.

The proportional-hazards model for survival data, also known as the Cox model, assumes that

$$\lambda(t \mid \boldsymbol{X}) = \lambda_0(t) \exp\Big(\sum_j X_j \beta_j\Big) \tag{8.15}$$

where $\lambda(t \mid \boldsymbol{X})$ is the hazard at time t given predictor values $\boldsymbol{X} = (X_1, \ldots, X_p)$, and $\lambda_0(t)$ is an arbitrary baseline hazard function. The hazard function is the instantaneous probability of failure for an individual at time t, given it has survived until time t:

$$\lambda(t \mid \boldsymbol{X}) = \lim_{\Delta t \to 0} P\{T \in (t, t + \Delta t) \mid T \geq t, \boldsymbol{X}\}.$$

Once again, it is natural to generalize (8.15) to an additive proportional hazards model

$$\lambda(t \mid \boldsymbol{X}) = \lambda_0(t) \exp\Big\{\sum_j f_j(X_j)\Big\}.$$

8.3.2 *Estimation*

There are a variety of techniques for estimating the parameters in the model (8.15). Some approaches assume a parametric form for $\lambda_0(t)$; in this case, the full sampling distribution and hence the likelihood can be written down. As in the case-control model with its matched set indicators, this baseline hazard function is often not of primary interest, and we do not wish to restrict or distort the model through its misspecification. Fortunately the parameter $\boldsymbol{\beta}$ in the proportional-hazards model (8.15) can be estimated without specification of $\lambda_0(t)$. The *partial likelihood*

$$L(\boldsymbol{\beta}) = \prod_{r \in D} \frac{\exp\left(\sum_{j \in D_r} \boldsymbol{\beta}^T \mathbf{x}^j\right)}{\left\{\sum_{j \in R_r} \exp\left(\boldsymbol{\beta}^T \mathbf{x}^j\right)\right\}^{d_r}} \tag{8.16}$$

is similar to the conditional likelihood in the previous section, and factors out the nuisance parameters. Each term in the product reflects the conditional probability of a failure at observed failure time t_r given all the individuals that are still in the study and at risk at time t_r. In equation (8.16), D is the set of indices of the failures, R_r is the set of indices of the individuals at risk at time $t_r - 0$, D_r is the set of indices of the failures at t_r, and we are using Peto's (1972) approximation for ties. We also assume that the censoring is noninformative, so that the construction of the partial likelihood is justified.

The generalized additive proportional-hazards model is

$$\lambda(t \mid \boldsymbol{X}) = \lambda_0(t) \exp\{\eta(\boldsymbol{X})\} \tag{8.17}$$

where $\eta(\boldsymbol{X}) = \sum_{j=1}^{p} f_j(X_j)$. Let $\eta_i = \eta(\mathbf{x}^i)$ and $\boldsymbol{\eta} = (\eta_1, \ldots, \eta_n)^T$. The partial likelihood corresponding to (8.17) is

$$L(f_1, \ldots, f_p) = \prod_{r \in D} \frac{\exp\left(\sum_{j \in D_r} \eta_j\right)}{\left\{\sum_{j \in R_r} \exp\left(\eta_j\right)\right\}^{d_r}}. \tag{8.18}$$

As before, we work with the log partial likelihood $l = \log L$, and add a penalty term of the form $\lambda_j \int \{f_j''(x)\}^2 \, dx$ for each nonlinear function. The minimization problem is almost the same as in the previous section; letting $\mathbf{u} = \partial l / \partial \boldsymbol{\eta}$ and $\mathbf{A} = -\partial^2 l / \partial \boldsymbol{\eta} \boldsymbol{\eta}^T$, we again obtain equations of the form (8.14) for the Newton-Raphson steps. Once again the weight matrices $\mathbf{A}$ are nondiagonal, and to solve the equations (8.14) directly, even using the Gauss-Seidel method, requires $O(n^3)$ computations. We use the same trick and replace $\mathbf{A}$ by a diagonal matrix $\mathbf{D}$, having the same diagonal elements as $\mathbf{A}$. On the average, the off-diagonal elements of $\mathbf{A}$ are smaller than the diagonal elements by an order of magnitude. In particular, if there is no censoring and there are no predictors in the model, then one can show that the diagonal elements of $\mathbf{A}$ average about two, while the nondiagonal elements average about $1/n$. Thus the algorithm that uses $\mathbf{D}$ should be not very different from the exact Newton-Raphson procedure.

Although we do not give details here (Exercise 8.7), one can also develop a specialized local-scoring algorithm as in the previous section. Then one can use the usual linear Newton-Raphson algorithm step to update the $\hat{\boldsymbol{\beta}}$s and the additive Gauss-Seidel step to update the nonlinear component.

The diagonal approximation slows the algorithm down, as in the matched case-control case. Our experience suggests that it doesn't slow it down substantially, and the extra computation involved with step length optimization is usually not warranted. One can also try convergence-acceleration techniques, but once again our limited experience suggests that this is not necessary.

8.3.3 *Further details of the computations*

We now give details on the computation of $\partial l/\partial\eta_i$ and $\partial^2 l/\partial\eta_i\partial\eta_{i'}$ for $\mathbf{u}$, $\mathbf{A}$ and $\mathbf{D}$. Let $C_i = \{k : i \in R_k\}$ denote the risk sets containing individual i, and $C_{ii'} = \{k : i, i' \in R_k\}$ the risk sets containing individuals i and i'. Straightforward calculations yield

$$\frac{\partial l}{\partial \eta_i} = \delta_i - \exp(\eta_i) \sum_{k\in C_i} \frac{d_k}{\sum_{j\in R_k} \exp(\eta_j)};$$

$$\frac{\partial^2 l}{\partial \eta_i^2} = -\exp(\eta_i) \sum_{k\in C_i} \frac{d_k}{\sum_{j\in R_k} \exp(\eta_j)} + \exp(2\eta_i) \sum_{k\in C_i} \frac{d_k}{\left\{\sum_{j\in R_k} \exp(\eta_j)\right\}^2},$$

and

$$\frac{\partial^2 l}{\partial \eta_i \partial \eta_{i'}} = -\exp(\eta_i)\exp(\eta_{i'}) \sum_{k\in C_{ii'}} \frac{d_k}{\left\{\sum_{j\in R_k} \exp(\eta_j)\right\}^2}; \quad (i \neq i').$$

There is a close relation between the quantity $\partial l/\partial\eta_i$ and the generalized residual for the proportional-hazards model. The generalized residual is defined by

$$\hat{e}_i = \hat{\Lambda}(t_i)\exp(\eta_i) \tag{8.19}$$

where

$$\hat{\Lambda}(t_i) = \sum_{k \in C_i} \frac{d_k}{\sum_{j \in R_k} \exp(\eta_j)}.$$

Thus $\partial l/\partial \eta_i = \delta_i - \hat{e}_i$. Hence if there is no censoring, $\hat{e}_i$ and $\partial l/\partial \eta_i$ are equivalent; with censoring, they are not quite the same if one uses the procedure of adding one or $\log 2$ to the $\hat{e}_i$ corresponding to censored observations. In addition, it is customary to plot not $\hat{e}_i$ but $\log(\hat{e}_i)$ versus a predictor.

We obtain approximate standard-error curves for the fitted functions by generalizing the arguments given for exponential family models in section 6.8. Similarly we obtain approximate degrees of freedom. Both hinge on the asymptotic approximation

$$\hat{\boldsymbol{\eta}} \approx \mathbf{S}\mathbf{z}_0 \tag{8.20}$$

(for a model with a single smooth term) where as before $\mathbf{S} = (\mathbf{A} + \lambda\mathbf{K})^{-1}\mathbf{A}$ is the cubic smoothing-spline operator matrix and $\mathbf{z}_0$ is the "true" adjusted dependent variable $\boldsymbol{\eta}_0 + \mathbf{A}_0^{-1}\mathbf{u}_0$. For an additive fit, $\mathbf{S}$ is replaced by $\mathbf{R}$, the weighted additive operator. Apart from the fact that $\mathbf{A}$ is not diagonal, the details are the same as in section 6.8.

8.3.4 *An example*

Kalbfleisch and Prentice (1980) analysed the results of a study designed to examine the genetic and viral factors that may influence the development of spontaneous leukemia in mice. The original data set contains 204 observations, with six predictors and both cancerous and noncancerous deaths recorded. Kalbfleisch and Prentice performed a number of analyses. Here we follow Hastie and Tibshirani (1990) and consider any death as the endpoint, and predictors `antibody` level (% gp 70 ppt), `virus` level (PFU/ml), `mhc` phenotype (1 or 2), `sex` (1 =male, 2 =female), and `coat` colour (1 or 2). After removing observations with missing values to allow comparisons to the analysis of Kalbfleisch and Prentice, the data set contained 175 observations. `Antibody` level and `virus` level have continuous values, although about half of the mice have a value of zero recorded for `antibody` level (levels less than 0.5% are considered undetectable, and were coded as 0).

Table 8.1 show the results of a number of models fitted to these data. Convergence is typically achieved in two to five iterations. Our analysis follows analyses "5" and "6" given by Kalbfleisch and Prentice (1980 p. 216). `Virus` level was thought to be highly related to `antibody` level, so it is of interest to consider models containing either one or both of these variables. Models (ii)–(viii) examine each variable separately, model (ix) has smooth terms in both `virus` level and `antibody` level, model (x) shows the full fit, and models (xi)–(xiv) give the results of a backward stepwise procedure described in Chapter 9, with a significance cutoff of 10%. The smooth terms were fitted using smoothing splines in the local-scoring algorithm with $df = 4$, and the final model chosen is (xi). We see that `virus` level is strongly significant, even with `antibody` level in the model. `Antibody` level is marginally significant, by itself or in combination with `virus` level, but does not appear in the final model (xi) chosen by the backward stepwise procedure. The coefficient of `coat` colour in model (xi) is 0.44, (corresponding to a relative risk of 1.6), with an estimated standard error of 0.15.

Of particular interest are the estimated functions for `antibody` level and `virus` level (Fig. 8.2). The left figure shows an increasing risk up to 55% `antibody` level followed by a sharp decrease, and the right figure shows the opposite effect at about 5,000 for `virus` level. It might seem from the range of the fitted functions that `antibody` level is a more important predictor than `virus` level. In fact the opposite is true (see Table 8.1); the exotic nonlinear behaviour of antibody level is not supported by many observations.

Kalbfleisch and Prentice split `virus` level into four categories: $\texttt{virus} < 3.98$, $3.98 \le \texttt{virus} < 10^4$, $10^4 \le \texttt{virus} < 10^5$, and $\texttt{virus} \ge 10^5$. The four-category `virus`-level variable produces (on its own) a value of 896.1 for $-2l$ on three degrees of freedom, compared with 892.4 on three degrees of freedom for the smooth `virus` level model. The jittered data sights for `virus` level in Fig. 8.2 suggest that there are three groups of interest: the two extremes and the interior. The uncharacteristic behaviour of the standard-error curves for `virus` level confirm this. They are wider in the interior than at the ends — due to the sparsity of data in

Table 8.1 *Analysis of deviance table for mice leukemia data*

	Model	$-2l(\mathbf{y};\boldsymbol{\eta})$	df^{err}	ΔDev	Δdf^{err}
(i)	`null`	925.9	175.0		
(ii)	`antibody`	922.7	174.0	3.2	1.0
(iii)	`s(antibody)`	917.6	171.2	8.3	3.8
(iv)	`virus`	897.1	174.0	28.8	1.0
(v)	`s(virus)`	892.3	171.4	33.6	3.6
(vi)	`sex`	920.7	174.0	5.2	1.0
(vii)	`coat`	923.1	174.0	2.8	1.0
(viii)	`mhc`	924.4	174.0	1.5	1.0
(ix)	`s(virus)+s(antibody)`	887.1	167.7		
(x)	`s(virus)+s(antibody)+sex+ coat+mhc`	877.1	164.8	10.0	2.9
(xi)	`s(virus)+coat`	887.7	170.4		
(xii)	`virus+coat`	893.2	173.0	5.5	2.5
(xiii)	`coat`	923.1	174.0	35.4	3.6
(xiv)	`s(virus)`	892.3	171.4	4.6	1.0

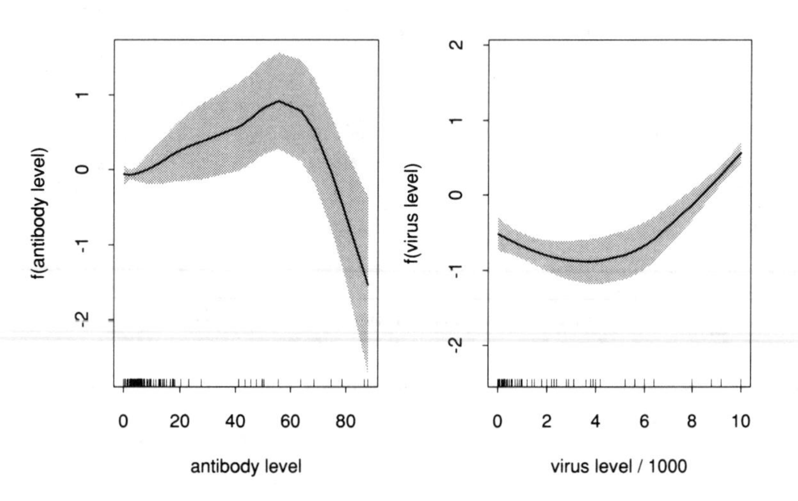

Fig. 8.2. *Fitted functions for* `antibody` *and* `virus` *level. Shaded regions are 2× standard-error bands. The functions are plotted on the same scale.*

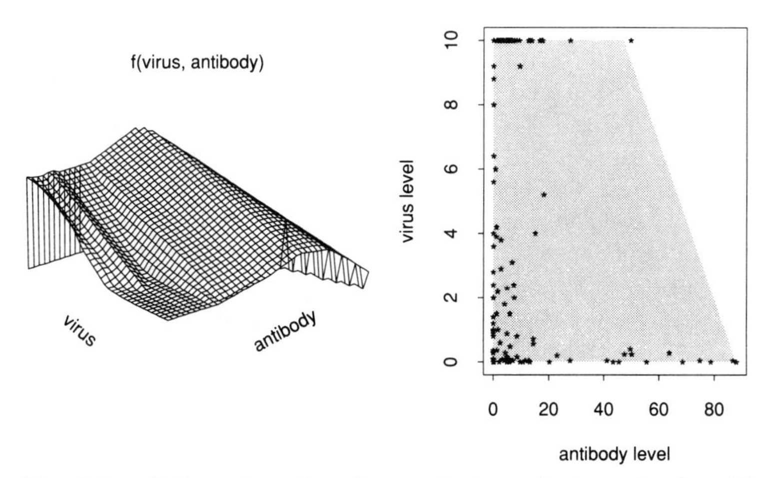

Fig. 8.3. *Estimated surface for* `antibody` *and* `virus` *levels. The scatterplot shows where the data lie, and the shaded region is their convex hull. The surface is interpolated and plotted over this region.*

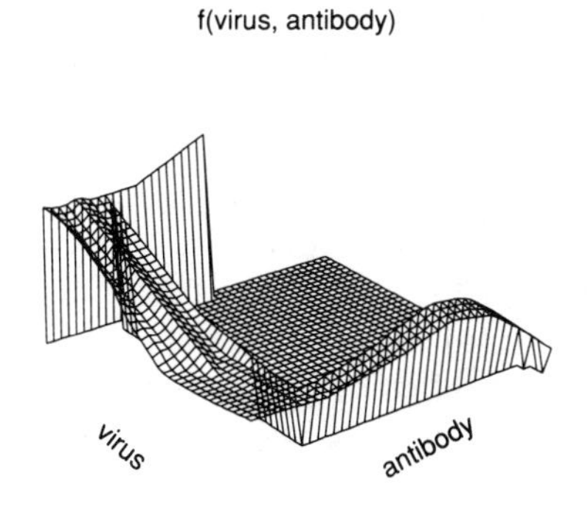

Fig. 8.4. *Estimated surface for* `antibody` *and* `virus levels` *as in Fig. 8.3, except plotted only in regions where data exist.*

the interior.

Kalbfleisch and Prentice also split `antibody` level into two categories (0 and > 0). We can try separating out the mice with zero for `antibody` level by adding an appropriate dummy variable and excluding the observations with `antibody` level zero when computing the smooth for that variable. This does not change the overall results substantially, but in the model containing only `antibody` level and `virus` level, the `antibody` level$=0$ group has a fitted value of about 0.2 units higher than the points to their right. This indicates that this group may be a high risk group that is different from the mice with some detectable `antibody`. However, `antibody` again does not appear in the final model chosen by the backward stepwise procedure.

Figure 8.3 shows a jointly estimated surface for `antibody` level and `virus` level, an attempt to investigate a possible interaction between the variables. The function η is now a general function of `antibody` level and `virus` level. A bivariate kernel smoother is used in the local-scoring algorithm to compute the surface. From the perspective plot we might deduce that the two effects are approximately additive, but we need to be careful. Figure 8.3 also shows the bivariate behaviour of the two predictors `virus` level and `antibody`. There is no data in the interior! The perspective surface plotting routine has approximated the surface over the region defined by the convex hull of the predictors (the shaded region in Fig. 8.3), and interpolated smoothly where there is no data. This is a natural procedure when there is sufficient data plotted over an unstructured region. No wonder the surface looks additive! Figure 8.4 shows the surface again, but with the region with no data set to zero. It is clear that there is not much bivariate information in the data, except at the two extreme values of `virus` level. An additive model has a good chance of fitting well.

An alternative way to model the interaction is to fit separate curves for `antibody` at different levels of `virus` level. This is more appropriate in this case, since we appear to have data for `antibody` level at the two extremes of `virus` level, but not much information between. We discuss the modelling of interactions in section 9.5.

8.4 The proportional-odds model

8.4.1 *Background*

The proportional-odds model has a similar flavour to the logistic model, the matched case-control model and the proportional-hazards model, yet it is different enough to require special treatment.

In this setting, the response variable Y takes on one of K ordered categories, as opposed to two in the usual logistic regression model. If there are no predictors, we can summarize the observations by a histogram of the counts in each category. Instead of a single proportion, as in logistic regression, we have a whole vector of proportions — one for each category. The proportional-odds model provides a way of describing how the category proportions redistribute themselves as the predictors vary.

A useful way of visualizing and generating the model is to think of an underlying *continuous* response variable, say Z. What we sample is Y, a categorization of Z into K contiguous intervals. If the categories are fixed, and the distribution of Z changes as the predictors change, then so does the distribution of Y (McCullagh and Nelder, 1989, Fig. 5.1).

The proportional-odds model is derived as follows. Suppose Z has distribution function $F\{z - \eta(\boldsymbol{X})\}$ where $\eta(\boldsymbol{X})$ is a location parameter depending on the predictors: typically $\eta(\boldsymbol{X}) = \boldsymbol{X\beta}$. Define the K categories of Z by the cutpoints $\alpha_1 < \cdots < \alpha_{K-1}$. Then

$$\begin{aligned} \gamma_k(\boldsymbol{X}) &= \mathrm{pr}(Y \le k \mid \boldsymbol{X}) \\ &= \mathrm{pr}(Z \le \alpha_k \mid \boldsymbol{X}) \\ &= F\{\alpha_k - \eta(\boldsymbol{X})\}. \end{aligned} \tag{8.21}$$

We need to choose F, and a variety of choices are available. The logistic distribution, $F(z) = \exp(z)/\{1+\exp(z)\}$ has computational advantages, since then

$$\log\left\{\frac{\mathrm{pr}(Y \le k \mid \boldsymbol{X})}{1 - \mathrm{pr}(Y \le k \mid \boldsymbol{X})}\right\} = \alpha_k - \eta(\boldsymbol{X}). \tag{8.22}$$

The name proportional-odds arises because the *odds-ratio* is independent of k:

$$\frac{\gamma_k(\boldsymbol{X}_1)\{1 - \gamma_k(\boldsymbol{X}_2)\}}{\{1 - \gamma_k(\boldsymbol{X}_1)\}\gamma_k(\boldsymbol{X}_2)} = \exp\{\eta(\mathbf{X}_2) - \eta(\mathbf{X}_1)\}.$$

Here we see the similarity to the proportional-hazards or Cox model. We emphasize that $\eta(\boldsymbol{X})$ does not depend on the response category k. This is only one of many possible models for polytomous-response data; McCullagh and Nelder (1989) describe others.

8.4.2 *Fitting the additive model*

The additive proportional-odds model is obtained simply by using an additive predictor $\eta(\boldsymbol{X}) = \sum_{j=1}^{p} f_j(X_j)$. Although the algorithm we present for this model has an intuitive flavour, its derivation is intricate. We outline the derivation here, and devote Exercises 8.10 to 8.12 to the details.

We think of our observations as a pair of vectors $(\mathbf{y}^i, \ \mathbf{x}^i)$, where $\mathbf{y}^i$ is a K-vector of frequencies for the ith observation. This notation allows $m_i = \sum_{k=1}^{K} y_{ik}$ to be one, and thus allows either *grouped* or *ungrouped* data. Both occur frequently in practice.

We concentrate first on the linear model, and develop an algorithm that extends easily to the additive model. The likelihood of the data has contributions $M_K(m_i, \mathbf{p}^i)$ where $p_{ik} = \gamma_{ik} - \gamma_{i\,k-1}$ and $M_K(m, \mathbf{p})$ denotes a multinomial distribution of m objects distributed amongst K categories with probabilities p_k. One can write this in vector notation as $\boldsymbol{\gamma}^i = \mathbf{L}\mathbf{p}^i$ where $\mathbf{L}$ is a $(K-1) \times K$ lower triangular matrix with ones on and below the diagonal, and constructs the vector of cumulative frequencies. Similarly $\mathbf{g}^i = \mathbf{L}\mathbf{y}^i$ denotes the vector of observed cumulative frequencies for the ith observation.

To estimate the linear model, we differentiate the log-likelihood, construct the score vector and information matrix, and use the Newton-Raphson iterations to find a solution to the score.

After some work (Exercise 8.10), we see that these iterations are equivalent to repeatedly solving the system

$$\begin{pmatrix} \sum_1^n \mathbf{W}_i & \sum_1^n \mathbf{W}_i \mathbf{1} \mathbf{x}^{i^T} \\ \sum_1^n \mathbf{x}^i \mathbf{1}^T \mathbf{W}_i & \sum_1^n (\mathbf{1}^T \mathbf{W}_i \mathbf{1}) \mathbf{x}^i \mathbf{x}^{i^T} \end{pmatrix} \begin{pmatrix} \boldsymbol{\alpha}^1 \\ \boldsymbol{\beta}^1 \end{pmatrix} = \begin{pmatrix} \sum_1^n \mathbf{W}_i \mathbf{z}^i \\ \sum_1^n \mathbf{x}^i (\mathbf{1}^T \mathbf{W}_i \mathbf{z}^i) \end{pmatrix} \quad (8.23)$$

for the $q = K - 1 + p$ parameters, where:

(i) $\mathbf{W}_i$ is a $(K-1)\times(K-1)$ weight matrix for each observation based on the current fitted values,

(ii) $\mathbf{z}^i$ is a (vector-valued) adjusted response defined by $\mathbf{z}^i = \boldsymbol{\alpha}^0 + \mathbf{1}\mathbf{x}^{i^T}\boldsymbol{\beta}^0 + \mathbf{C}_i^{-1}(\mathbf{g}^i - \boldsymbol{\gamma}^i)$,

(iii) $\boldsymbol{\gamma}^i$ are the fitted cumulative probabilities based on $\boldsymbol{\alpha}^0$ and $\boldsymbol{\beta}^0$, and

(iv) $\mathbf{C}_i$ is diagonal with kth entry $\gamma_{ik}(1-\gamma_{ik})$.

Although this system is itself a repeated regression, we can simplify it even further. Intuitively, we would like to reduce the estimation of the scalar $\eta(\boldsymbol{X})$ to a univariate response regression problem, for then we should be able to easily generalize the procedure to additive models.

Previous experience with solving systems such as (8.23) suggests the Gauss-Seidel type iterations:

$\boldsymbol{\alpha}$ step: Let $\mathbf{z}^i(\alpha) = \mathbf{z}^i - \mathbf{1}\mathbf{x}^{i^T}\boldsymbol{\beta}^0 = \boldsymbol{\alpha}^0 + \mathbf{C}_i^{-1}(\mathbf{g}^i - \boldsymbol{\gamma}^i)$, an adjusted dependent vector for the $\boldsymbol{\alpha}$s. Then

$$\boldsymbol{\alpha}^1 = \left(\sum_{i=1}^{n}\mathbf{W}_i\right)^{-1}\sum_{i=1}^{n}\mathbf{W}_i\mathbf{z}^i(\alpha),$$

a weighted average of the $\mathbf{z}(\alpha)$s.

η step: Let $z_i(\eta) = \eta_i^0 + \mathbf{1}^T\mathbf{W}_i\mathbf{C}_i^{-1}(\mathbf{g}^i - \boldsymbol{\gamma}^i)/\mathbf{1}^T\mathbf{W}_i\mathbf{1}$, a (scalar) adjusted dependent variable for η, where $\eta_i^0 = \mathbf{x}^{i^T}\boldsymbol{\beta}^0$. Obtain $\boldsymbol{\beta}^1$ by linearly regressing $z_i(\eta)$ on the predictors $\mathbf{x}^i$ with weights $w_i = \mathbf{1}^T\mathbf{W}_i\mathbf{1}$.

We leave it as Exercise 8.11 to show that the iteration of these two steps converges to the solution of (8.23).

The algorithm has some intuitively appealing features. The αs are like constants or intercepts, so we expect to compute them as an average. From these equations we see that if there are no predictors, all the $\mathbf{A}_i$ are the same, and $\boldsymbol{\alpha}$ is simply the logit of the average cumulative frequency. It is also convenient that the regression coefficients are found from a scalar regression. One can also show that when there are only two categories, the whole procedure is the usual iterative reweighted least-squares for solving the score equations (and the Gauss-Seidel step is redundant). Have we gone to all this trouble to simplify the rather messy Newton-Raphson algorithm purely for aesthetic reasons? Not really when

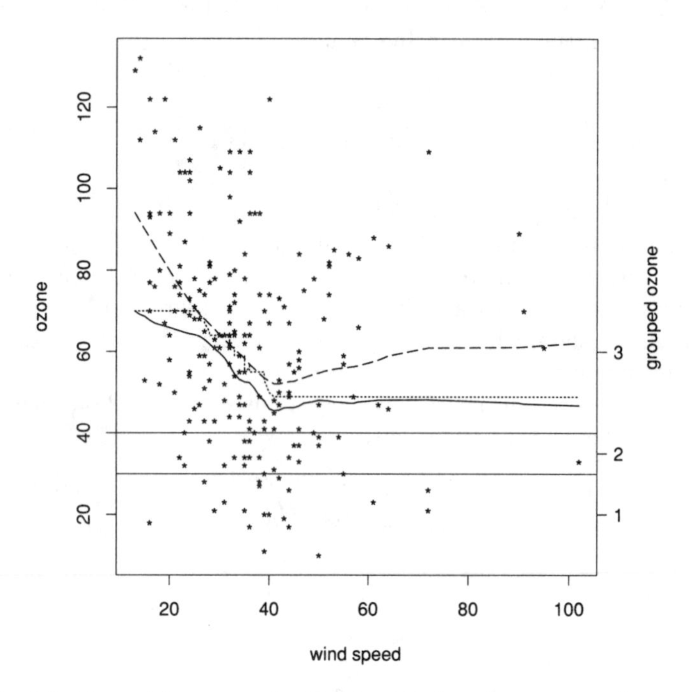

Fig. 8.5. *The solid curve is the function fitted to the categorized version of* `ozone` *by the nonparametric proportional-odds procedure, and estimates the median of the underlying continuous* `ozone` *as a function of* `wind speed`. *The horizontal lines indicate the categorization of* `ozone`. *The dotted curve is the running median of the original data and nearly coincides with the solid curve. The dashed curve is a locally-weighted running-line fit to the original data.*

we notice that the two steps above immediately suggest an additive-model algorithm. Indeed, we simply replace the *linear* η step by the *additive* η step:

η step: Let $z_i(\eta) = \eta_i^0 + \mathbf{1}^T\mathbf{A}_i\mathbf{C}_i^{-1}(\mathbf{g}^i - \gamma^i)/\mathbf{1}^T\mathbf{A}_i\mathbf{1}$ where $\eta_i^0 = \sum_{j=1}^p f_j^0(x_{ij})$. Obtain f_j^1, $j = 1, \ldots, p$, by fitting a weighted additive model of $z_i(\eta)$ onto x_{ij} with weights $w_i = \mathbf{1}^T\mathbf{A}_i\mathbf{1}$.

8.4.3 *Illustration*

Figure 8.5 is a plot of two of the air-pollution variables. The response is `ozone` concentration and the predictor is `wind speed`. Included in the figure is a locally-weighted running-line smooth which shows the relationship between the two. What if the data are categorized before we see it? This is not an uncommon practice, especially for survey data. Figure 8.5 also shows the nonparametric proportional-odds fit (solid curve). Usually such a curve is location and scale free, but we have chosen this example well (and have only two cutpoints). We match the two estimated cutpoints to the values 30 and 40 used to cut the original data, and hence fix the scale and location. This allows us to see how well the proportional-odds model performs in modelling the underlying continuous response variable (which is what we are doing in this case). In order to judge it, we superimpose the running median of the original data, and note the strong similarities between the two.

To what extent does our model estimate the median regression of the underlying variable? If there are a large number of categories, then the logit assumption does not play much of a role, and we are essentially modelling the distribution of the errors nonparametrically. The logit is simply a monotone transformation of the category labels to a new set of labels, estimated from the data. The stronger assumption is that the shape of the distribution does not change as the predictor changes. Consequently the regression, which thus depends only on the ranks of the categories for the response, can be thought of as estimating (up to a constant shift) any chosen quantile of the underlying distribution.

Another way of viewing the model is in terms of the Box-Cox transformation model. The traditional Box-Cox procedure attempts to simultaneously find a transformation of the response Y to Z, such that Z has a Gaussian distribution with error variances independent of X and the regression of Z on X is linear (or additive). Box-Cox uses maximum likelihood on the original scale of Y. Here we seek transformations to a new variable Z with a logistic distribution (Gaussian via the probit link is also possible), with error variances independent of X and the regression of Z on X is linear (or additive). The only difference is that we don't transform the category labels to *values* of Z, but rather to *intervals* of Z. Once again, maximum likelihood estimation is performed on the original scale, using the distribution induced by this model.

8.5 Seasonal decomposition of time series

Although smoothing techniques originated in the study of time series data, these applications have received little attention so far in this book. In this section we describe a new technique for decomposing a time series into three components: *seasonal*, *trend* and *remainder*. We include it not only to interest our time series readers, but also for its novel use of backfitting and the subsequent analysis of the estimating equations. This section is based on Cleveland *et al.* (1990), and we present their decomposition of some carbon dioxide data.

8.5.1 *The STL procedure*

The data are a realization of a time series Y_t, measured at n times $t = 1, \ldots, n$. The goal is to form a decomposition

$$Y_t = T_t + S_t + R_t \tag{8.24}$$

consisting of a trend component T_t, a seasonal component S_t and the remainder R_t. In our terminology, *time* is the only predictor, and the additive model consists of two different time effects plus a residual. In the example in Fig. 8.6, Y_t is a series of monthly averages of atmospheric CO_2 measurements made at Mauna Loa Observatory in Hawaii (Keeling, Bacastow and Whorf, 1982). There are 348 observations, from January 1959 to December 1987. The STL procedure ("Seasonal and Trend decomposition using Loess") uses a backfitting algorithm to estimate the trend and seasonal components, with an appropriate smoother for each term. Indeed, it is the construction of these smoothers, and their respective bandwidths, that define the trend and seasonal components. We now discuss each smoother in turn.

The seasonal smoother $\mathcal{S}_S$ is rather special, so we describe it first. Denote by n_S the period of the seasonal cycle. In the example the natural cycle length is $n_S = 12$ corresponding to the months of the year. We describe the smoother for such a monthly series; the general case is clear from the description. The data to be smoothed

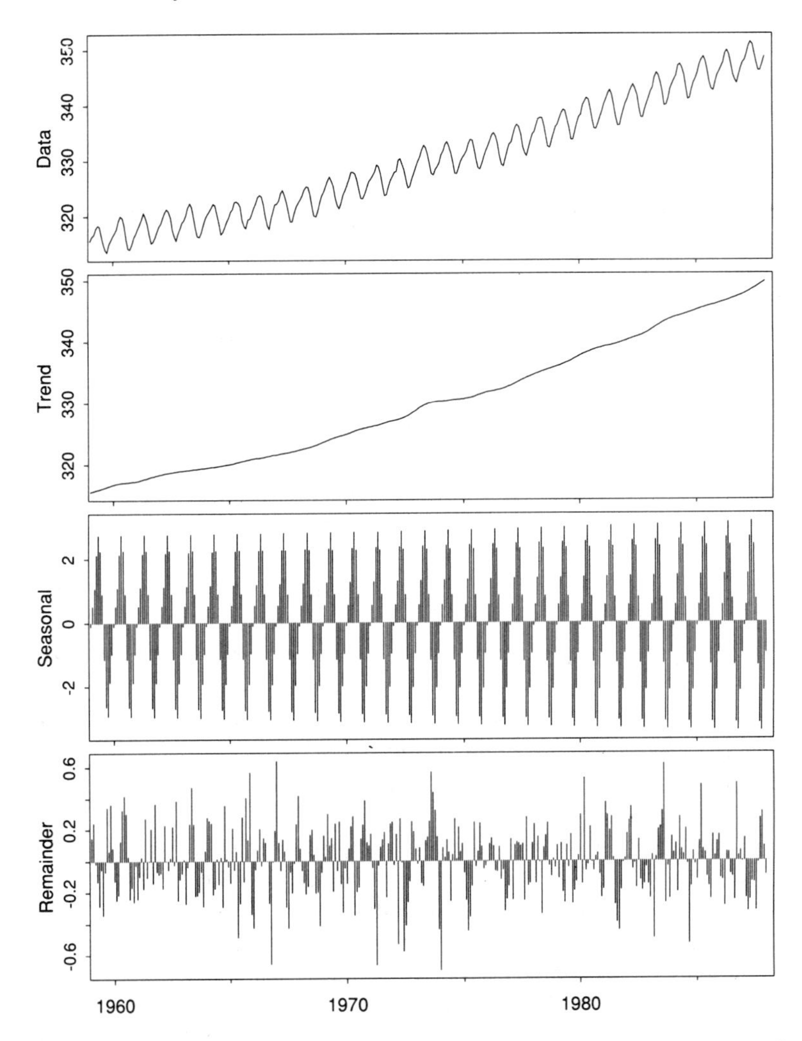

Fig. 8.6. *A STL decomposition plot of monthly* CO_2 *data. The top panel is the data* Y_t *plotted against time. Next is the fitted trend component, followed by the seasonal component, and finally the remainder. Note the difference in vertical scales for the lower two panels.*

by $\mathcal{S}_S$ (partial residuals within the backfitting iterations) is divided into 12 monthly subseries. There is a subseries of 29 January measurements, 29 February measurements and so on. Each of the subseries is then smoothed separately using a regression smoother, as a function of year. The 12 separate smoothed subseries are then recomposed. This results in a seasonal effect that is not necessarily smooth from one month to the next, but for any given month the effect varies smoothly over the years.

As described so far, the seasonal smoother is perfectly capable of picking up both long term trend effects as well as seasonal effects. For example, if the data presents with values Y_t linear in time, each of the subseries is linear as well, and the whole linear effect is absorbed by the 12 pieces comprising the seasonal effect. To avoid this aliasing, the final step in the construction of $\mathcal{S}_S$ is to remove any long-term trend that may have crept in, by using the trend smoother described next or some other low-pass filter $\mathcal{S}_L$. If $\mathcal{S}_C$ represents the recomposed cyclical smoother without the detrending, then $\mathcal{S}_S$ is defined to be

$$\mathcal{S}_S = (I - \mathcal{S}_L) \circ \mathcal{S}_C. \tag{8.25}$$

The third panel in Fig. 8.6 is a seasonal component computed and recomposed in this way. Notice that the amplitude of the seasonal effect is increasing with time, and with the long term trend or level of CO_2. This observation has of course been made before (Keeling *et al.*, 1982; Cleveland, Freeny and Graedel, 1983).

Figure 8.7 shows the separate subseries data, and the smooth curves that describe them. Here the smoothing is performed using *loess* (section 2.11), which fits locally-weighted polynomials (linear in this case). As can be seen, a large bandwidth is used, resulting in a nearly linear fit to each of the subseries.

The *trend* smoother $\mathcal{S}_T$ can be any regression smoother. Some care is needed in selecting the smoothing parameter, as we see in the next section, to avoid aliasing with the seasonal component. *Loess* is used for all the smoothing in these examples. The smoothing parameter for *loess* is the number q of nearest neighbours, which is particularly intuitive for evenly spaced time series data.

The backfitting algorithm alternates between $\mathcal{S}_T$ and $\mathcal{S}_S$ until convergence. The STL procedure described in Cleveland *et al.*

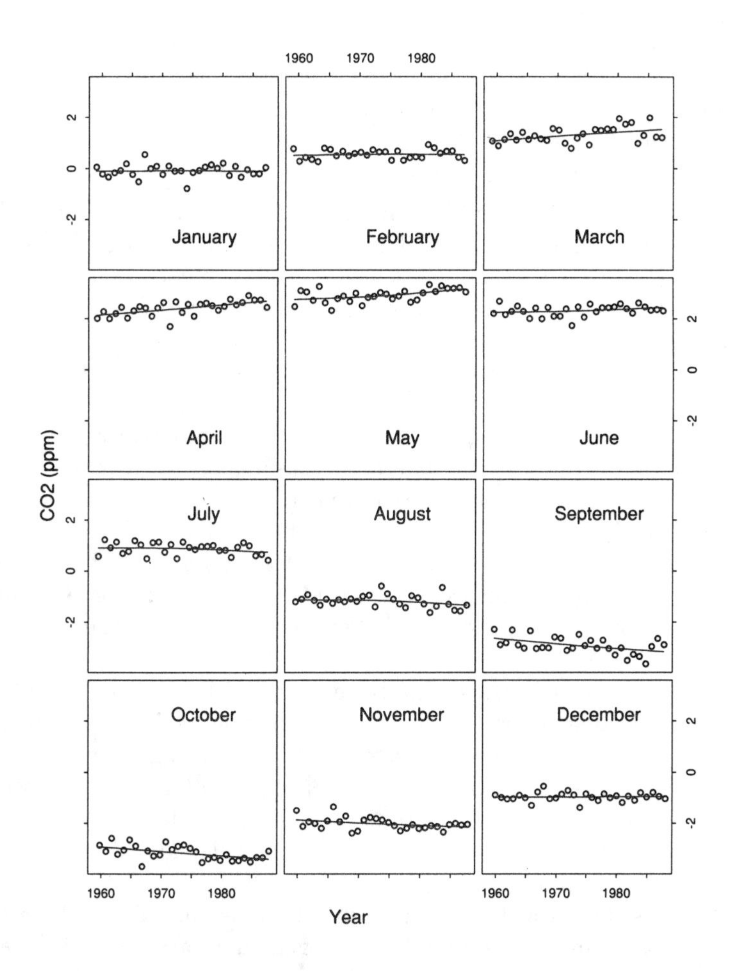

Fig. 8.7. *Seasonal subseries plots. Each panel shows the fitted subseries curve, together with partial residuals. The average slopes of each of the 12 subseries curves themselves average about zero over the 12 plots, as one would expect once the long term trend effect is removed from the seasonal component.*

allows an outer robustness loop which down-weights points with large residuals. They give a more detailed version of the seasonal smoother, and also give guidelines on selecting the various smoothing parameters, clearly an important task if we hope to separate the two time effects.

Although the STL procedure may appear *ad hoc* to some, its modularity produces some interesting features:

(i) The smoothers can be modified to suit the occasion. For example, the seasonal smoother described above allows for nonstationary seasonal behaviour; one can construct a simpler stationary version by using the average alone of each subseries to represent the month effect.

(ii) "Trend" is somewhat of a misnomer, since there may be a number of different trends of interest, each at different frequencies. As Cleveland *et al.* point out, the trend smoother $\mathcal{S}_T$ is really there to allow extraction of the seasonal component. Once the seasonal component is removed from Y_t, the remainder (plus trend) can be smoothed with different bandwidths to highlight different features. For example, by removing the almost linear component from the trend component in Fig. 8.6, they expose the wiggles as real effects correlated with the southern oscillation.

(iii) Other components can be extracted, by simply including additional operators in the backfitting loop. Once again, special care should be taken to avoid aliasing. For example, in monthly economic data, a special *trading day* regression term can be included to account for the different numbers of business days in each month; other *calendar* effects can be treated similarly.

(iv) Missing data do not cause problems as they do for Fourier-based time series filters. The missing data are simply ignored, and replaced by values fitted by the smoothers, as needed.

(v) The nature of the estimates encourage graphical inspection rather than formal model based hypothesis tests. Often these latter tests are based on spectral density estimates and thus take place in the frequency domain. It is far simpler for nonspecialists to understand effects in the time domain.

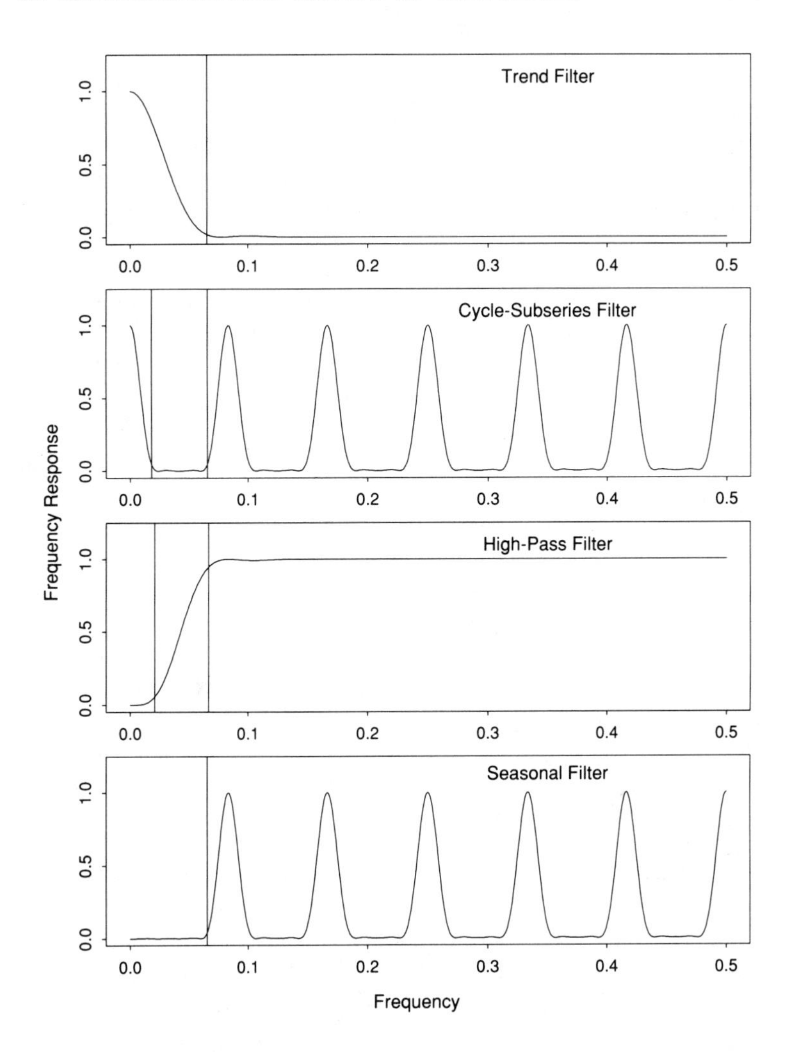

Fig. 8.8. *Eigen-analysis for simplified circulant smoothers. The top panel gives the power transfer function (squared eigenvalues) for* $\mathcal{S}_T$*, the second panel for* $\mathcal{S}_C$*, the third for* $I - \mathcal{S}_L$*, and the fourth for* $\mathcal{S}_S = (I - \mathcal{S}_L) \circ \mathcal{S}_C$.

8.5.2 *Eigen-analysis and bandwidth selection*

In Chapter 5 we gave explicit expressions in equation (5.25) for the solution to the additive model estimating equations for two smoothers. Suppose the smoother matrices $\mathbf{S}_T$ and $\mathbf{S}_S$ correspond to the smoothers $\mathcal{S}_T$ and $\mathcal{S}_S$ for smoothing the time series $\mathbf{y}$. Then the operators for the trend and seasonal components are

$$\begin{aligned} \mathbf{R}_T &= \mathbf{S}_T(\mathbf{I} - \mathbf{S}_S\mathbf{S}_T)^{-1}(\mathbf{I} - \mathbf{S}_S) \\ \mathbf{R}_S &= \mathbf{S}_S(\mathbf{I} - \mathbf{S}_T\mathbf{S}_S)^{-1}(\mathbf{I} - \mathbf{S}_T) \end{aligned} \tag{8.26}$$

provided $\|\mathbf{S}_S\mathbf{S}_T\| < 1$ and $\|\mathbf{S}_T\mathbf{S}_S\| < 1$. Since both $\mathcal{S}_T$ and $\mathcal{S}_S$ are defined in terms of the same predictor *time*, and the time values are evenly spaced, there is some hope that these equations might simplify. In particular we need to arrange that $\mathbf{S}_T$ and $\mathbf{S}_S$ to be symmetric and share the same eigenspace. The rows of $\mathbf{S}_T$ and $\mathbf{S}_S$ are in the order of $\mathbf{y}$, which is in time order, so the special structure of the smoother matrices is not disturbed by permutation matrices as is usually the case. To achieve the desired simplification, we need to make the additional assumption that the data are circular, i.e., $Y_{n+t} = Y_t$. In practice this means that $\mathbf{S}_T$, $\mathbf{S}_C$ and $\mathbf{S}_L$ are all circulant matrices. A circulant (smoother) matrix $\mathbf{S}$ has the same set of smoother weights in each row, shifted by one each time. This in turn implies that $\mathbf{S}$ has a Fourier eigendecomposition, and thus the smoothers above share the same eigenspace. The eigenvalues are the Fourier coefficients corresponding to the discrete Fourier transform of the weight sequence (section 3.7; Exercise 8.14).

The properties of the smoothers and the converged backfitting solutions are all captured entirely by the corresponding compositions of their Fourier coefficients. For example, if λ_{Ti} and λ_{Si} are the ith eigenvalues of $\mathbf{S}_T$ and $\mathbf{S}_S$, then the ith eigenvalue of the trend and seasonal operators are (Exercise 5.16)

$$\lambda(\mathbf{R}_T)_i = \frac{\lambda_{Ti}(1 - \lambda_{Si})}{1 - \lambda_{Ti}\lambda_{Si}}$$

and

$$\lambda(\mathbf{R}_S)_i = \frac{\lambda_{Si}(1 - \lambda_{Ti})}{1 - \lambda_{Si}\lambda_{Ti}}. \tag{8.27}$$

The rate of decay of the eigenvalue sequence for a regression smoother is determined by the bandwidth (see Exercise 3.6, 3.7 and section 9.3.6, where this is made explicit for smoothing splines). One can use this information to select the various bandwidths of the smoothers to ensure little or no overlap in the *power transfer functions* (squared eigenvalues) of the trend and seasonal operators. Figure 8.8, taken from Cleveland *et al.*, illustrates the idea, showing the frequency response functions for $\mathcal{S}_T$, $\mathcal{S}_C$, $I - \mathcal{S}_L$ and $\mathcal{S}_S = (I - \mathcal{S}_L) \circ \mathcal{S}_C$. Here the bandwidths are carefully chosen so as to ensure very little overlap for $\mathcal{S}_T$ and $\mathcal{S}_S$. From (8.26) and (8.27) it is clear that if there is no overlap at all, the backfitting solutions are immediately available from the individual fits; when there is overlap, they indicate which components of the terms are shared (aliased), and to what extent. Cleveland *et al.* give some clever heuristics based on these considerations for selecting the various bandwidths.

8.6 Bibliographic notes

Breslow and Day (1980) gave a lucid account of the techniques associated with the analysis of matched case-control data. Pregibon (1982) discussed diagnostic procedures for this model. Hastie and Pregibon (1988) discussed alternative algorithms for the linear case, and derived the specialized local-scoring algorithm for the additive model. Jørgenson (1984) studied the delta method. Breslow (1987) discussed applications of generalized additive models in epidemiological settings.

The proportional-hazards model was introduced by Cox (1972), and has enjoyed much coverage in the literature. Kalbfleisch and Prentice (1980) gave a more recent account; see also Cox and Oakes (1984), Lawless (1982) or Miller (1981). O'Sullivan (1988) gave an algorithm for the proportional-hazards model with one predictor, based on a conjugate gradient method. The additive model algorithm presented here was first given in Hastie and Tibshirani (1986), although details of the approximation were not; these were provided later in Hastie and Tibshirani (1990). Kalbfleisch and Prentice described generalized residuals for the proportional-hazards model, as do Crowley and Storer (1983). Brasher (1989) studied a number of definitions of residuals, including one derived from the penalized partial-likelihood approach. Gentleman (1988) described a local-likelihood approach (section 6.13) using the full likelihood in conjunction with the EM-algorithm (Clayton and Cuzick, 1985). Sleeper and Harrington (1989) used B-splines in an additive Cox proportional-hazards model.

McCullagh (1980) developed the proportional-odds model, which was also discussed in detail in McCullagh and Nelder (1989). The additive extension given here is taken from Hastie and Tibshirani (1987a). Hastie, Botha and Schnitzler (1989) described an application.

The STL procedure of Cleveland, Cleveland, McRae and Terpenning (1990) is the successor of a similar earlier procedure named SABL (Cleveland, Devlin and Terpenning, 1982). Both of these were motivated by the even earlier X-11 system (Shiskin, Young and Musgrave, 1967; Dagum, 1978) still used today by the U.S. Census Bureau for the seasonal adjustment of time series. On a historical note, X-11 is one of the earliest known programs to have used the backfitting algorithm.

8.7 Further results and exercises 8

8.1 Derive the expression for the conditional probability (8.5). Think of randomly assigning the case and control labels to observations with the given predictors, and use Bayes theorem to write down the probabilities. The conditional likelihood is the ratio of the probability for the observed assignments to the sum of probabilities for all possible assignments.

8.2 Verify equations (8.7) and (8.8) for the score and information of the conditional likelihood.

8.3 The IRLS algorithm (8.9) for the linear matched case-control model involves a nondiagonal weight matrix consisting of the blocks $\mathbf{A}_k$ on the diagonal. Show that these can be decomposed into a simpler form $\mathbf{A}_k = (\mathbf{I} - \mathbf{1}\boldsymbol{\mu}_k^T)^T \mathbf{U}_k (\mathbf{I} - \mathbf{1}\boldsymbol{\mu}_k^T)$, where $\mathbf{1}$ is a column of R+1 ones. Use this decomposition to derive a simpler diagonally weighted IRLS algorithm. Why does this trick not work for additive models?

8.4 An alternative derivation of the algorithm for the linear case-control model is possible. Suppose Y has a Poisson distribution with mean μ, and let $\log(\mu) = \alpha_k + x_1\beta_1 + \cdots + x_p\beta_p$ for the

kth matched set. Derive the score and information for the $K+p$ parameters in the model. Show that the part of the Newton-Raphson step corresponding to $\boldsymbol{\beta}$ is the same as (8.9). Suggest applications.

8.5 The Poisson likelihood of the previous exercise can also be used to derive a nonparametric Newton-Raphson algorithm.

(i) Go through the machinery for one predictor, and show that the Newton-Raphson update for $\mathbf{f}$ coincides with that produced by the conditional likelihood.

(ii) Suggest an alternating algorithm for the nonparametric model using the Poisson approach.

(iii) The algorithm you probably suggested in step (ii) is extremely slow. Suggest why.

8.6 There are strong analogies between the matched case-control model with its conditional likelihood, and the proportional-hazards model with its partial likelihood. Use these to derive a "Poisson" algorithm for the proportional-hazards model, in the linear case.
[Whitehead, 1980]

8.7 Outline an efficient modified backfitting algorithm for the additive proportional-hazards model.

8.8 Although the proportional-hazards model used in section 8.3 is the most popular, other formulations are also used. Derive the local-scoring procedure for the additive relative-risk model $\lambda(t \mid \boldsymbol{X}) = \lambda_0(t)\{1 + \sum f_j(X_j)\}$.

8.9 If one assumes a parametric form for the baseline hazard in the proportional-hazards model, the full likelihood can be used for estimating the regression parameters and the parameters of the hazard function. Suggest a semi-parametric version in which the hazard is modelled by a smooth function. Derive an algorithm for estimating this model
[O'Sullivan, 1988]

8.10 Using the multinomial likelihood and the notation given in section 8.4.2, derive equation (8.23), the IRLS equations for estimating the linear proportional-odds model. Show that the derivation above can be achieved using the quasi-likelihood equations for $\mathbf{g}$.

8.11 We present an alternating algorithm for solving the system (8.23); outline a proof that this alternating algorithm converges to a solution of (8.23).

8.12 Derive the additive version of the proportional-odds algorithm.

8.13 Make rigorous the claims at the end of section 8.4 regarding the proportional-odds and transformation models. Suggest what might happen as the number of categories grows.

8.14 Suppose $\mathbf{S}$ is an $n \times n$ ($n = 2k + 1$) circulant matrix with ijth element a_{i-j}, where a_i, $i = -k, \ldots, k$ is a sequence of real numbers. Show that the eigenvalues of $\mathbf{S}$ are the Fourier coefficients of the discrete Fourier transform of a_i, and that the eigenvectors are the corresponding vectors of evaluated cosinusoid terms.

[Grenander and Szego, 1958]

8.15 Fit an additive proportional-hazards model to the Stanford heart-transplant data. (These data are listed in Andrews and Herzberg, 1985.) Compare the resulting fit to a standard (parametric) proportional-hazards fit using linear and quadratic terms for the covariate `age`, and a linear term for `T5 mismatch score`.

CHAPTER 9

Further topics

9.1 Introduction

In this chapter we investigate some special topics associated with additive models and nonparametric regression. These include resistant fitting, parametric additive model fitting using regression splines and ridge regression, model selection, and exploration of interactions.

The object of resistant fitting is to ensure that the estimates are not unduly influenced by a small number of data points. We discuss resistant fitting for exponential-family generalized additive models, but in principle it can be applied to any of the models of the previous chapters. Although we allow parametric components in an additive model, in this chapter we discuss techniques that make exclusive use of them. In particular, additive regression splines enjoy all the advantages of the linear regression model, with the added potential of allowing flexible functional forms. In order to achieve the level of flexibility one demands of a semi-parametric technique, we investigate adaptive knot-selection strategies. This is closely associated with model selection, which is also discussed here.

Throughout the text we have emphasized that components of an additive model can be bivariate or multivariate, to capture interactions between predictors. Here we discuss interactions in more detail, and some adaptive techniques for identifying them.

9.2 Resistant fitting

The goal of resistant fitting methods is to reduce the influence of outlying points on the estimated model. For the Gaussian linear model, *M-estimation* (maximum likelihood-type estimation) is probably the best known technique. This works by replacing the usual sum of squares by a tapered least-squares criterion of the form

$$m(\boldsymbol{\beta}) = \sum_{i=1}^{n} \rho\Big(\frac{y_i - \boldsymbol{\beta}^T \mathbf{x}^i}{\hat{\sigma}}\Big). \tag{9.1}$$

For example, the ρ function corresponding to Huber's (1964) proposal is

$$\rho(r) = \begin{cases} r^2/2, & \text{for } |\, r \,| \leq k; \\ k^2/2 + k \,|\, r - k \,| & \text{for } |\, r \,| > k. \end{cases}$$

Hence $\rho(r)$ is quadratic in $[-k, k]$ and linear outside of that interval, and data points with large residual values have less influence than they have in a least-squares fit. The choice $k = 1.345$ results in 95% efficiency when the data are Gaussian.

In (9.1), $\hat{\sigma}$ is an estimate of the dispersion parameter. Simultaneous estimation of $\boldsymbol{\beta}$ and σ is possible but the more common approach (and that taken here) is to obtain a rough estimate of σ from an initial least-squares fit. One recommendation is to use the median absolute deviation of the residuals divided by 0.67. This estimate is resistant and approximately unbiased for the standard deviation when the data are Gaussian. With this value of $\hat{\sigma}$, one then minimizes (9.1) over $\boldsymbol{\beta}$, usually by an iterative procedure.

It is reasonably straightforward to extend this technique to the class of generalized additive models. Let's first look at the Gaussian model in some detail.

9.2.1 *Resistant fitting of additive models*

An obvious and simple approach is to add smoothness penalties to the tapered criterion, as is done in the previous chapters:

$$\sum_{i=1}^{n} \hat{\sigma}^2 \rho \left\{ \frac{y_i - \sum_{j=1}^{p} f_j(x_{ij})}{\hat{\sigma}} \right\} + \tfrac{1}{2} \sum_{j=1}^{p} \lambda_j \int \{f_j''(t)\}^2 \, dt, \tag{9.2}$$

where we are absorbing the constant term into the functions. The criterion is similar in flavour to a penalized likelihood, as in section 6.5.2, and its solution is appropriately named a penalized M-estimate. As long as ρ is convex, we can use arguments similar to those used in Chapter 6 to establish that a solution always exists and is a sum of cubic splines. This in turn allows us to write down the equivalent finite dimensional form

$$\sum_{i=1}^{n} \hat{\sigma}^2 \rho \left\{ \frac{y_i - \sum_{j=1}^{p} f_{ij}}{\hat{\sigma}} \right\} + \tfrac{1}{2} \sum_{j=1}^{p} \lambda_j \mathbf{f}_j^T \mathbf{K}_j \mathbf{f}_j. \tag{9.3}$$

The score equations have the form

$$-\hat{\sigma}\boldsymbol{\psi} + \lambda_j \mathbf{K}_j \mathbf{f}_j = \mathbf{0}; \qquad j = 1, \ldots, p \tag{9.4}$$

where $\boldsymbol{\psi}$ is an n-vector with ith element $\psi(r_i/\hat{\sigma})$ with $r_i = y_i - \sum_{j=1}^{p} f_j(x_{ij})$, and ψ is the derivative of ρ:

$$\psi(r) = \begin{cases} r & \text{for } |\, r \,| \leq k; \\ k & \text{for } |\, r \,| > k. \end{cases}$$

Equations (9.4) are nonlinear in the $\mathbf{f}_j$, and so an iterative procedure is needed to solve them.

The obvious choice might appear to be a Newton-Raphson procedure, since we have had success with it in the likelihood models. However $\rho'' = \psi'$ is discontinuous and this leads to an algorithm which gives observations weights of zero or one depending on the size of the scaled residuals (Exercise 9.1). The method of iterative weighting is a better choice, popular for robust linear regression. We can rewrite the elements of the leftmost term in (9.4) as

$$\begin{aligned} \hat{\sigma}\psi(r_i/\hat{\sigma}) &= \frac{\psi(r_i/\hat{\sigma})}{r_i/\hat{\sigma}} \Big\{ y_i - \sum_{j=1}^{p} f_j(x_{ij}) \Big\} \\ &= w_i \Big\{ y_i - \sum_{j=1}^{p} f_j(x_{ij}) \Big\} \end{aligned}$$

where w_i are the weights

$$w_i = \frac{\psi(r_i/\hat{\sigma})}{r_i/\hat{\sigma}}.$$

Equations (9.4) now simplify to the usual set of estimating equations for additive spline models, except that the smoothers are weighted splines with weights w_i that depend on the residuals. Because of this dependence, we iteratively solve the weighted estimating equations until the criterion (9.3) converges (Exercise 9.3). Once again, we have derived a simple algorithm like local-scoring using the spline methodology. Alternatively we can use any weighted smoother in place of the smoothing splines. The justification is along the same lines as the Hilbert space justification for local-scoring (Exercise 9.2).

Let's examine the effectiveness of this approach on a simulated example (an advantage of using simulated examples is that you know the truth!). We generated 100 observations from the model

$$y = \frac{2}{3}\sin(1.3x) - \frac{3}{10}z^2 + \varepsilon$$

where x and z have a standard bivariate Gaussian distribution with correlation 0.4, and $\varepsilon = (1-\omega)N(0,1)+\omega N(0,10)$, with $\omega = 1$ with probability 0.05 and zero otherwise (a 5% contamination model).

Figure 9.1 shows the two generating functions with the response superimposed. In the implementation used to fit these data, a robust linear fit is computed as a set of starting functions. Using the dispersion parameter thus derived (and weights), a weighted additive model is fitted. This is used to recompute the dispersion estimate, and the algorithm above is then iterated with this dispersion estimate. The circles in Fig. 9.1 are those points that receive weights less than 0.75 in the final iteration of the algorithm. Figure 9.2 shows two pairs of fitted functions, as well as the generating functions. The dotted curves are the nonresistant additive-model estimates for these data, whereas the dashed curves are the resistant estimates. The nonresistant fitting appears to respond to the particular pattern of outliers present in these data, while the resistant algorithm is unaffected. The resistant fitted functions are very similar to those produced for the uncontaminated version of this model (not shown).

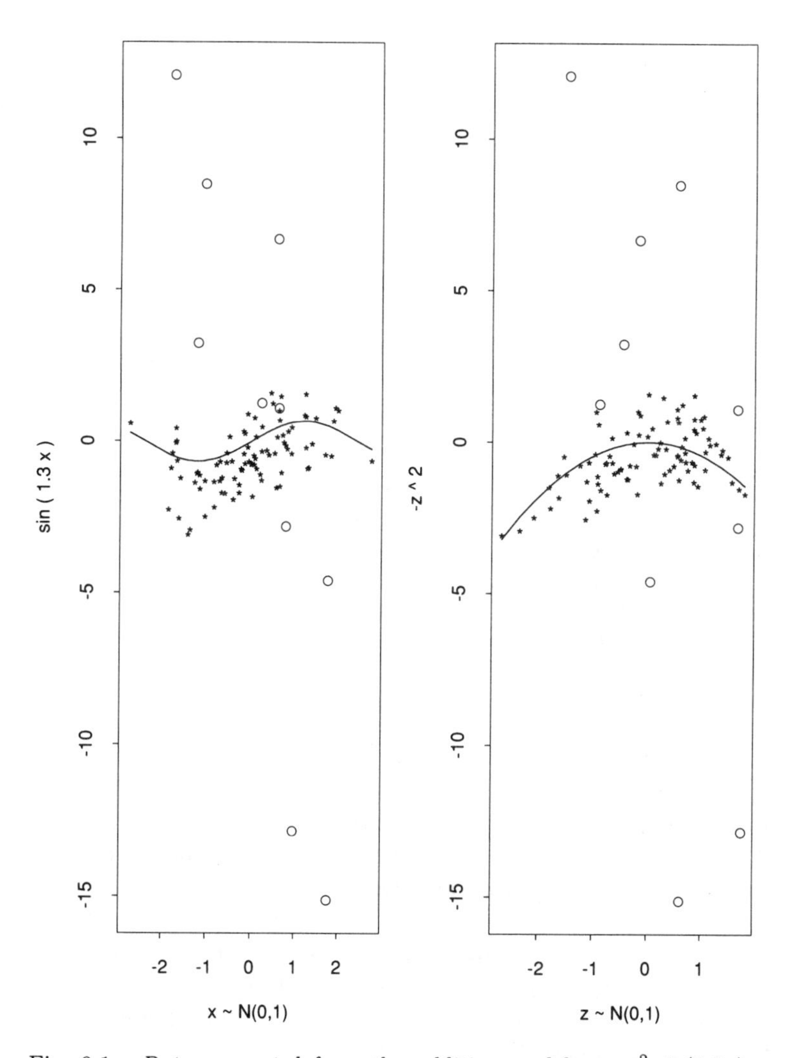

Fig. 9.1. *Data generated from the additive model* $y = \frac{2}{3}\sin(1.3x) - \frac{3}{10}z^2 + \varepsilon$ *with contaminated noise. The plots represent the two generating functions with the response (not partial residuals) superimposed. The plotted circles are those points that receive weights less than 0.75 in the resistant additive model algorithm.*

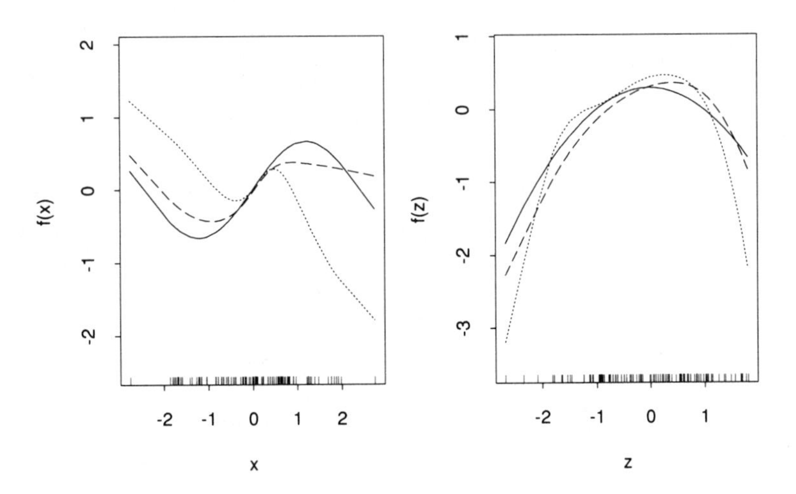

Fig. 9.2. *The solid curves are the generating functions. The dotted curves were obtained using a nonresistant additive model algorithm; the estimate of the sine curve has an average slope that is way off the target (average slope 0), which is not surprising if we inspect the data. The dashed curves were produced by the resistant algorithm; these functions are very similar to the nonresistant additive-model fitted to the uncontaminated version of this model.*

9.2.2 *Resistant fitting of generalized additive models*

It is not quite so obvious how to extend (9.1) to incorporate non-Gaussian models such as the binomial-logistic or Poisson models. Following Pregibon (1982), one way to achieve this is by viewing the deviance contributions $D(y_i; \hat{\mu}_i)$ as the analogue of the squared residuals $(y_i - \hat{\mu}_i)^2$. As in Chapter 6, we assume that there is a known link function $g(\mu)$ that relates the mean μ to the additive predictor, that is $g(\mu_i) = \alpha + \sum_j f_j(x_{ij})$. To simplify matters we re-express the function $\rho(r)$ as a function ω with argument $t = r^2$:

$$\omega(t) = \begin{cases} t/2, & \text{for } t \leq k^2; \\ k\sqrt{t} - k^2/2 & \text{for } t > k^2. \end{cases}$$

Along the lines of the previous section, we can write down a penalized criterion for the resistant fitting of generalized additive models:

$$\sum_{i=1}^{n} \omega\{D(y_i;\mu_i)\} + \tfrac{1}{2}\sum_{j=1}^{p} \lambda_j \int \{f_j''(t)\}^2\, dt. \tag{9.5}$$

Let $w_i = 2\partial\omega(t)/\partial t$ evaluated at $t = D(y_i;\mu_i)$:

$$w_i = \begin{cases} 1, & \text{for } D(y_i;\mu_i) \le k^2; \\ k/\sqrt{D(y_i;\mu_i)} & \text{for } D(y_i;\mu_i) > k^2, \end{cases}$$

and let $\mathbf{W}$ represent the diagonal matrix of these w_is. Reducing to a finite dimensional basis in the usual way, the following score equations define the generalized additive M-estimate:

$$-\mathbf{W}(\mathbf{y} - \boldsymbol{\mu}) + \lambda_j \mathbf{K}_j \mathbf{f}_j = \mathbf{0}; \qquad j = 1, \ldots, p, \tag{9.6}$$

where for simplicity we have used the canonical link, so that $\partial D(y_i;\mu_i)/\partial f_{ij} = -2(y_i - \mu_i)$.

Once again there are several approaches for solving this system of equations. Exercise 9.4 explores the Newton-Raphson and Fisher scoring approaches; here we derive an algorithm similar to the previous section using penalized iterative reweighted least-squares. We can rewrite (9.6) as

$$\mathbf{W}\mathbf{A}(\boldsymbol{\eta} - \mathbf{A}^{-1}(\mathbf{y} - \boldsymbol{\mu}) - \boldsymbol{\eta}) + \lambda_j \mathbf{K}_j \mathbf{f}_j = \mathbf{0}; \qquad j = 1, \ldots, p, \tag{9.7}$$

where $\mathbf{A}$ is the diagonal matrix with the usual weights $a_{ii} = (\partial\mu/\partial\eta)_i^2 V_i^{-1}$. Writing $\mathbf{z} = \boldsymbol{\eta}^0 + (\mathbf{A}^0)^{-1}(\mathbf{y} - \boldsymbol{\mu}^0)$, a little more algebra leads to

$$\mathbf{f}_j^1 = \mathbf{S}_j\Big(\mathbf{z} - \sum_{k \ne j} \mathbf{f}_k^1\Big); \qquad j = 1, \ldots, p, \tag{9.8}$$

where $\mathbf{S}_j$ is a weighted smoothing spline with weights $w_i a_i$. We have derived an algorithm similar to the local-scoring algorithm; the only difference is the weights $w_i a_i$ used in place of the usual weights a_i.

Note that the above procedure does not make sense for a model with a free dispersion parameter such as the Gaussian model. In that instance $D(y_i;\mu_i)$ should be replaced by $D(y_i;\mu_i)/\hat\sigma^2$, where $\hat\sigma$ is a resistant estimate of dispersion, such as the median deviance contribution.

9.2.3 *Illustration*

We illustrate this procedure by applying it to some data on vasoconstriction of the skin, given by Finney (1971), and reanalysed by Pregibon (1982). The response is presence (1) or absence (0) of vasoconstriction of the skin in 39 subjects, and the predictors are `volume` of inspired air, and the `rate` of inhalation. Following Pregibon, we consider binary logistic models for these data. Pregibon fitted linear logistic models in `log(volume)` and `log(rate)`, using both standard and resistant maximum likelihood. When weighted linear regression is used in the inner loop of the resistant additive model fitting procedure just described, the technique is equivalent to that proposed by Pregibon. The resistant fitting procedure for the linear logistic model gives approximately-unit weights to all of the points except two, which receive weights of 0.44 and 0.47.

Figure 9.3 (top panels) shows the estimated 0.1, 0.5, and 0.9 probability contours from the linear logistic model plotted against `volume` and `rate`, along with the raw response values. The down-weighted points are the ones in the south-west corner; these points are seen to have less effect on the estimated 0.1 contour in the resistant linear fit. The bottom panels in Fig. 9.3 show the nonresistant and resistant additive fits. The raw (not logged) predictors are used; the resulting transforms are somewhat logarithmic in shape. As in the linear fit, the resistant additive fit down-weights the two ones in the south-west corner. After closer scrutiny we find that the log transformation is not critical in the linear analysis.

The deviances for the linear models on the log and nonlogged scales are not very different. The nonparametric fit for `volume` is approximately linear (not shown), while the fit for `rate` is given in Fig. 9.4. The solid curve was produced by the nonresistant additive model algorithm, the broken curve by the resistant algorithm. We include the partial residuals for this variable (from the nonresistant fit) in the plot. The two outlying points are those that are down-weighted in the resistant algorithm; their influence is evident in the figure.

There is a rather curious twist to these data. If the two offending points are removed entirely, there is complete linear separation between the zeros and ones and the maximum likelihood estimates are infinite. Fortunately (or perhaps misleadingly), the resistant

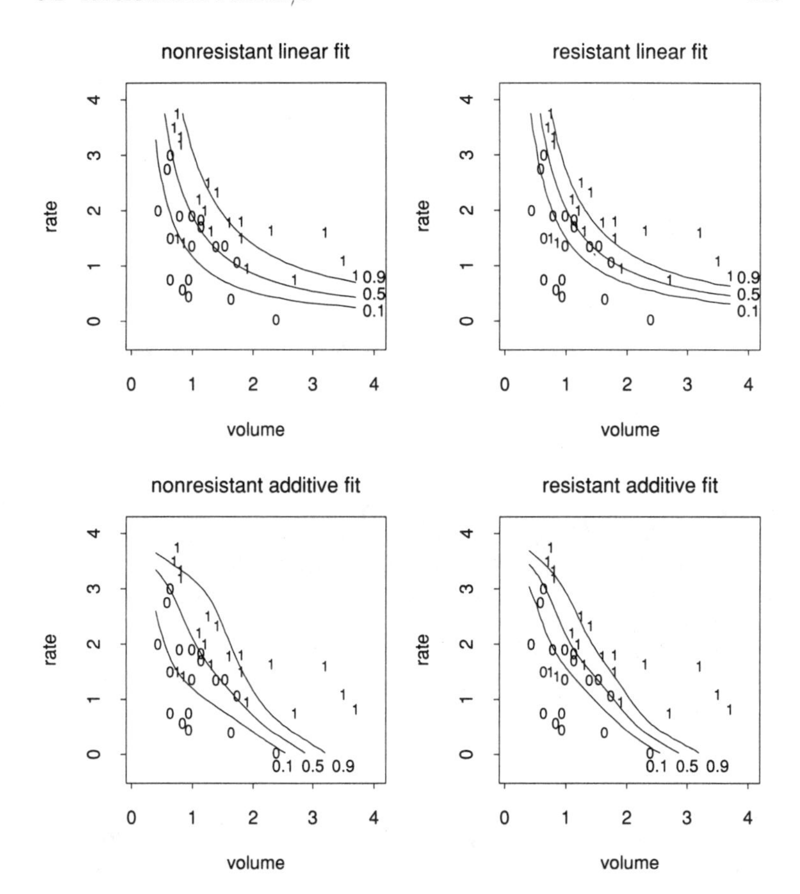

Fig. 9.3. *Estimated 0.1, 0.5, and 0.9 probability contours, as a function of* `volume` *and* `rate`, *for the vasoconstriction data. The curved contours in the top two panels reflect the log transformation of* `volume` *and* `rate` *used in the linear model; the additive model approximates these curves despite the small number of points.*

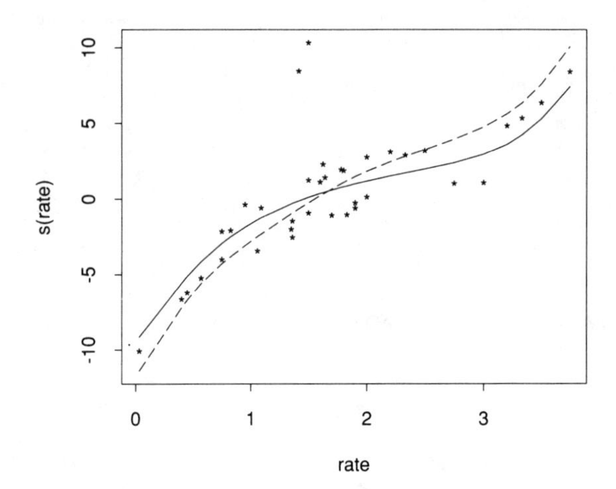

Fig. 9.4. *The solid curve is the nonresistant fitted function for* `rate`. *The points are the partial residuals from this fit. The broken curve is the resistant fit for this term, in which the two outlying residuals are down-weighted.*

algorithms give them a small nonzero weight and so the estimates are finite.

9.2.4 *Influence and resistance*

In standard linear modelling, the M-estimation technique does not completely solve the problem of influential points. It deals with data points with large residuals but does not take into account the position of a point in the predictor space. Thus a point can have large influence because its predictor values are far from the centre of the predictor space, even though it does not result in an unusually large residual. For this reason, alternatives to M-estimation have been proposed, including bounded-influence regression and least median-of-squares regression. The former augments the M-estimation criterion with a factor that reflects influence due to position in predictor space; the latter takes a completely different tack, seeking to minimize the median squared residual, rather than the mean squared residual as in least-squares regression. References are given in the bibliographic notes.

It is not clear whether one needs to be concerned about this further kind of influence in the additive model. The use of smoothers rather than global least-squares fitting should tend to reduce such effects, especially in the interior of the range of a predictor; fitted functions on the boundary, however, can suffer more severely from influential points. Often a useful strategy is to smoothly transform the predictors prior to fitting the additive model, where the transformation is chosen to pull in the tails. This still produces a smooth function of the original untransformed predictor.

When there are two or more predictors, resistance problems can become more severe. If a point or group of points correspond to a set of joint outliers for a pair of variables, the additive terms can complement each other, resulting in wild tail behaviour. This is an example of *concurvity*, for which there is no simple solution. An initial transformation of the predictors might be helpful in these situations; this is a relatively new area and requires further study.

9.3 Parametric additive models

One focus of the research effort into additive models is concerned with the properties of the backfitting algorithm. This algorithm is attractive because it builds up a composite high dimensional model from many univariate or low dimensional regression methods. Since these modules are rather arbitrary, the method is extremely flexible. However, we usually want to compute more than just the fitted functions; in particular, degrees of freedom and pointwise standard-error curves are a useful supplement, and the diagonal of the overall operator matrix is needed for cross-validation and diagnostics. These are at least an order of magnitude more expensive to compute. To date, the only exact and efficient approach we know of is to apply the algorithm n times to the columns of the identity matrix, as described in section 5.4.4. This allows one to build up the relevant operator matrices which are then available for other computations as well.

The difficulty is that the smoother matrices involved have full dimension n and are not projection matrices. An alternative and natural approach is to use a flexible parametric method to fit each of the terms in an additive model, thus allowing the entire fit to be parametric. The smoother matrices are projection matrices with dimension much less than n. This in turn allows the explicit computation of degrees of freedom, standard errors, and all the auxiliary information that has become standard in linear modelling.

One can argue that all of the methods that we have dealt with so far are parametric, since they all tend to concentrate on finding the fitted functions at the observed data sites, resulting in p variables $\times$ n fitted values $= np$ parameters. This stretches the definition of parametric too far, since the effective dimension of these estimates is much lower.

The next section is a discussion of regression splines, which provide a way to model terms flexibly but strictly in a low dimensional parametric form. In order to achieve this flexibility some form of adaptive knot selection is required.

9.3.1 *Regression splines*

The additive predictor for an (additive) regression spline model can be represented in the form

$$\begin{aligned} \eta_i(\boldsymbol{X}) &= \beta_0 + \sum_{j=1}^{p} \mathbf{b}_j^T(X_j)\boldsymbol{\beta}_j \\ &= \beta_0 + \sum_{j=1}^{p}\sum_{l=1}^{k_j} b_{lj}(X_j)\beta_{lj} \end{aligned} \tag{9.9}$$

where each of the $\mathbf{b}_j$s is a *vector* of k_j basis functions associated with predictor X_j. For example, for each predictor we can use a set of B-splines with an associated knot sequence. B-splines are described in some detail in Chapter 2. If we evaluate these functions at each of the n realized values of the predictor X_j we obtain a regression matrix $\mathbf{B}_j$ for the jth variable. Associated with the k_j columns of $\mathbf{B}_j$, or functions in $\mathbf{b}_j$, is a vector of k_j parameters $\boldsymbol{\beta}_j$.

Overall there are $k = 1 + \sum k_j$ parameters, and the regression matrices and parameter vectors can be concatenated to form one big linear predictor $\boldsymbol{\eta} = \mathbf{B}\boldsymbol{\beta}$. Once the basis vectors are determined, the additive model becomes a large linear model, and can be fitted using the usual linear least-squares technology. Similarly a *generalized* additive model becomes a large *generalized* linear model.

This is all very convenient, as long as we can supply the basis functions in a flexible manner. If the choice is B-splines, we need to select the number and locations of the knots. Of course, any set of basis functions can be used. Traditional choices include polynomial terms, sines and cosines, and other parametric transformations.

9.3.2 *Simple knot-selection schemes for regression splines*

Probably the simplest method for selecting knots is a generalization of the cardinal spline strategy for univariate spline problems. We select the number of knots p_j for each predictor and then place them uniformly over their individual ranges. So if cubic B-splines are used, p_j knots correspond to a cardinality of $k_j = p_j + 4$ basis functions for the jth predictor. The only remaining decision is how many knots to use per variable. This can be decided by fixing the degrees of freedom in advance, and therefore the number of knots, or else by using some criterion such as cross-validation or the C_p statistic to automatically drive the choice. This latter approach is used by Atilgan (1988), who uses the *AIC* (Akaike Information Criterion; see section 6.8.3), a close relative to C_p. If there are many variables, one has to decide whether to optimize separately over each k_j, or set them all equal.

Stone and Koo (1985) used a slightly different fixed knot-selection strategy which seems to make efficient use of available degrees of freedom. They place a knot at each of the fifth largest and smallest values of X_j, and then three knots evenly spaced in between. They then use the natural B-spline basis for this set of knots, or piecewise-cubic polynomial in the four interior partitions, and a linear function in the two outside partitions. This results in five degrees of freedom, or four if we ignore the constant.

9.3.3 *A simulated example*

To crystallize these ideas, let's look at a small simulation. Figure 9.5 shows the two fitted functions for the model

$$y_i = f_1(x_i) + f_2(z_i) + \varepsilon_i; \qquad i = 1, \ldots, 100$$

$$f_1(x) = \begin{cases} -2x, & \text{for } x < 0.6; \\ -1.2 & \text{otherwise,} \end{cases}$$

$$f_2(z) = \cos(2.5\pi z)/(1 + 3z^2),$$

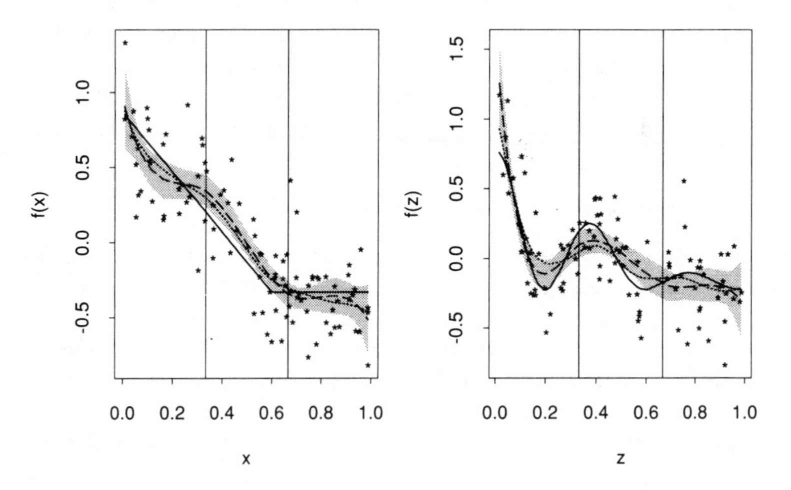

Fig. 9.5. *Additive model fits to simulated data. The solid curves are the generating functions. The dashed curves were fitted using cardinal B-splines in an additive model with two interior knots placed at the vertical lines. The dotted curve was fitted using a locally-weighted running-line smoother, with the same df of five for each term to match the B-splines. The shaded regions are estimated 2× standard-error curves for the spline model, and the points are the partial residuals for the spline model.*

with x_i, z_i generated independently from a $U(0,1)$ distribution, and ε_i from a $N(0, 0.25)$ distribution. This represents an approximate signal to noise ratio of two to one. We used cardinal B-splines for each variable, with two interior knots each. A single B-spline with two interior knots has six associated parameters: four for each partition less three at each interior knot due to the constraints. However, the constant is in the span of these functions, so we use only five of the six basis functions, after adjusting for the constant. This gives a total of 5+5+1=11 *df* in the model; the knots are placed at the locations 1/3 and 2/3 for both predictors.

The estimated functions are $\hat{f}_1(x) = \mathbf{b}_1^T(x)\hat{\boldsymbol{\beta}}_1$ and $\hat{f}_2(z) = \mathbf{b}_2^T(z)\hat{\boldsymbol{\beta}}_2$, where $\hat{\boldsymbol{\beta}} = (\hat{\beta}_0, \hat{\boldsymbol{\beta}}_1, \hat{\boldsymbol{\beta}}_2)$ is the least-squares estimate of the 11 element coefficient vector. Their evaluations at the observed sample points are the fitted vectors $\hat{\mathbf{f}}_1 = \mathbf{B}_1\hat{\boldsymbol{\beta}}_1$ and $\hat{\mathbf{f}}_2 = \mathbf{B}_2\hat{\boldsymbol{\beta}}_2$. Computing the pointwise standard errors is straightforward as well.

Suppose

$$\begin{aligned}\boldsymbol{\Sigma} &= \operatorname{cov}(\hat{\boldsymbol{\beta}})\\ &= (\mathbf{B}^T\mathbf{B})^{-1}\sigma^2.\end{aligned}$$

Then

$$\begin{aligned}\operatorname{cov}(\hat{\mathbf{f}}_1) &= \operatorname{cov}(\mathbf{B}_1\hat{\boldsymbol{\beta}}_1),\\ &= \mathbf{B}_1 \operatorname{cov}(\hat{\boldsymbol{\beta}}_1)\,\mathbf{B}_1^T,\\ &= \mathbf{B}_1\boldsymbol{\Sigma}_{11}\mathbf{B}_1^T,\end{aligned}$$

where $\boldsymbol{\Sigma}_{11}$ is the appropriate submatrix of $\boldsymbol{\Sigma}$. At an arbitrary point $\operatorname{var}[\hat{f}_1(x)] = \mathbf{b}_1^T(x)\boldsymbol{\Sigma}_{11}\mathbf{b}_1(x)$. As usual we can estimate σ^2 via the residual sum of squares. The diagonal of $\operatorname{cov}(\hat{\mathbf{f}}_1)$ estimates the pointwise variances. Figure 9.5 displays the fitted functions (broken lines), as well as the upper and lower $2\times$ standard-error bands. Also displayed in Fig. 9.5 are the generating functions (solid curve), and the functions estimated using the backfitting algorithm with a locally-weighted running-line smoother (dotted curve). The smoothing parameters are set to match the *df* of five per variable.

Both models do a reasonable job, although both exhibit a fair amount of bias. The B-splines seem to introduce spurious wiggles more often than the running-line smoother. The choice of knots happens to be favourable for the spline model, so it has been successful here. It seems as long as an appropriate knot sequence is used, the regression spline model is simple and satisfactory. We return to this example after the next section.

9.3.4 *Adaptive knot-selection strategies*

More recently, Friedman and Silverman (1989) gave an algorithm for optimizing over the *number and location* of the knots in an adaptive way. We briefly outline their method, first for a single predictor. The key idea is to use the piecewise-linear basis functions $(x - \xi_k)_+$ rather than cubic functions for locating the knots. These basis functions are illustrated in Fig. 9.6. Every observed predictor value is a candidate knot site. Once the knot positions are determined, the piecewise-linear functions are converted into piecewise-cubic functions by essentially rounding the corners at each knot. These are then used at the chosen knots to compute the piecewise-cubic fit.

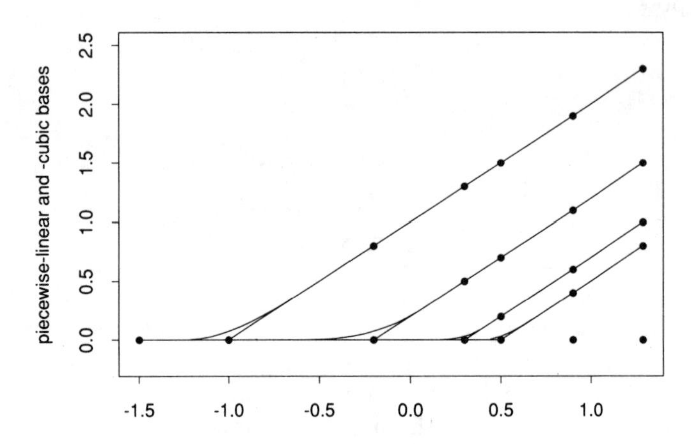

Fig. 9.6. *Part of a sequence of piecewise linear splines used by Friedman and Silverman, evaluated at the data points. The nonzero contributions of adjacent linear splines are a constant distance apart, allowing for $O(1)$ updating of the least-squares calculations. Once the knots are selected, they are replaced by piecewise cubic functions with similar characteristics.*

Why use piecewise-linear functions? The reason is that updating formulae are available and *all* n candidate knot sites can be evaluated in $O(n)$ operations. Suppose that k knots are already selected, and that we are using a criterion such as GCV to decide when to stop. Suppose also that we have evaluated the fit using x_i as the $(k+1)$th knot. If we remove x_i and replace it with x_{i+1}, then each evaluation of the basis function beyond the point x_{i+1} changes by the same constant amount. This in turn allows us to adjust the least-squares fit in $O(1)$ operations. This is done for each candidate knot, moving from left to right, and the one that reduces the criterion the most is included. The procedure is stopped when the criterion starts to increase.

For multiple predictors, the procedure is the same, except that each variable is scanned separately for the best knot. The best knot (amongst all p predictors) is then augmented to the set. Friedman and Silverman also employ a knot deletion strategy which involves first finding a rich set of knots, and then deleting them until the criterion is minimized. They demonstrated their procedure

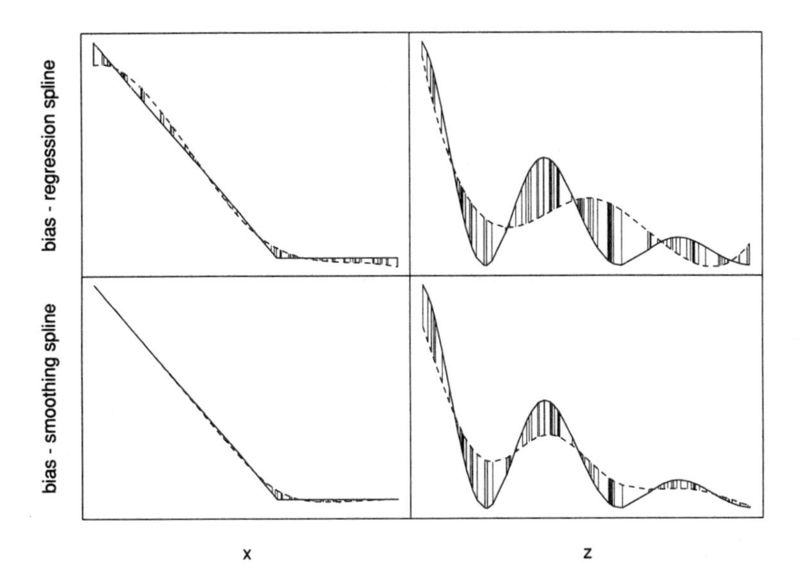

Fig. 9.7. *The bias of a regression spline smoother with one interior knot compared to a smoothing spline, when used in an additive model with two terms. The solid curves are the generating functions, and in the simulation of Fig. 9.5, Gaussian noise is added to their sum. The broken curves are what the additive model algorithms are estimating.*

successfully on a number of examples. The overall effect for a single variable is that of a variable bandwidth smoother. The big gains are for additive models since the procedure automatically selects a bandwidth for each term in the model. We return to this technique when we discuss model selection in section 9.4.

9.3.5 *Discussion of regression splines*

The advantages of using parametric basis functions are clear, and B-splines are a reasonable choice. The additive model fitting algorithm is simply a large linear least-squares problem and all the usual auxiliary information is readily available. Unless the number of knots chosen is large, this presents no real computational problem.

There are, unfortunately, some negative aspects to regression splines. We focus our criticisms first on the fixed knot strategies.

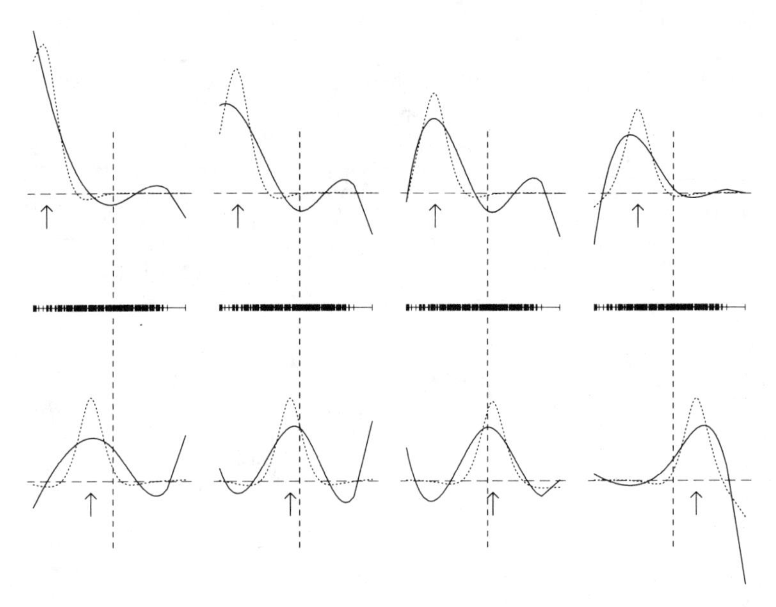

Fig. 9.8. *A comparison of the equivalent kernels of a cubic regression spline with one interior knot and a cubic smoothing spline with five df. The smoothing variable is* `Daggert Pressure Gradient` *from the Los Angeles ozone data, and the data sites are indicated by the rug plots in the centre of the figure. The one interior knot is indicated by the broken vertical line, and the arrow indicates the target point for each kernel. The dotted kernels represent smoothing splines, the solid kernels represent regression splines.*

In order to keep the degrees of freedom down, we typically want to use a small number of interior knots, such as one or two (resulting in four or five *df*). In these situations, however, an unfortunate knot placement leads to misleading results.

Figure 9.7 shows the estimation bias when we use *one* interior knot for each variable in the simulated example of Fig. 9.5. The bias is especially large for the second term. We have included the bias for smoothing splines used in a backfitting algorithm, where the smoothing parameter was chosen to give four degrees of freedom per term. Although the latter smoother is also biased, it does not tend to be as misleading about where the local extrema in the functions are located. In summary, the local smoothers trim down the hills and fill in the valleys, but they don't move them somewhere else! The adaptive knot placing algorithm presumably alleviates some of these problems.

The strange local behaviour of regression splines is also evident if we examine the equivalent kernel at various locations, especially when the number of knots is small. Figure 9.8 compares these to the equivalent kernel of smoothing splines, where we have once again matched for *df*. There is one interior knot placed at the median. The regression spline does not exhibit the local behaviour we expect of a smoother. This behaviour does of course improve as we increase the number of knots. By analogy to digital filters (section 3.7) the asymmetric nature of the equivalent kernel means that the regression-spline smoother will cause a phase shift in the output, thus moving local extrema as noted above.

Another problem with cardinal B-splines is that as we increase the degrees of freedom from say five to six, the entire set of knots changes, and hence the space of fits changes in a jumpy fashion. This is in contrast to the neighbourhood smoothers and smoothing splines, where the fit changes smoothly as we vary the smoothing parameter.

Although the adaptive techniques appear quite promising, we do not have a great deal of experience with them. One reservation is that, with all the selection going on, the shapes of the fitted functions are liable to exhibit excess variability. A similar selection variability is encountered in all-subsets regression. Fortunately, we don't mind too much if knot i is chosen rather than knot $i + 1$, if they belong to the same variable, since the fitted function for that variable should look much the same. It is the competition for knots between variables we worry about, since the selection criterion is not concerned with individual function, but rather the overall fit. Hastie (1989) demonstrates some of this variability in a discussion of the Friedman-Silverman paper. The simulation study in section 10.2 also supports this claim.

The adaptive techniques have not yet been fully developed for generalized additive models. Since computing the fit when a knot is added typically involves iteration, some approximations are needed. One approach (untried) is to use the adjusted dependent variable and weights based on the current K knots, and find the $(K + 1)$th knot by weighted least-squares. One can use an appropriate weighted version of the selection criterion to select the next knot for inclusion. Once the optimal $(K+1)$th knot is located, it can be incorporated into the fit, and only then is the nonlinear fitting algorithm iterated until convergence. This is repeated until a criterion appropriate on the deviance scale, such as *AIC*, fails to decrease. This is an area of current research.

9.3.6 *Generalized ridge regression — pseudo additive models*

A parametric approach closer in spirit to the additive smoothing-spline model is based on generalized ridge-regression. Once again, let's focus on the univariate problem first.

Suppose $\mathbf{S} = (\mathbf{I} + \lambda\mathbf{K})^{-1}$ is a smoothing-spline matrix, based on a set of n unique values of the predictor. Recall the eigendecomposition (3.24) of section 3.7: $\mathbf{S} = \mathbf{U}\mathbf{D}_\psi\mathbf{U}^T = \sum_{k=1}^n \psi_k \mathbf{u}_k \mathbf{u}_k^T$, and the associated Fig. 3.6. Here $\mathbf{D}_\psi$ is the diagonal matrix of eigenvalues ψ_k of $\mathbf{S}$. The smoother damps the components of $\mathbf{y}$ along its eigenvectors by differing amounts; the higher the order, the more damping. The first two eigenvalues are one, corresponding to constant and linear functions; these components are passed through the smoother.

If we parametrize the fitted vector $\mathbf{f}$ with respect to the basis defined by $\mathbf{U}$ (the Demmler-Reinsch basis) so that $\mathbf{f} = \mathbf{U}\boldsymbol{\beta}$, then the smoothing spline minimizes

$$\|\mathbf{y} - \mathbf{U}\boldsymbol{\beta}\|^2 + \lambda\boldsymbol{\beta}^T\mathbf{D}_\theta\boldsymbol{\beta}, \tag{9.10}$$

where $\mathbf{D}_\theta$ is a diagonal matrix of eigenvalues θ_k of $\mathbf{K}$, with corresponding eigenvectors $\mathbf{u}_k$. Note that θ_k is independent of the smoothing parameter λ, whereas $\psi_k = 1/(1 + \lambda\theta_k)$ depends on λ.

At least three insights emerge from these manipulations:

(i) The eigenvectors represent a family of (evaluated) functions that increase in complexity as their order increases.

(ii) The eigenvalues of $\mathbf{K}$ represent a sequence of relative penalties that are applied to the functions in the ridge criterion (9.10). The overall level of penalization is determined by λ.

(iii) From the eigenvalue plots in Fig. 3.6, the dimension of the space of fits is effectively much less than n for moderate amounts of smoothing.

One can design a smoother by supplying a set of basis functions $p_k(x)$, $k = 1, \ldots, K$, and a corresponding sequence of penalties δ_k. We can either supply the functions themselves, or else an $n \times K$

basis matrix $\mathbf{P}$ of their evaluations. In any event, we typically need far fewer functions than n, perhaps $K = 8$ or 10 for most problems. Let $\mathbf{D}_\delta$ be the $K \times K$ diagonal matrix of penalties. We can then define the smooth to be the minimizer of $\|\mathbf{y} - \mathbf{P}\boldsymbol{\beta}\|^2 + \lambda\boldsymbol{\beta}^T\mathbf{D}_\delta\boldsymbol{\beta}$, which is $\hat{\mathbf{f}} = \mathbf{P}\mathbf{D}_{1/(1+\lambda\delta)}\mathbf{P}^T\mathbf{y}$.

In this univariate case there is no compelling reason for using an approximation of this kind, since the smoothing spline is cheap enough to compute. Nevertheless the approximations give us cheap access to all the features of the smoother matrix that are not always available (such as its eigen-decomposition).

The big winnings come when we design similar smoothers $\{\mathbf{P}_j, \mathbf{D}_{\delta_j}\}$ for each of the terms in an additive model. In this case, $\mathbf{f}_j = \mathbf{P}_j\boldsymbol{\beta}_j$, and $\mathbf{f}_+ = \mathbf{P}\boldsymbol{\beta}$, where $\mathbf{P} = (\mathbf{1}, \mathbf{P}_1, \ldots, \mathbf{P}_p)$, and the additive fit is defined as the solution to $\|\mathbf{y} - \mathbf{P}\boldsymbol{\beta}\|^2 + \boldsymbol{\beta}^T\mathbf{D}\boldsymbol{\beta}$ where $\mathbf{D} = \text{diag}(\lambda_1\mathbf{D}_{\delta_1}, \ldots, \lambda_p\mathbf{D}_{\delta_p})$. The solution is a generalized ridge regression.

The advantages of this approach include:

(i) This parametric fit is close in spirit to the smoothing-spline additive fit, and has properties in common. We can study issues such as concurvity more easily.

(ii) The fit can be computed without backfitting.

(iii) Many quantities of interest can be computed cheaply, such as the effective operator matrix $\mathbf{R}$ satisfying $\mathbf{f}_+ = \mathbf{R}\mathbf{y}$, and the component operators $\mathbf{f}_j = \mathbf{R}_j\mathbf{y}$. These can be used for computing standard errors and diagnostics.

(iv) If weights are present, the problem is simply a weighted ridge regression.

(v) We can parametrize each of the terms with a single smoothing parameter, given the basis. In particular, we can specify the df_j for each term and solve for λ_j.

(vi) Used within the generalized additive-model framework, the local-scoring iterations become a series of iteratively-reweighted ridge regressions.

We have to choose a basis and penalty sequence for each term. Hastie (1988) developed a simple algorithm for approximating the first k eigenvectors and eigenvalues of the smoothing-spline matrix itself. His algorithm is equivalent to the QR algorithm for powering up sets of vectors towards an eigenspace of a square matrix. Each full step takes $O(n)$ operations and typically only one or two steps are needed to achieve the required accuracy. In this case, the additive spline solutions and the ridge regression approximations are typically indistinguishable. Other bases are currently being explored, and we refer to them in general as *pseudo smoothers* and *pseudo additive models.*

9.3.7 *Illustration: diagnostics for additive models*

The example here serves two purposes. It outlines an approach to diagnostics for additive models, and demonstrates the usefulness of the pseudo additive models of the previous section.

Much has been written on diagnostics and influence measures for smoothing splines and related problems. Here we generalize these ideas to additive models.

Typically we require the effective smoothing matrices $\mathbf{R}_j$ and $\mathbf{R}$, or at least their diagonals, where for example $\hat{\mathbf{f}}_j = \mathbf{R}_j\mathbf{y}$. As an illustration, we concentrate on versions of Cook's (1977) distance appropriate for additive Gaussian models. These results can be extended to include other generalized additive models.

Consider first a version of Cook's distance appropriate for the fitted values $\hat{\mathbf{f}}_+$:

$$D(\hat{\mathbf{f}}_+)_i = \frac{\left\|\hat{\mathbf{f}}_+ - \hat{\mathbf{f}}_+^{(i)}\right\|^2}{df^{\,\mathrm{var}}\hat{\sigma}^2}; \qquad i = 1, \ldots, n.$$

The notation $\hat{\mathbf{f}}_+^{(i)}$ represents the vector of fitted values with the ith point removed. $D(\hat{\mathbf{f}}_+)_i$ measures the total change in the fitted additive predictor before and after the removal. The constants in the denominator arise from

$$\begin{aligned} E\left\|\hat{\mathbf{f}}_+ - E(\hat{\mathbf{f}}_+)\right\|^2 &= \mathrm{tr}(\mathbf{R}\mathbf{R}^T)\sigma^2 \\ &= df^{\,\mathrm{var}}\sigma^2, \end{aligned}$$

also seen to be the sum of the variances of the fits (section 5.4.5). Using the standard deletion formulae, we obtain

$$D(\hat{\mathbf{f}}_+)_i = \frac{e_i^2 \sum_{k=1}^n R_{ik}^2}{(1 - R_{ii})^2 df^{\,\mathrm{var}}\hat{\sigma}^2}, \tag{9.11}$$

where e_i is the ith residual and R_{ij} an element of $\mathbf{R}$. Probably more interesting is the Cook's distance for the individual functions, defined as

$$D(\hat{\mathbf{f}}_j)_i = \frac{\left\|\hat{\mathbf{f}}_j - \hat{\mathbf{f}}_j^{(i)}\right\|^2}{df_j^{\text{var}}\hat{\sigma}^2} \quad \forall i, j,$$

with computing formula

$$D(\hat{\mathbf{f}}_j)_i = \frac{e_i^2 \sum_{k=1}^n (\mathbf{R}_j)_{ik}^2}{\{1 - (\mathbf{R}_j)_{ii}\}^2 df_j^{\text{var}}\hat{\sigma}^2} \tag{9.12}$$

and $df_j^{\text{var}} = \text{tr}(\mathbf{R}_j\mathbf{R}_j)^T$.

In order to perform these computations, we need the entire matrices $\mathbf{R}$ and $\mathbf{R}_j$. We describe an algorithm in section 5.4.4 for computing these matrices exactly, but this requires repeating the backfitting algorithm n times. Here we use instead the pseudo additive model fit described in the previous section.

We demonstrate these ideas on data from a study by Bruntz *et al.* (1974) of the dependence of `ozone` on some meteorological variables on 111 days from May to September of 1973 at sites in the New York metropolitan region. As in Cleveland *et al.* (1988), we work with the cube root of `ozone`. For illustration we focus on `wind speed` and `temperature`, and Fig. 9.9 shows the fitted pseudo additive model together with the partial residuals and pointwise standard-error curves.

Figure 9.10 gives boxplots of $D(\hat{\mathbf{f}}_+)_i$ and $D(\hat{\mathbf{f}}_j)_i$, and also shows a scatterplot of the two predictors indicating the points above the threshold in the boxplot. This threshold was arbitrarily chosen simply to illustrate the effect of deleting the most influential points (the distribution theory of Cook's distance in this setting has yet to be developed). The large dots in Fig. 9.9 are the residuals of the observations that exceed the threshold. The model is then fitted with these points removed, and the results are shown in Fig. 9.11, with the deleted points indicated by hollow circles. The effect is clear, and does not surprise us too much since the areas most affected are on the boundaries. The width of the standard-error curves has not changed much.

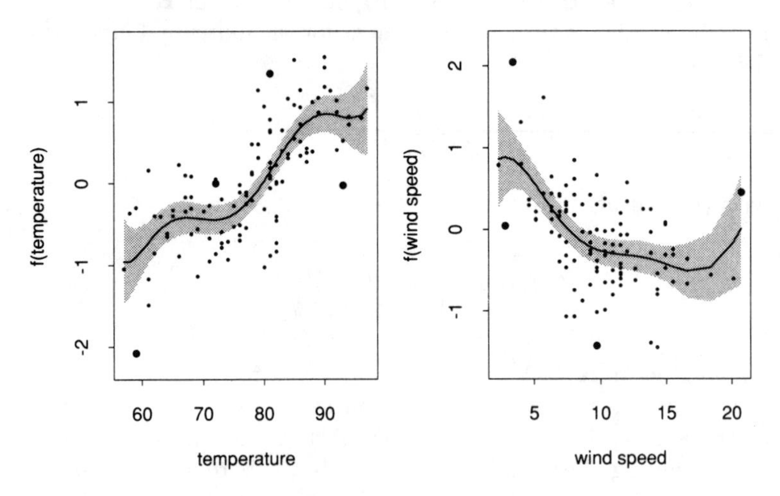

Fig. 9.9. *The pseudo additive model fit for the New York ozone data, together with partial residuals and pointwise 2× standard-error curves. The partial residuals plotted as large dots are those flagged as having large Cook's distance in Fig. 9.10.*

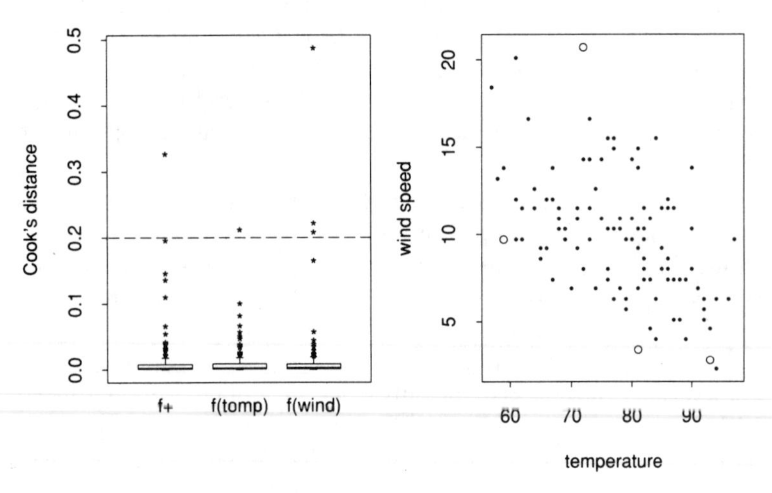

Fig. 9.10. *The panel on the left shows boxplots of the Cook's distances for the additive predictor and the individual functions in Fig. 9.9. The horizontal line indicates the threshold chosen (arbitrarily). On the right is a scatterplot of the two predictors, the thresholded points shown with circles.*

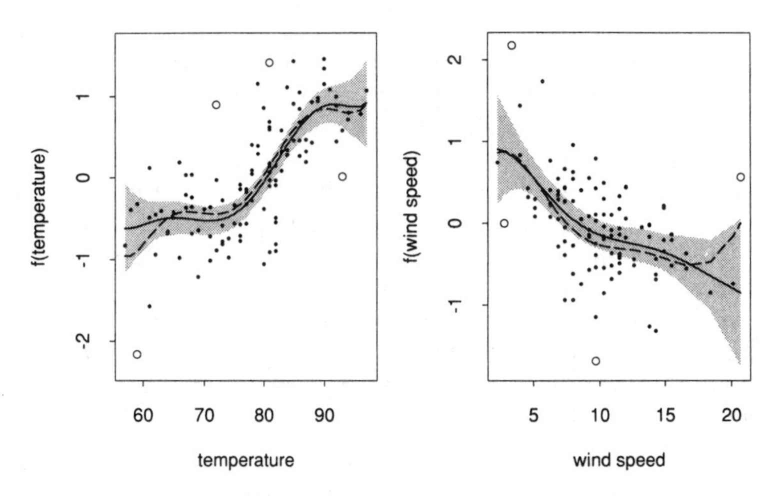

Fig. 9.11. *The pseudo additive model fit (solid curve) for the New York ozone data, computed without the four influential points. The fitted values for these four points are included, however, as well as their residuals. The fitted curves from Fig. 9.9 (broken curves) are also included.*

9.4 Model selection techniques

There are a variety of different ways to approach model selection for generalized additive models. They range from the manual interactive (and therefore more limiting) methods to rather adventurous automatic methods. When nonparametric smoothers are used to fit some terms, model selection develops some new flavours. We not only need to select which terms to include in the model, but also how smooth they should be.

A simple approach to model selection is the manual one we use in earlier chapters. One fits a logical sequence of models, and examines the deviances and differences in deviances that arise. To evaluate these differences, one uses approximate χ^2 tests (or F-tests), based on the approximate *df*. The problem of selecting the amount of smoothing for a given term is side-stepped by fixing its *df*. One tries fitting a smooth term with approximately four *df* say, and then tests to see if a linear term does as well. Of course, one also tests the effect of omitting the variable as well.

The *df* of a term is usefully viewed as a continuous scale for combining variable selection and smoothing parameter selection. For all types of terms $df = 0$ means the term is deleted. For linear terms or single-parameter terms, $df = 1$. For all of the shrinking smoothers that we consider, $df \geq 1$ (since linear fits represent one extreme end of their scale), and in all cases, the larger the degrees of freedom, the rougher the fit. Higher-order parametric terms also fit into this framework, except that df increases in integer steps.

In summary, we define a rather course regimen of fits, say $df \in \{0, 1, 4\}$ for each term, and operate in a manual stepwise mode to select a model.

In the following sections we describe other more streamlined approaches that automate this selection in a natural way. These are all illustrated in Chapter 10 where they are compared in an analysis of the Los Angeles air-pollution data.

9.4.1 *Backward and forward stepwise selection techniques*

There are two requirements needed to implement this approach:

(i) For each possible term in the model, we need a regimen or set of methods which describes the fits of different complexity that we wish to consider for that term. One can then think of a *model space* of all possible models, which is the product space of the regimens for each term. In our manual strategy above, each regimen had three elements, defined in terms of df. The product space for p terms has 3^p different models.

(ii) We also need a criterion to compare the fits of two different models. We use a χ^2 or F-test.

Here is a suggested approach. Start with some model, for example the linear model. Test each term in the model for a possible upgrade or downgrade of its status in terms of the regimen of available degrees of freedom. In our example above there are only three possibilities per term. So if we start with a linear term, we evaluate the effect of it being smooth with $df = 4$, as well as the effect of it being omitted. Do this for all the variables, and pick the best modification. This is repeated until the best modification is "not significant".

There are a number of variations/options available:

(i) The regimens may consist of more than three choices. For example we can try both $df = 4$ and $df = 7$ in addition to zero or one. In fact, there can be an arbitrary list of fitting methods, including omitting the variable. For example, if a variable is categorical, we might consider collapsing some categories, and thereby reduce the complexity of the fit for that term. This collapsed fit might then be a candidate for that term, as well as the full categorical fit. Although not strictly necessary, it is more convenient if there is a natural ordering on the elements of a regimen (such as that imposed by *df*).

(ii) The initial model can be any prespecified member of the product space of regimens for each term. Although it often makes more sense to start with the largest model, in some cases it might be overparametrized and never attainable.

(iii) Our present software allows choices such as backward only, forward only or both, and the ability to force terms in the model.

(iv) The criterion for selection can be one of the prediction criteria, such as AIC or CV, in which case we can always make the choice corresponding to the largest decrease.

This technique performs well in practice, but is rather slow. This is not surprising since before a term is changed, $2(p-1)$ different models have to be fitted (for regimens of size three). Of course the backfitting algorithm can get off to a good start from the previous fit. In section 9.4.3 we describe a more continuous and efficient version of this stepwise approach that combines selection and backfitting.

9.4.2 *Adaptive regression splines: TURBO*

The real strengths of the adaptive spline methodology (called TURBO by Friedman and Silverman) of section 9.3.4 lie in its ability to both select which terms to include and the amount of smoothing for those included in an efficient way. Predictors for which no knots are chosen are not in the model. For those that are in, the relative amount of smoothing is determined automatically by the number of knots per predictor. Furthermore, the amount of smoothing can vary over the range of a variable.

When all of the basis functions are selected, they are grouped by predictor and an additive model is defined in terms of the composite functions of each predictor. We have sketched how the same techniques can be used for generalized additive models. We compare TURBO to a number of alternatives in Chapter 10, in an analysis of the Los Angeles ozone data.

9.4.3 *Adaptive backfitting: BRUTO*

One can think of simply optimizing a criterion such as *GCV* (generalized cross-validation) over all p smoothing parameters λ_j in a p-term additive model. Gu and Wahba (1988) described an efficient algorithm for doing just that with smoothing splines. Unfortunately their algorithm still requires $O(n^3)$ computations, and so will not be in routine use for a while.

Here we outline an approximate method that has the same flavour as the adaptive regression splines, but, like backfitting, is modular and allows any smoother to be used. The algorithm (called BRUTO) combines backfitting and smoothing parameter selection.

GCV for an additive model is defined as

$$GCV(\lambda_1,\ldots\lambda_p) = \frac{\sum_{i=1}^n \{y_i - \sum_j \hat{f}_{j,\lambda_j}(x_{ij})\}^2}{n\{1 - \operatorname{tr}\mathbf{R}(\lambda_1,\ldots,\lambda_p)/n\}^2}. \tag{9.13}$$

$\mathbf{R}(\lambda_1,\ldots,\lambda_p)$ is the additive-fit operator for the given values of the smoothing parameters, and the terms $\hat{f}_{j,\lambda_j}$ denote the fitted functions. The computational difficulties are mainly due to the computation of the denominator.

BRUTO attempts to minimize the *GCV*-like statistic:

$$GCV^B(\lambda_1,\ldots\lambda_p) = \frac{\frac{1}{n}\sum_{i=1}^n \{y_i - \sum_j \hat{f}_{j,\lambda_j}(x_{ij})\}^2}{\left(1 - [1 + \sum_1^p\{\operatorname{tr}\mathbf{S}_j(\lambda_j) - 1\}]/n\right)^2} \tag{9.14}$$

The difference between this and *GCV* lies in the denominator, where GCV^B uses $1+\sum_j\{\operatorname{tr}\mathbf{S}_j(\lambda_j)-1\}$ in place of $\operatorname{tr}\mathbf{R}(\lambda_1,\ldots,\lambda_p)$. This approximation requires $O(n)$ computations, rather than $O(n^3)$, and is discussed in section 5.4.5.

The minimization of GCV^B is carried out one parameter at a time, by applying the appropriate smoother to the corresponding partial residuals, but selecting the smoothing parameter to minimize this (global) criterion. This step simply requires a univariate *GCV* optimizer, slightly modified to include the degrees of freedom for other terms in the model. Having obtained the best candidate for each of the p available terms, we incorporate the update corresponding to the minimum GCV^B. Actually we need only consider $p - 1$ updates at each step, since the update of the last step cannot reduce the criterion. This is continued until GCV^B converges, which it must do, since each step is a decrease. Figure 9.12 displays an example of BRUTO's convergence for the Los Angeles air-pollution data, analysed in Chapter 10. The plotting characters indicate which of the nine predictors was adjusted at each step. The final model included predictors 2, 4, 5, 6, 8 and 9. Notice, for example, that variable number four is the first to be chosen, with just over five *df*. It is selected three times after that, and each time its contribution is diminished by a small amount.

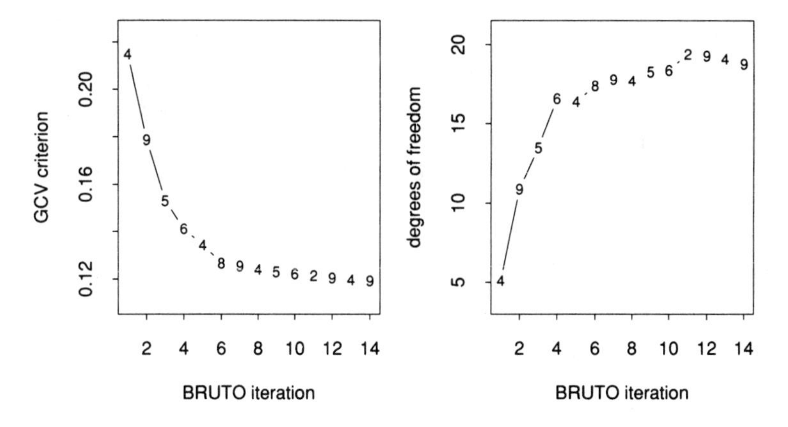

Fig. 9.12. *Left: Convergence of BRUTO's criterion* GCV^B *when fitted to the Los Angeles ozone data. The plotting characters indicate which of the nine predictors is updated. Right: Total degrees of freedom in the model* $1 + \sum_j df_j$ *as a function of iteration number.*

A further modification to the smoothers allows the null fit (mean) or the linear fit to be candidates. This means that a valid step can be to remove a variable from a model, or make it linear.

BRUTO (inspired by TURBO) is also a recent arrival, and has yet to be tested thoroughly. Our experience has been very favourable. It allows a continuous regimen of fits for each term, and combines the backfitting algorithm with the model selection algorithm. Since each iteration requires $p - 1$ smooths, it is still a relatively cheap procedure for model selection.

9.5 Modelling interactions

Most of our attention throughout this book is concentrated on models that are component-wise additive, that is

$$E(Y \mid X_1, \ldots, X_p) = \alpha + \sum_{j=1}^{p} f_j(X_j) \tag{9.15}$$

where each of the f_j are univariate functions. This is an extension of the usual linear regression model. We explore some variations on this theme, namely models for other kinds of regression data (Chapters 4 and 8), and methods that provide transformations of the response Y (Chapter 7). We also note in Chapter 4 that the additive model paradigm and associated backfitting algorithm allows terms that are more complex than simple component-wise additive functions. For example, one might add a term like $f(X_1, X_2)$ to model (9.15) and estimate it with a surface smoother in a backfitting algorithm.

In this section we outline several approaches to modelling interactions. Much of this work is still under development by us and others.

9.5.1 *Simple interactions*

Rather than model an interaction term $f(X_1, X_2)$ in a general way, one can resort to a variety of more parsimonious representations:

(i) Model simple functions of the predictors, e.g. $g(X_1 X_2)$.

(ii) Fit one degree-of-freedom terms of the form $\gamma \cdot h_1(X_1) \cdot h_2(X_2)$ to the residuals $Y - \hat{\eta}$, where $\hat{\eta} = \hat{\alpha} + \hat{h}_1(X_1) + \hat{h}_2(X_2)$ is the fitted additive model.

(iii) As in (ii), except estimate the h_j and γ simultaneously. For example, consider the model $\eta(X_1, X_2) = \mu + \beta_1 h_1(X_1) + \beta_2 h_2(X_2) + \gamma \cdot h_1(X_1) \cdot h_2(X_2)$, with h_j monotone, $Eh_j(X_j) = 0$, and $\|h_j(X_j)\| = 1$. This model is useful in analysing data from drug studies, where h_j represents the metameter or scale on which the drug acts.

Clearly there are many more varieties. The particular form used may depend on the nature of the data as in (iii), or else might be chosen to seek a simple interpretation.

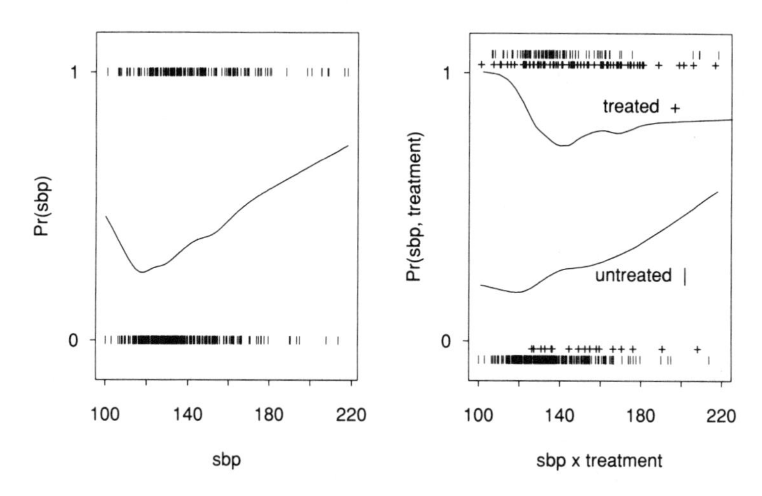

Fig. 9.13. *A nonparametric estimate of the prevalence of heart attacks as a function of* `sbp` *has a puzzling U shape. When we model the interaction of* `sbp` *with* `treatment` *for high blood pressure, this U shape is explained. The treated group (with much higher prevalence for heart attacks) have lower blood pressure as a result of the treatment.*

9.5.2 *Interactions resulting in separate curves*

A natural form of interaction, and one we can easily plot, is that between a categorical predictor and a continuous one. For example, in the heart attack data of Fig. 4.7 the fitted prevalence curve for `sbp` exhibits a U shape. It turns out that in these retrospective data some of the men had been on treatment for high blood pressure *before* their blood pressure readings were taken, but *after* their heart attack. Fitting the interaction between this dummy variable `treatment` and `sbp` amounts to fitting two separate curves in this case. Figure 9.13 shows both fits, on the prevalence scale. Note once again that this is a retrospective case-control sample, so the overall level of the prevalence is exaggerated.

More often there are other terms in the additive model, as is actually the case here, and these interactions have to be fitted in a slightly different fashion. One method is to have two terms for `sbp` in the model, one for the `treated` group and one for the `untreated` group. When smoothing to fit a `sbp` term for the `treated` group in the backfitting algorithm, only those partial residuals belonging to the `treated` group are smoothed against `sbp`. Fitted values of zero are returned for this function for observations belonging to the `untreated` group. The reverse is done when fitting the `sbp` curve for the untreated group. This procedure can be motivated formally using the usual penalized likelihood argument (Exercise 9.13).

9.5.3 *Hierarchical models*

So far our concept of an interaction has been rather vague; any term involving more than one variable might be called an interaction. These terms are fitted jointly with the other terms in the model. Here we proceed more systematically.

Suppose that we have fitted a component-wise additive model of the form (9.15) and wish to explore lack of fit. Perhaps one of the simplest possible departures from this model is a pairwise interaction between two of the variables, say X_1 and X_2. We can model this as

$$E(Y \mid X_1, \ldots, X_p) = \alpha + \sum_1^p f_j(X_j) + f_{12}(X_1, X_2). \tag{9.16}$$

As it is, the components of this model are not identifiable, because, for example, we can add $h(X_1)$ to $f_1(X_1)$ and subtract it away from $f_{12}(X_1, X_2)$ without changing the overall model. We evidently need to place restrictions on $f_{12}(X_1, X_2)$; a convenient choice is

$$E\{f_{12}(X_1, X_2) \mid X_j\} = 0, \qquad j = 1, 2. \tag{9.17}$$

These restrictions are analogous to those put on each $f_j(X_j)$, namely $E\{f_j(X_j)\} = 0$, and are analogous to the usual constraints for second order interactions in a multiway analysis of variance.

Now let's see how to estimate these components, first in the population case. Assume that Y has been centered so that $E(Y) = 0$. The simplest approach is to ignore the constraint on the interaction term, and fit the model as usual, using backfitting to solve the estimating equations

$$\begin{aligned} f_j(X_j) &= E\Big\{Y - \sum_{k \neq j} f_k(X_k) - f_{12}(X_1, X_2) \mid X_j\Big\} \quad \forall j, \\ f_{12}(X_1, X_2) &= E\Big\{Y - \sum_{j=1}^p f_j(X_j) \mid X_1, X_2\Big\}. \end{aligned} \tag{9.18}$$

In order to enforce the constraints, we need to remove the component in $\hat{f}_{12}$ that is additive in X_1 and X_2. We do this in the obvious way, by fitting an additive model $h_1(X_1) + h_2(X_2)$ to $\hat{f}_{12}$. Then $\hat{f}_{12}^*(X_1, X_2) = \hat{f}_{12}(X_1, X_2) - \hat{h}_1(X_1) - \hat{h}_2(X_2)$ satisfies the constraints (9.17), and with $\hat{f}_j^*(X_j) = \hat{f}_j(X_j) + \hat{h}_j(X_j), \; j = 1, 2$; the additive predictor defined by (9.18) is unchanged by this operation (Exercise 9.14). Notice that we can leave the components f_1 and f_2 out of (9.16) and (9.18), since they are completely aliased with f_{12} and can be recovered later. An important fact emerges from this procedure: the fitted model is independent of the constraints on the interaction term. The only purpose of the constraint is to sensibly define what part of a function f_{12} is *main effect* and what part is *interaction.*

The model (9.16) is *hierarchical*, which means that if a term like $f_{12}(X_1, X_2)$ is present in the model, then the *sub-terms* that it contains, namely $f_1(X_1)$ and $f_2(X_2)$, are in the model as well. We have seen that in the population setting, the fitted model is the same whether we start with a hierarchical model or with the nonhierarchical model that omits the terms f_1 and f_2.

Now let's consider the data version of these procedures. If the terms in the additive model are all defined by projections onto linear subspaces, then the procedure above can be applied by replacing conditional expectations by projection operators. The simplest example is again multiway analysis of variance; for a more complicated example using tensor product splines see section 9.5.7.

For nonprojection smoothers, such as smoothing splines, some differences arise. The hierarchical and nonhierarchical models now lead to different fitted surfaces. The hierarchical model is probably preferable because it captures smooth main effects with fewer degrees of freedom than the nonhierarchical model. Note that if we replace the expectation in (9.17) by a shrinking smoother, the condition (9.17) is never satisfied unless $f_{12} = 0$. Hence it is not clear how to impose suitable identifiability conditions on the additive and interaction components. In practice we can simply carry out the program described for the population case, and fit an additive model to the fitted interaction term. Then we can remove it and consider the remainder to be pure interaction for plotting and inspection purposes. We can also evaluate the interaction by observing the decrease in the residual sum of squares, relative to the decrease in df^{err}.

9.5.4 *Examining residuals for interactions*

In practice an even simpler but approximate approach is to examine the residuals for interactions, rather than refitting the additive part each time. A natural first step is to represent the residuals graphically to alert one to the presence of interactions; see Fig. 6.4 in Chapter 6. The residuals can then be modelled using surface smoothers, or additive models comprised of surface smoothers, or more complicated high dimensional smoothers such as regression trees (section 9.5.6). Smoothing residuals with, say, a surface smoother, can be viewed as the first step in a backfitting algorithm for fitting the entire model. As such the procedure has the same flavour as a score test, or a diagnostic for the interaction. Once an interaction is identified, the entire model can be iterated to convergence.

The above discussion pertains to the simple (Gaussian) additive model. For other models, such as the generalized additive models and the models of Chapter 8, a similar approach is possible. Write the model as $\eta = \eta_A + \eta_I$, where η_A is the fitted additive model and η_I the interaction part to be estimated. We fix η_A as an *offset*, and then only the terms in η_I are estimated by local scoring. This is analogous to working with residuals. Once the important interactions are identified, the entire local-scoring procedure is iterated to convergence.

9.5.5 *Illustration*

For an example of modelling interactions, let's return to the `ozone` data from section 9.3.7. Here we include the third variable `radiation`. We follow our strategy above, and begin with an additive model in the three variables (111 observations).

We then fit separately all three pairwise surfaces to the residuals from the additive fit using a locally-weighted surface smoother. The residual sum of squares for the additive model is 20.7 with 99.6 df^{err}. The surface $f(\texttt{wind}, \texttt{temperature})$ dropped the RSS to 17.1 with an additional 4.4 df. Using a crude F-test, this drop is significant ($p = 0.004$), and is the only significant pairwise interaction among the three. The full three-dimensional surface is not

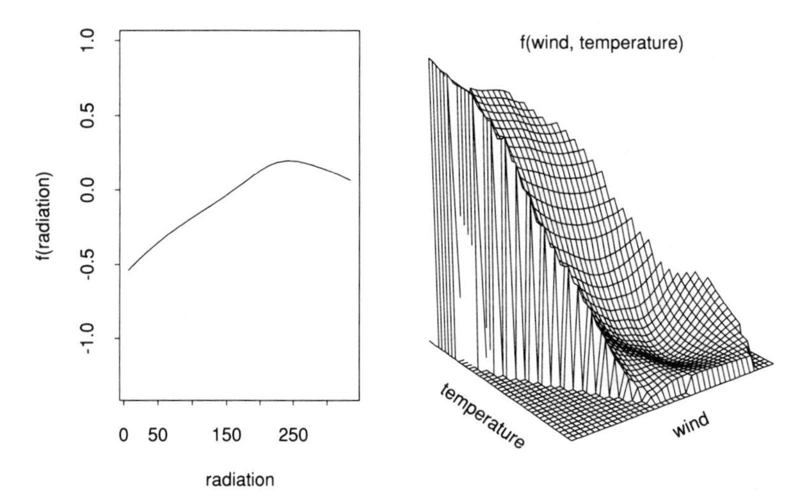

Fig. 9.14. *The fitted functions for the model* $\texttt{ozone}^{\frac{1}{3}} = f(\texttt{radiation}) + f(\texttt{wind}, \texttt{temperature}) + \varepsilon$. *The heights of the figures are scaled to reflect the relative importance of the terms.*

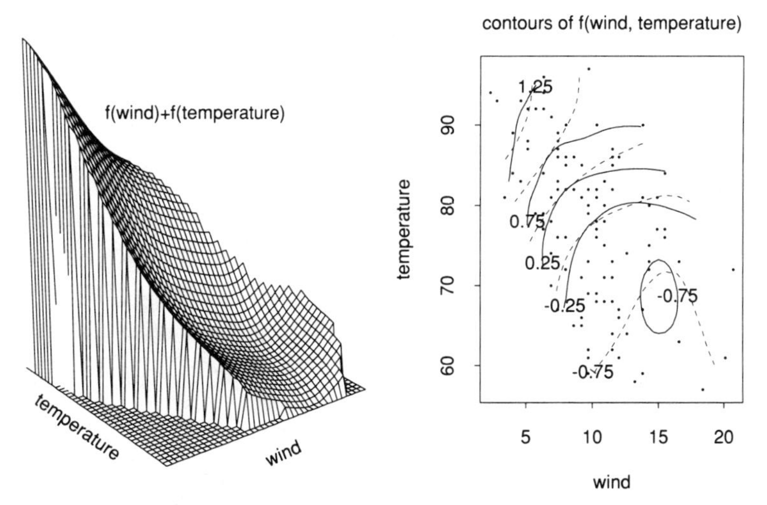

Fig. 9.15. *The surface on the left is the additive surface* $f(\texttt{wind}) + f(\texttt{temperature})$ *corresponding to* $f(\texttt{wind}, \texttt{temperature})$ *above. The contour plot on the right is another method for displaying surfaces; in it we compare the two-dimensional surface (solid contours) to the additive surface (dashed contours). The predictor sites are also indicated.*

significant either. We can also test to see if the interaction is due to the influential observations found in the previous section for these data; this turns out not to be the case. Figures 9.14, 9.15 and 9.16 represent this interaction in a variety of ways. Figure 9.14 shows the main effect curve for `radiation` and combines the main effects and interactions for `wind` and `temperature` in a perspective plot. The vertical scale of the plots represents the relative ranges of the functions accurately. Nevertheless, perspective plots are difficult to digest quantitatively. Figure 9.15 (left) shows the additive version of the surface, and the contour plot (right) compares the two. The contour plot allows the predictors to appear in the plot as well, and finer comparisons are possible.

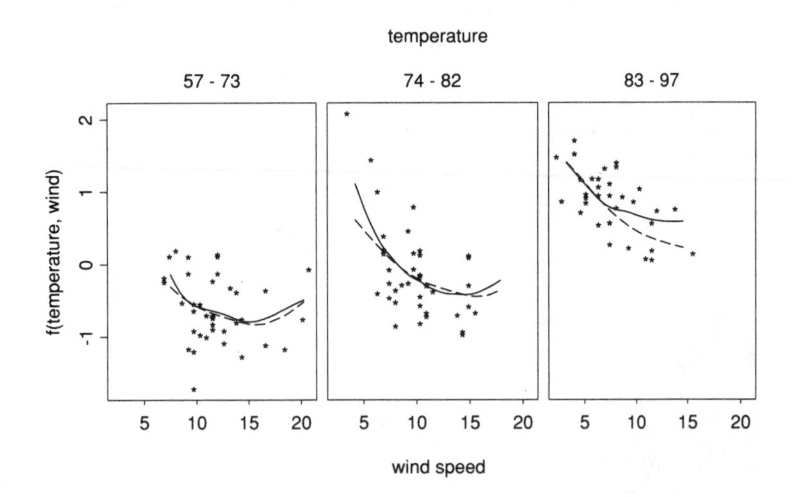

Fig. 9.16. *The conditioned plot is a useful way of representing and comparing surfaces. The points are partial residuals for the 2-D surface, also conditioned on the values of* `temperature`*. The solid curves represent the interactive surface, the broken curves the additive surface.*

Finally, Fig. 9.16 represents the surfaces by conditioning on specific values of one of the variables, in this case `temperature`. We have conditioned on the first, third and fifth sextile of `temperature`, and compare the additive and two-dimensional surfaces. The partial residuals are those from the two-dimensional surface, and are all those partial residuals whose values of `temperature` lie in the three regions defined by the second and fourth sextile. Although the surface significantly improves the fit, our interpretations do not change much. As `temperature` increases, the effect of `wind` changes from a U shape to a monotone function. However, because `wind` and `temperature` are negatively correlated, the additive model is able to do a reasonable job in depicting this relationship. Different parts of the curve for `wind` serve different regions of `temperature`.

The preceding methods are useful for exploring bivariate interactions, but what about higher order interactions? As we have seen, surfaces are difficult to estimate and display in higher dimensions. An alternative, described next, is to fit a regression tree to the residuals from the additive model.

9.5.6 *Regression trees*

In this section we give a very brief description of the tree-based approach to regression. This approach is effective when there is significant interaction structure in the predictors. Further details can be found in the references given in the bibliographic notes.

As usual we have a response Y and predictors $\boldsymbol{X} = (X_1, \ldots, X_p)$. We consider splitting the data into two parts, along any of the predictors, so that the two resulting groups are most homogeneous with respect to the response. Specifically, we examine all splitting points along all predictors, and choose the one that produces the smallest total within-group variance in the two groups. The data are then split into two parts, and the process is repeated on each part. At each stage, all split points along all predictors are considered, so that a predictor can be used for splitting more than once. If instead the response Y is a dichotomous 0-1 variable, then the deviance or a similar quantity is used to measure homogeneity of the groups.

This repeated binary-style splitting can be conveniently summarized by a binary tree, with the leaves, or terminal nodes, representing the subgroups. The tree can be used for predicting the response of a new observation by determining in which terminal node its predictors lie. The mean of the response values in a terminal node is the estimate of the regression function, which is used to make the prediction.

Figure 9.17 shows a regression tree fitted to the New York `ozone` data. For simplicity and comparison to the other surfaces in this

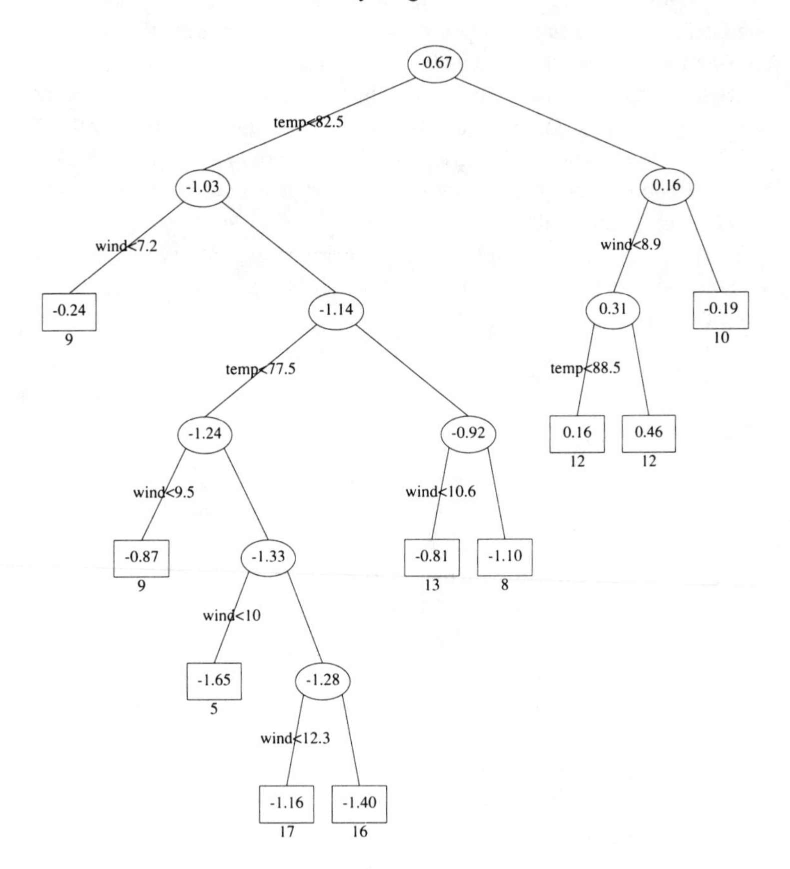

Fig. 9.17. *The interaction effect of* `temperature` *and* `wind` *speed modelled by a regression tree. The tree is limited to 10 terminal nodes for this demonstration. The mean response is inscribed at each node, and the number of occupants is given below each terminal node.*

section we first removed a smooth effect for `radiation`, and fitted the tree to the partial residuals as a function of `temperature` and `wind`. The tree has 10 terminal nodes, each indicated by a square with the mean response value of its occupants inscribed. For many predictors, such a tree is a useful way to render the regression model. Since we are using only two predictors, we can represent the fit as a piecewise-constant surface, as in Fig. 9.18. The fitted surface is qualitatively similar to that in Fig. 9.14. If the actual regression surface is smooth, then a regression tree requires many cuts (and therefore degrees of freedom) to approximate it. There is some evidence of that occurring here. The additive model with a smooth term in `radiation` and a regression tree with 10 *df* in `temperature` and `wind` has a residual sum of squares of 19.2 compared to 17.1 for the fit in Fig. 9.14 with the same *df*.

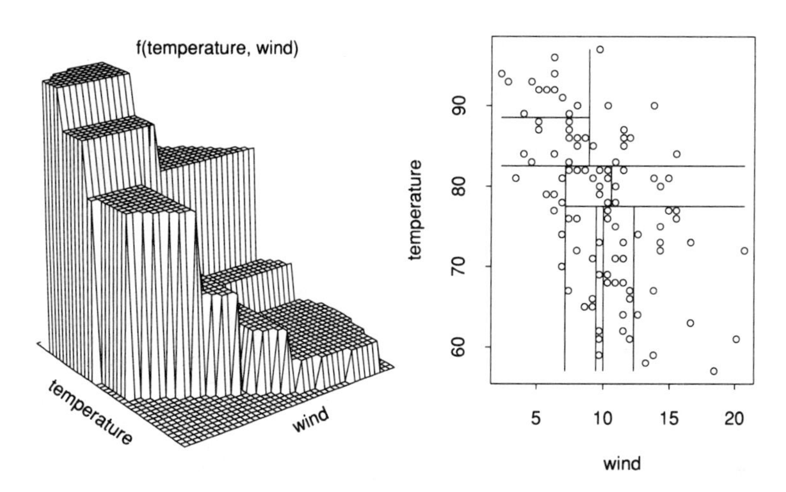

Fig. 9.18. *The regression tree represented as a piecewise-constant surface. This is only possible for one or two predictors, while the binary tree representation can be used for any number. The figure on the right shows how the predictor space is recursively partitioned by the tree-growing algorithm.*

The splitting process can be terminated when no further splits can be found to significantly improve the homogeneity of the subgroups. However, such a *greedy* strategy can miss effective splits further down in the tree by stopping too soon. A preferable strategy is to build a very large tree, then *prune* it back to a reasonable size. Cross-validation is used to guide the pruning, and the final tree chosen is that subtree of the large tree having the smallest estimated prediction error. While the process of building and pruning the tree may appear to be computationally formidable, an effective algorithm has been developed and implemented in the CART program (Classification and Regression Trees, Breiman *et al.* 1984). See the bibliographic notes for details.

A regression tree can be expressed more formally as a regression function

$$E(Y \mid \boldsymbol{X}) = \sum_{k=1}^{K} c_k I(\boldsymbol{X} \in R_k)$$

where the R_ks are regions in predictor space determined by the splitting, and the c_ks are coefficients that are estimated by the mean value of Y in the region in R_k. Figure 9.18 shows the partition obtained for the ozone data. In fact, we used this representation in fitting the tree represented in Figs 9.17 and 9.18. The cutpoints are identified by modelling the residuals from a simple smooth term in `radiation`; once the regions R_k defined by the cutpoints are found, the joint fit is obtained by a semi-parametric fit to the model

$$\texttt{ozone} = \sum_{k=1}^{10} c_k I(\texttt{temperature}, \texttt{wind} \in R_k) + f(\texttt{radiation}) + \varepsilon.$$

An additional feature of the CART methodology is the ability to make effective use of observations that are missing in at least one predictor value. These observations are not discarded (as is often the case in regression modelling) but instead each nonmissing predictor is used in determining the value of a split on that predictor. So-called *surrogate variables* are constructed to estimate the strata membership for observations missing on any predictor.

The piecewise-constant nature of a regression tree surface is unattractive, and can be extremely inefficient if the underlying surface is quite smooth. However the simple binary tree representation makes the model easy to understand for nonstatisticians. CART is probably most useful for *classification* of binary or categorical-response data: details can be found in Breiman *et al.* (1984). Next we describe some more recent methodology for fitting piecewise-smooth regression surfaces.

9.5.7 *Multivariate adaptive regression splines: MARS*

Friedman (1990) generalized the adaptive additive regression spline procedure to include multivariate tensor-spline bases, calling it "MARS" for Multivariate Adaptive Regression Splines. A tensor spline basis function for two predictors, for example, is a product of two one dimensional basis functions — one for each predictor. If both functions are truncated linear functions, the result is a surface that is zero in three quadrants and hyperbolic in the fourth.

MARS builds up its tensor-product basis in an adaptive way, and in so doing adds an important dimension to TURBO. Not only are variables with their smoothing parameters selected, but the interaction terms are selected as well, and their amount of smoothing. Let's look briefly how this ambitious task is tackled.

Suppose there are K basis functions in the model, which can be tensor products representing any order of interaction. The constant function is always included. A new basis function is added by searching among all one-dimensional bases, using updating as before. The tensor product of this candidate with each of the K bases in the set is formed, and the fit is computed each time. The best tensor product amongst all bases defines the $(K+1)$th tensor basis. There are some fine but important details:

(i) Each tensor product basis includes at most one univariate basis from each predictor.

(ii) The bases are included in pairs: $(x_j - t)_+$ and $(t - x_j)_+$ where t is the knot. Although this leads to redundant bases, it makes the procedure palindromically equivariant to sign changes. As a consequence the class of surfaces includes the piecewise-constant surfaces resulting from regression trees.

(iii) Typically every kth distinct value on a predictor is considered as a knot site, where k is a chosen parameter.

(iv) The *GCV* criterion for basis inclusion (and deletion) charges a *cost* per basis function (default of 3) to account for the degrees of freedom used in the search.

(v) The basis selection is followed by a similar basis deletion procedure, at which time redundant basis functions are removed.

After the tensor product bases are selected, they can also be converted to piecewise-cubic bases as in the additive case. The final fit is then computed using these smooth bases. Once the final model is fitted, all the main effect bases for each variable are grouped to represent the main effects, all second order bases for each pair are grouped to form second order interaction surfaces and so on (Exercise 9.14).

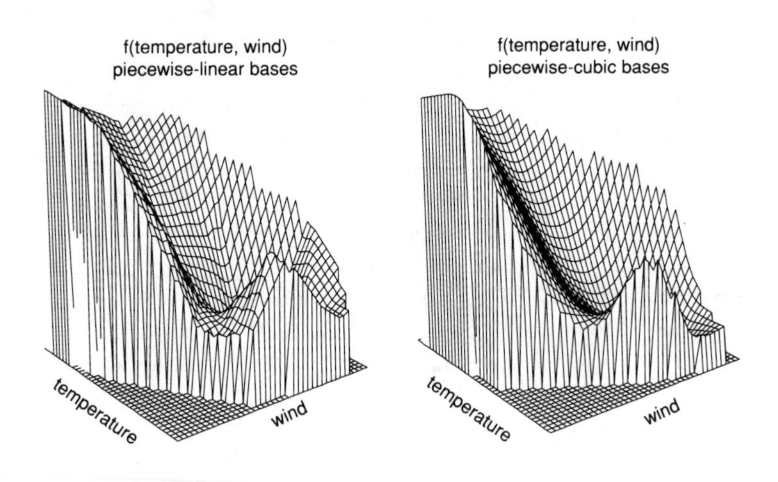

Fig. 9.19. *The MARS fit for the term* $f($**temperature**, **wind**$)$ *in the model* $f($**radiation**$) + f($**temperature**, **wind**$)$. *The figure on the left is the surface composed of tensor-product bases with piecewise-linear components. The surface on the right uses the same knots as the figure on the left, but the piecewise-linear bases are converted into piecewise-cubic functions.*

The MARS procedure has yet to endure the tests of time; it nevertheless seems extremely promising as an automatic high dimensional smoother. It also serves well as a continuous generalization of the regression-tree technology.

We applied the MARS algorithm to the `ozone` data of the previous sections. The surface selected by MARS has the form $\eta = f(\texttt{radiation}) + f(\texttt{radiation}, \texttt{temp}) + f(\texttt{wind}, \texttt{temp})$. The implementation we use (June, 1989) does not try and enforce hierarchical structure, which makes it more difficult to interpret the results. The fitted values are similar to those produced in the previous section. For more direct comparison, we forced MARS to allow only univariate terms in the predictor `radiation`; as expected, it included a surface for `temp` and `wind` as well. Figure 9.19 shows the surfaces produced by MARS using both the piecewise-linear and piecewise-cubic components for the tensor product bases. They are not very different, and both resemble the surfaces found earlier, apart from the bump in the foreground.

9.6 Bibliographic notes

Huber (1964) established the foundations of robustness, with detailed discussion of robustness in regression. Launer and Wilkinson (1979) is a useful collection of topics by different authors on robustness. In particular, the article by Huber on robust smoothing splines is one of the first formal motivations for robust smoothing. Huber does remark, however, that the technique had probably been used for a while before. Cleveland's (1979) description of locally-weighted running-line smoothing includes iterative estimation of robustness weights. In fact, our robust additive model algorithm in section 9.1 reduces exactly to Cleveland's smoother if there is one variable and the locally-weighted running-line smoother is used for smoothing (Exercise 9.2). Our treatment of resistant generalized additive models is an easy extension of Pregibon's (1979, 1982) work on generalized linear models. Mo (1989) studied existence and uniqueness of solutions to the resistant additive model, their rates of convergence, and some numerical properties of the problem. Bounded influence regression was introduced by Krasker and Welsch (1982).

Many people have used regression splines in the context of additive models. We referred to the work of Stone and Koo (1985), and Atilgan (1988). Breiman (1989a and 1989b) reported on his experiences, and undoubtedly regression splines have been used in this context for a long time. Breiman (1989a) described a methodology for fitting additive models with regression splines that is similar in flavour to Friedman and Silverman's (1989) TURBO approach. He applied a backward stepwise procedure to a set of power basis functions, using dynamic programming and cross-validation to choose the number and position of the knots. Both Breiman and Friedman and Silverman attributed the stepwise knot-selection idea to Smith (1982). In real and simulated examples, it performed comparably to TURBO and outperformed the ACE algorithm on some small, noisy data sets. Unfortunately, we did not receive his manuscript in time to include his methodology in the second case study of Chapter 10. The BRUTO procedure is due to Hastie (1989). Sleeper and Harrington (1989) used regression splines in the additive proportional hazards model. Figure 9.8 is taken from Hastie and Tibshirani (1988).

Eubank (1984, 1985) described diagnostics for smoothing splines, and Green and Yandell (1985) discussed similar ideas for the semi-parametric model. Connections between smoothing splines and ridge regression have been known for a long time (Titterington, 1985). Eubank and Gunst (1986) and Walker and Birch (1988) discussed diagnostics for ridge regression problems. Hastie (1988) introduced pseudo additive models as a computational vehicle and an approximation for exploring the nature of additive fits using *shrinking* smoothers. The conditioned plot was described by Cleveland and Devlin (1988).

Gu and Wahba (1988) and Chen, Gu and Wahba (1989) described automatic bandwidth selection algorithms for additive and interactive spline models. The interactive spline model discussed in Wahba (1986) was initiated by the Bayesian model of Barry (1986); see also the rejoinder of Buja *et al.* (1989).

Key references for classification and regression trees are Morgan and Sonquist (1963) and Breiman, Friedman, Olshen and Stone (1984). The latter authors are also responsible for the *CART* software implementation. The classification tree in Figs 9.17 and 9.18 was produced by the *tree.tools* functions (Becker, Clark and Pregibon, 1989), in the S language (Becker, Chambers, and Wilks, 1988).

Breiman (1989) proposed fitting interaction terms of a similar type to those in section 9.5.3; his algorithm is different, however, and attempts to estimate all the terms simultaneously without the constraints on the interactions. Method (iii) in section 9.5.1 for modelling interactions is due to Gennings and Hastie (1990).

9.7 Further results and exercises 9

9.1 Derive the Newton-Raphson algorithm for solving the penalized M-estimation score equation (9.4).

9.2 In Chapter 6 we give a population motivation for the local-scoring algorithm. Provide an equivalent motivation for a resistant additive model algorithm suitable for any smoother. Show that in the case of one predictor, this reduces to Cleveland's (1979) smoother if locally-weighted running-lines are used.

9.3 Outline a proof of convergence for the modified weights algorithm for solving the penalized M-estimation problem (for a single predictor). Following Huber (1981, page 184) show that each step results in a decrease in the criterion. Extend this result to cover the additive-model algorithm.

9.4 Derive the Newton-Raphson and Fisher scoring algorithms for solving (9.6) for the linear model in the case of a canonical link function. Even though the Fisher scoring procedure does not simplify the Hessian, suggest an approximation. Using the approximation, derive an algorithm for the additive model analogue.

9.5 Once a rich set of knots have been selected, TURBO eliminates knots using a backwards-deletion strategy. Suggest a more continuous approach for regularizing the potentially rough regression surface. Describe an appropriate algorithm [hint: how do smoothing splines achieve this?].

9.6 Given a basis $\mathbf{U}$ and a vector of penalties $\boldsymbol{\delta}$, construct a smoother with a free parameter λ to control the amount of smoothing. How do you guarantee that the smoother reproduces functions constant and linear in x?

9.7 Suppose $\mathbf{U}_1$ and $\mathbf{U}_2$ are orthonormal $n \times k$ basis matrices representing $\mathbf{x}_1$ and $\mathbf{x}_2$, with associated penalty vectors $\boldsymbol{\psi}_1$ and $\boldsymbol{\psi}_2$. Write down the estimating equations for a pseudo additive fit to $\mathbf{y}$. Show how these estimating equations make the concurvity results of Chapter 5 transparent.

9.8 Derive the expressions (9.11) and (9.12) in section 9.3.7 for computing the Cook's distances for additive models.

9.9 Derive the estimating equations for solving (9.10) in section 9.3.6. Suggest a method for solving these equations using standard least-squares methodology.

9.10 Describe how to use the pseudo additive model for nonparametric logistic regression. Give the criterion optimized, derive the Newton-Raphson update, and discuss the computations.

9.11 Describe an algorithm for using TURBO and BRUTO in the context of generalized additive models.

9.12 Describe underlying models for which the regression tree approach is (a) effective, and (b) ineffective. Do the same for an additive model. Are the two methodologies similar or complementary? Suppose both methods are applied to the same data set. How would one compare the predictive power of each of the methods?

9.13 Write down the model used in section 9.5.2 that incorporates an interaction between a continuous predictor and a binary predictor. Using a penalized likelihood estimator, give a rigorous motivation for the estimation procedure outlined there.

9.14 In section 9.5.3 we outline a method for fitting an additive model with interaction terms, subject to constraints. Discuss when this post-fitting adjustment is equivalent to the constrained minimization in L_2. Generalize the procedure to parametric linear models for data, using the MARS regression model (section 9.5.7) as an example.

CHAPTER 10

Case studies

10.1 Introduction

In this chapter we analyse two sets of data using some of the methodology described in this book. Practical experience is important in the development of new statistical tools. Through such experience, strengths and weaknesses of methodology are exposed, and important directions for further research and development are discovered. Both for our sake and the reader's, the examples in this chapter and other chapters in the book were not carefully selected to *show off* the tools described in this book. Rather, we required examples that show honestly the strengths of additive modelling as well the inherent difficulties.

The two examples also represent different modes of analysis. The first has a small number of observations on three variables, and the goal of the analysis is to understand the association of the predictors with the binary response. We need to use our flexible models with some caution. After fitting a variety of nonparametric models, we reparametrize the components of the model in terms of simple one or two parameter families of functions. These functions are suggested by the nonparametric fit, and are a useful summary of the estimated model.

In the second example we focus on prediction from an additive model. Although description has been our main emphasis throughout this book, measures of prediction are useful because they provide a quantitative basis for comparing various approaches to model building. We assess a number of estimation methods, including nonparametric, stepwise, parametric and linear methods, in terms of their predictive ability for a (large) real data set. The results give an idea of whether the gains from some of the "fancy" techniques described here are worthwhile, relative to the possible increase in variability that they might cause.

10.2 Kyphosis in laminectomy patients

Bell *et al.* (1989) studied multiple level thoracic and lumbar laminectomy, a corrective spinal surgery commonly performed in children for tumour and congenital or developmental abnormalities such as syrinx, diastematomyelia and tethered cord. The incidence of postoperative deformity is not known. The purpose of the study is to delineate the true incidence and nature of spinal deformities following this surgery and to assess the importance of age at time of surgery, as well as the effect of the number and location of vertebrae levels decompressed. The data in the study consists of retrospective measurements on 83 patients, one of the largest studies of this procedure to date.

The specific outcome of interest here is the presence (1) or absence (0) of kyphosis, defined to be a forward flexion of the spine of at least 40 degrees from vertical. The available predictors are `age` in months at time of the operation, the starting and ending range of vertebrae levels involved in the operation (`start` and `end`) and the number of levels involved (`number`). These last predictors are related by `number` = `end` − `start` + 1. The goal of the analysis is to identify risk factors for `kyphosis`, and a natural approach is to model the prevalence of `kyphosis` as a function of the predictors. In order to investigate this relationship, we fit a number of generalized additive logistic models. By the results of Chapter 5, the exact linear dependence between the level variables cause an exact concurvity in any additive model involving all three of them. We therefore want to include only two of the three level variables in the model. The medical investigator felt a priori that `number` and `start` would be more interpretable, so we use these in the analysis. For `start`, the range 1–12 corresponds to the thoracic vertebrae, while 13–17 are the lumbar vertebrae; this dichotomy was felt to be an important one.

There are 65 zeros and 18 ones for the response `kyphosis`. This is not a large sample, especially for binary data; we have to bear in mind the warnings of section 6.10 about overinterpreting additive logistic fits to binary data. Despite the small sample size, this study is the largest and considered to be one of the most important of its kind for studying kyphosis.

Figure 10.1 shows the pairwise scatterplots of all three predictors, with the binary `kyphosis` encoded in the plot symbol. The

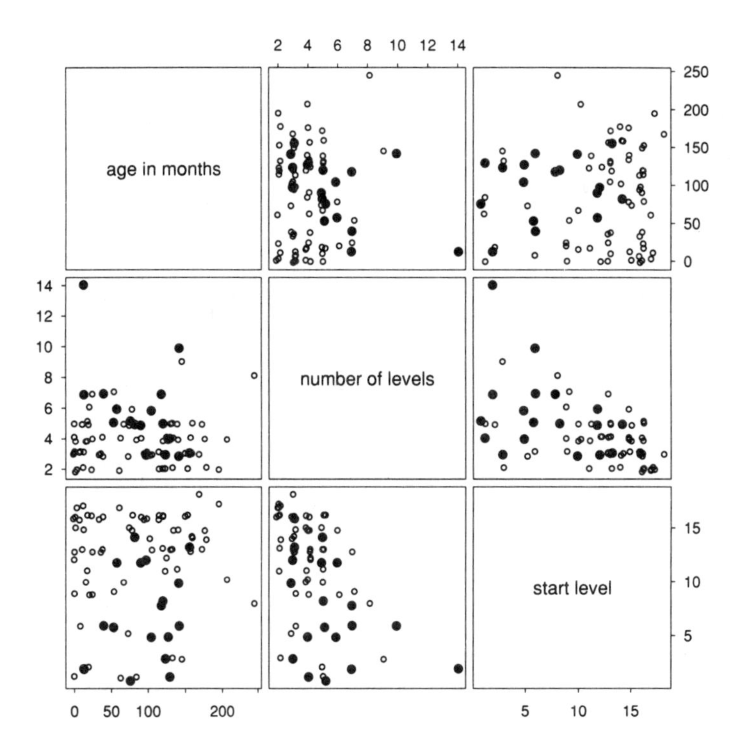

Fig. 10.1. *A scatterplot matrix of the three predictors, with the binary value of* **kyphosis** *encoded in the plotting symbol (black dots are* **kyphosis** *present). The positions of the points have been randomly jittered to break ties.*

positions of the points are slightly jittered to break ties; without this the bivariate densities perceived from the plots would be misleading. By inspecting the plots we immediately see that all three predictors are potentially important for predicting `kyphosis`, since there are noticeable clumps of *pure* black and white regions. Apart from four observations, the data appear to lie in a fairly uniformly filled convex region.

The boxplots in Fig. 10.2 show a strong location shift for all of the variables except `age`; the `age` effect turns out to be roughly quadratic with average effect zero.

The next step in our analysis is to fit an additive logistic model to

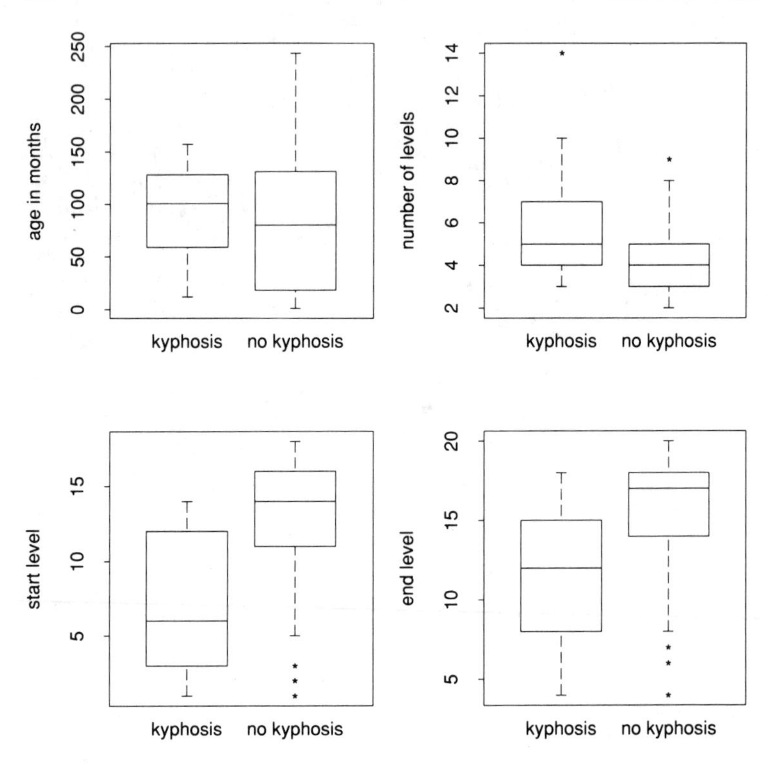

Fig. 10.2. *Boxplots by* `kyphosis` *for all 83 observations.*

the three predictors. Each term is fit using a smoothing spline with a nominal three degrees of freedom, that is $df_j = \text{tr}(\mathbf{S}_j) - 1 = 3$. Although this might seem a modest amount of flexibility, we prefer to be cautious with such a small data set. The df^{err} of the fitted model is based on the weighted additive-fit operator from the final local-scoring iteration. The additive fit is displayed in Fig. 10.3, and has some striking nonlinear features. The plots are deliberately not adorned with auxiliary information to achieve this effect. The fitted functions are represented in the same vertical units, although they all cover roughly the same range. We recommend this uniform plotting, otherwise even relatively unimportant effects can be made to look dramatic by blowing them up in a plot. The vertical labels in the plots show how the terms are fitted. For example, the label

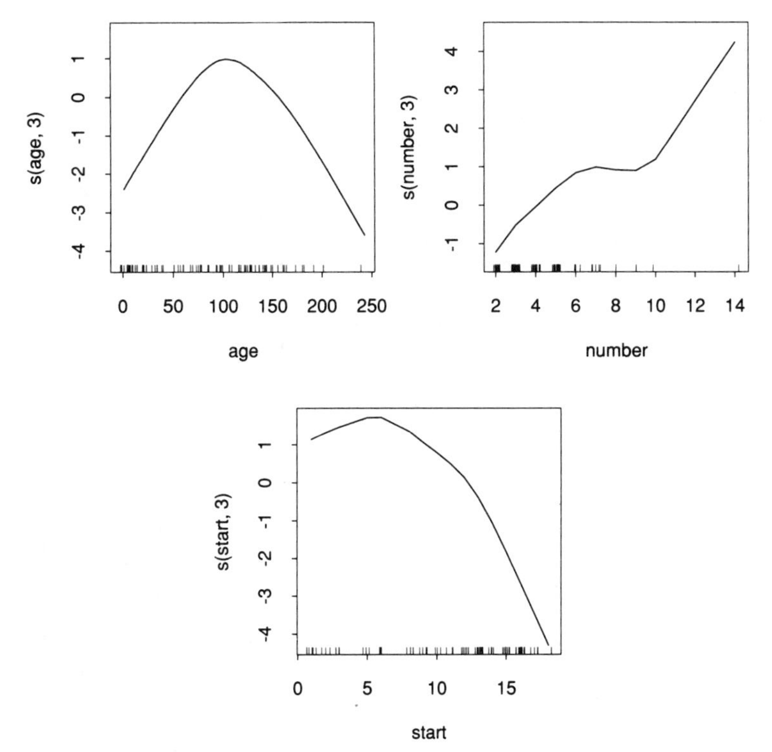

Fig. 10.3. *An additive logistic fit using all three predictors, with a nominal 3 df per term. The vertical labels in the plots are taken from the formula language used in the S software for fitting these models.*

`s(age,3)` means that a smoothing spline is used, with a nominal value of 3 *df*. This is the notation used in the formula language of the S software for fitting generalized linear and additive models (Chambers, ed. 1990), described in Appendix C.

The data sites are indicated at the base of the plots, and give the first warning about possible spurious effects. For example, in the plot for `number`, there is only one observation at 14. We see that about half the vertical range of the function is determined by this one point, which extends 50% beyond the range of the remaining points on the horizontal axis. It is also likely that this point has a large effect on the function estimates at neighbouring values of `number`. The fitted functions are displayed by joining

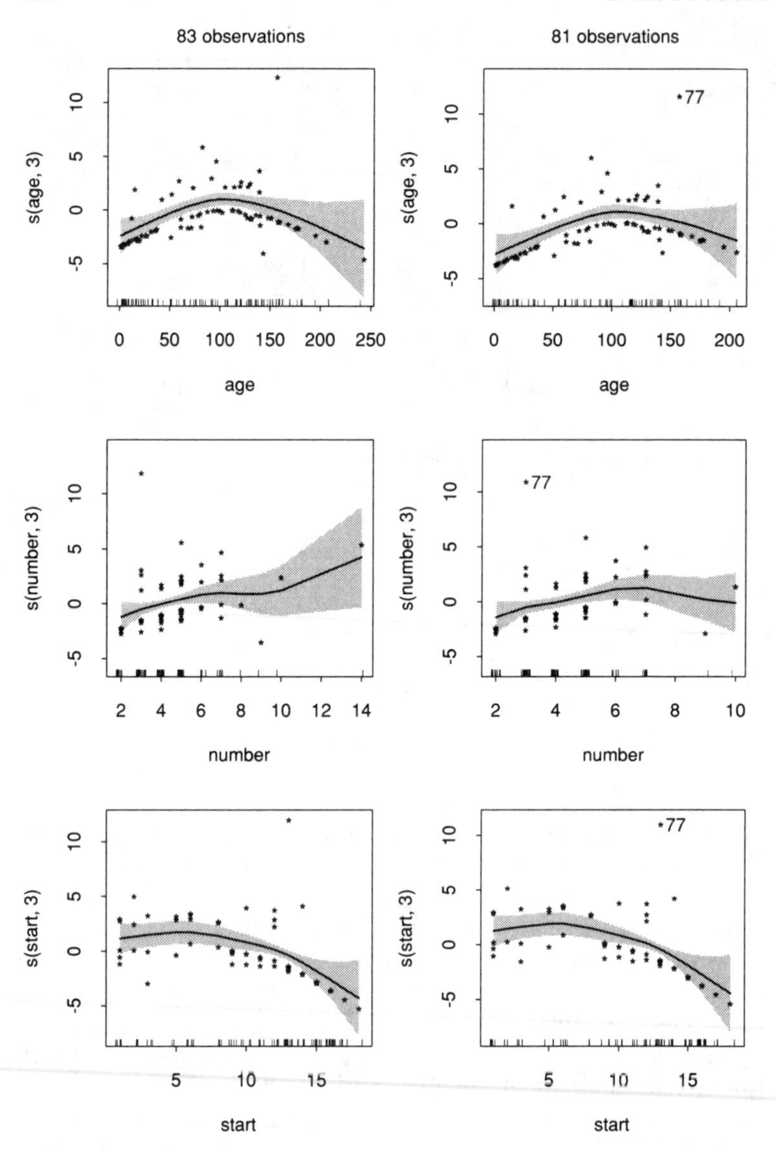

Fig. 10.4. *The left column is the same additive fit as in Fig. 10.3, but with partial residuals and pointwise standard errors included. The right column shows the same model, but fitted to the data excluding the observations with* `number` $= 14$ *and* `age` $= 243$*. The residual labelled* 77 *received a weight of* 0.55 *in the resistant fit of the additive model.*

the fitted values with straight lines. Although this is visually less pleasing than the smooth cubic-spline, it also draws attention to the regions of sparse data. The left column of plots in Fig. 10.4 shows the same fit, but this time with partial residuals and pointwise $2\times$ standard-error bands included. The pointwise standard-error bands are particularly wide in the region of `number` $= 14$. We shall remove this observation from this and all subsequent models, as well as the observation with `age`$= 243$ months. Although this latter observation would not necessarily be considered a severe outlier in `age`, it is seen to be a bivariate outlier in the pairwise scatterplots in Fig. 10.1.

The patterns in the residuals in the plots for `age` and `start` in the left column of Fig. 10.4 warn us of the pure-region effect (section 6.10). The sequence of negative residuals in the right hand region of each of these plots suggests that the fitted effects are biased there. The fitted functions in the right hand column of plots in Fig. 10.4 represent the model with the two observations removed. The nature of the `number` effect changes with the removal of these points, showing a negative slope in the right hand range rather than the positive slope in the figure on the left.

The deviance for the fitted model is 45.4 with $df^{\text{err}} = 69.6$. One of the residuals seems particularly large, so we refitted the model using the resistant algorithm for fitting generalized additive models described in section 9.2. The fitted functions hardly changed at all (not shown). Eight observations receive resistance weights less than one, shown in the top left panel of Fig. 10.5. Observation number 77 (numbered in the data set of 81 observations) receives a weight of 0.55, while the next lowest is observation number 11 at 0.69. Observation 77 also has the largest residual in the plots in the right column of Fig. 10.4. We delete observation 77 from the data (leaving 80 observations), and refit the nonresistant additive model. The deviance now drops to 38.8 with $df^{\text{err}} = 68.9$, rather a large change! Notice, however, that the fitted function for age has become even more extreme, since this region of age has become a purer region of 0s (observation 77 has `kyphosis`$= 1$). The large residual is simply a result of a one in a region of very low fitted probability, and as such is a potentially valuable observation. If the sample size was larger, we may have been able to decide whether or not it is an outlier; under the present circumstances we leave it in and pursue the analysis with all 81 observations.

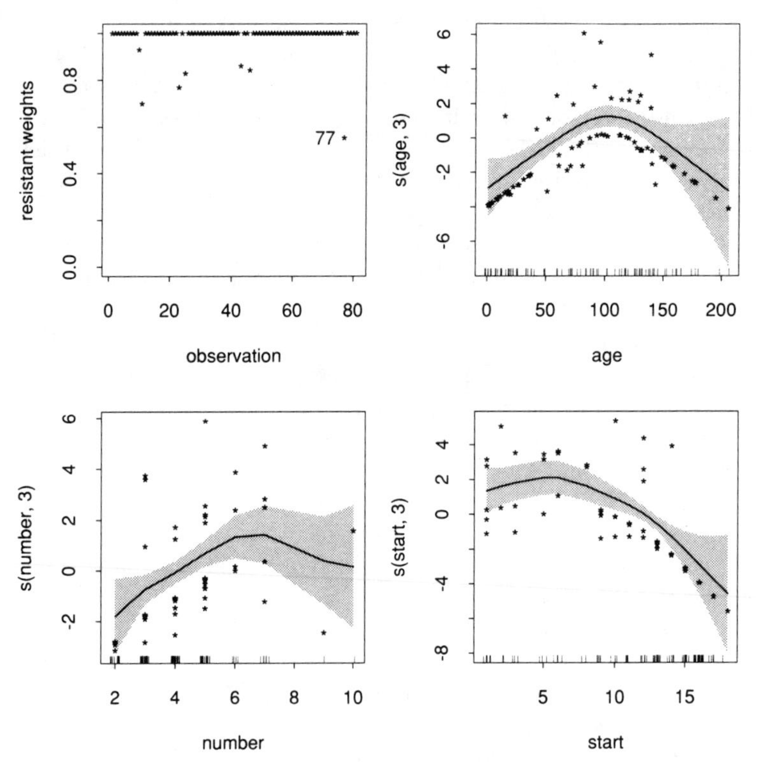

Fig. 10.5. *The top left plot shows the final weights assigned in the resistant logistic additive model fit to the three predictors. The remaining plots show the nonresistant fit to the same model, but with observation 77 removed.*

The next step in our analysis is to establish which effects, if any, are important. Table 10.1 summarizes the results from fitting a number of logistic additive models. For each model we show the deviance of the fit and the residual df^{err}, defined in section 6.8.3 as an asymptotic quadratic approximation to the mean residual deviance. The change in deviance when each variable is dropped out, or forced to have a linear effect, is also shown (labelled ΔDev) to indicate the strengths of the effects. Similarly Δdf^{err}, the expected change in deviance, is shown for each model.

Our software returns with a value of $df_j = \text{tr}(\mathbf{S}_j) - 1$ for each term in the final weighted additive fit; df_j can be used as a rough

Table 10.1 *Analysis of deviance table for preliminary fits to the kyphosis data*

	Model	*Dev*	df^{err}	ΔDev	Δdf^{err}
(i)	`null` (83 observations)	86.8	82.0		
(ii)	`s(age)+s(number)+s(start)`†	47.4	71.7		
(iii)	`null` (81 observations)	83.2	80.0		
(iv)	`s(age)+s(number)+s(start)`	45.4	69.6		
(v)	resistant fit to (iv)	43.8	69.7		
(vi)	model (iv) with obs. 77 deleted	38.8	68.9		
(iv)	`s(age)+s(number)+s(start)`	45.4	69.6		
(vii)	`age+s(number)+s(start)`	52.7	71.9	7.3	2.3
(viii)	`s(number)+s(start)`	57.4	72.8	12.0	3.2
(ix)	`s(age)+number+s(start)`	48.5	72.0	3.1	2.4
(x)	`s(age)+s(start)`	50.6	73.0	5.2	3.4
(xi)	`s(age)+s(number)+start`	50.8	72.0	5.4	2.3
(xii)	`s(age)+s(number)`	62.2	73.1	16.8	3.5

† The 3 in terms like `s(age,3)` has been suppressed for readability.

guide to the difference Δdf^{err} for models with and without the jth term, although it tends to underestimate this difference. In this analysis we have computed the df^{err} exactly; see Appendix B for details of various approximations to these quantities.

Both `age` and `start` seem to be important, while `number` does not attain the nominal significance levels. We are using $\chi^2_{\theta,0.95}$ to judge effects, where $\theta = \Delta df^{\text{err}}$ for the pair of models. The curve for `age` is nonmonotone, with a maximum risk occurring at about 110 months. The curve for `start` is roughly horizontal until about 10, and then it starts to decrease.

In order to simplify the model in a systematic way, we apply the backward/forward stepwise selection strategy outlined in section 9.4.1. Starting with the full model, the procedure checks at each stage whether any of the current variables can be dropped or simplified from a smooth fit to a linear fit. The minimum change in deviance is computed for dropping or *linearizing* any variable. The variable corresponding to the minimum is then dropped or linearized if the increase is less than a pre-specified threshold. Similarly, the current model is examined to check whether the addition of a variable or the relaxing of a linearity constraint results in

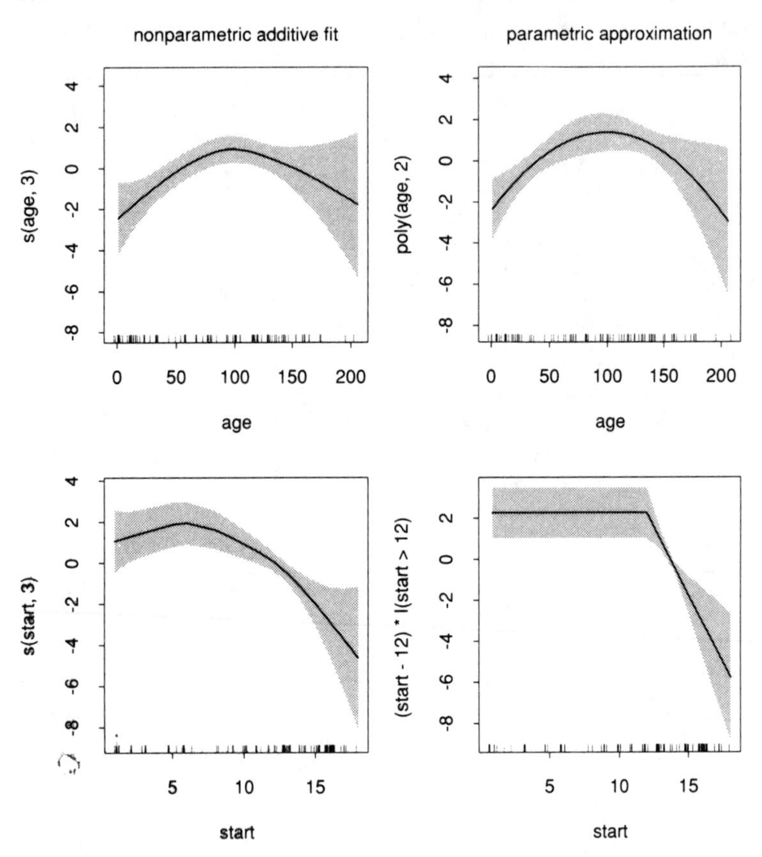

Fig. 10.6. *The left column displays the fitted functions for the final model* **s(age,3)+s(start,3)**, *together with asymptotic* 2× *pointwise standard-error curves. The right column displays the parametric approximation* **poly(age,2)+ (start-12)*I(start>12)**.

a change in deviance larger than a pre-specified threshold. The threshold value used was the 95% point of the appropriate χ^2 distribution. For these stepwise calculations we do not compute Δdf^{err} exactly each time, but use the approximations based on df_j mentioned earlier and described in Appendix B. We do not show the sequence of chosen models, but simply report that model (x) in Table 10.1 was selected, consisting of smooth terms for **age** and **start**.

Although the model is rather simple at this stage, it can be simplified even further. Both the fitted functions suggest simple parametric transformations. The fitted function for `age` is well approximated by a quadratic term. There are several possibilities for `start` which take into account the natural division into the two groups `start` $\leq$ 12 (thoracic vertebrae) versus `start` $>$ 12 (lumbar vertebrae). We express these once again using the formula language of the S modelling software:

- `I(start>12)` amounts to fitting a dummy variable to represent the two groups, or in other words approximating the function by a step function.
- `(start-12)*I(start>12)` will result in a constant for the `thoracic` group, joined continuously to a linear fit for the `lumbar` group.
- `bs(start,knot=12,degree=1)` fits a linear B-spline with one interior knot at `start` = 12. This is a piecewise linear fit in the two groups, but joined continuously at the knot.
- `bs(start,knot=12,degree=3)` fits a cubic B-spline with one interior knot at `start` = 12. The cubic splines are joined with two continuous derivatives at the knot.

Table 10.2 *Analysis of deviance table for the final fit to the kyphosis data, and several parametric approximations*

	Model	*Dev*	df^{err}	ΔDev	Δdf^{err}
(i)	`s(age,3)+s(start,3)`	50.6	73.0		
(ii)	`poly(age,2)+s(start,3)`	50.7	74.4	0.1	1.4
(iii)	`poly(age,2)+I(start>12)`	54.5	77.0	3.9	4.0
(iv)	`poly(age,2)+(start-12)*I(start>12)`	52.0	77.0	1.4	4.0
(v)	`poly(age,2)+bs(start,knot=12,d=1)`	51.6	76.0	1.0	3.0
(vi)	`poly(age,2)+bs(start,knot=12,d=3)`	50.0	74.0		

Table 10.2 summarizes the fits of these various approximations to the additive fit selected by the stepwise procedure. Model (iv) in the table performs the best, since it is as parsimonious as (iii) but captures the functional form suggested by the nonparametric fit. Figure 10.6 displays both the nonparametric fit in `age` and `start`, as well as the parametric approximation. Notice the rather strange form of the pointwise standard-error bands for the

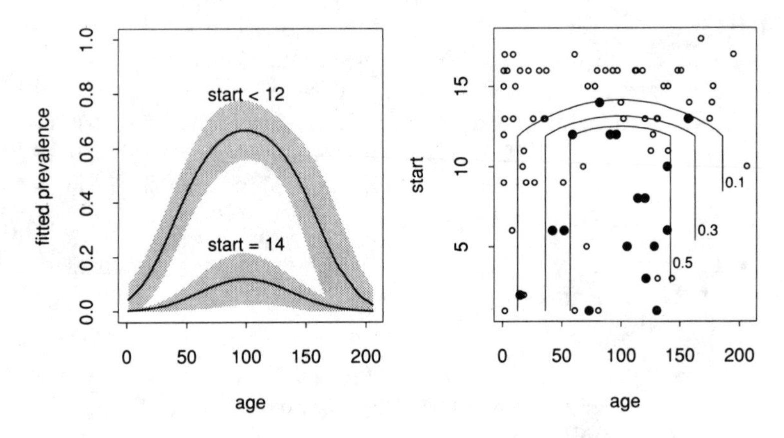

Fig. 10.7. *The left plot shows the fitted prevalence of* `kyphosis` *as a function of* `age`*, conditioned on* `start<12` *and* `start=14`*. Pointwise standard errors are indicated by the shaded regions. The contour plot shows three contours of the fitted prevalence surface, with the observations superimposed.*

term `(start-12)*I(start>12)`, which is a consequence of the continuity constraint.

So far we have not concerned ourselves with the possibility of interactions. With the relatively small number of observations, our ability to detect departures from nonadditivity is limited. Nonetheless, we tried adding pairwise products to the model, and also modelling the residuals with a regression tree. No significant interactions were found.

Figure 10.7 shows the fitted prevalence of `kyphosis` as a function of `age`, for `start<12` and `start=14`. From a medical point of view, this figure best summarizes the results of the analysis. It provides a clear quantification of the strong effects of both `age` and `start` on the prevalence of `kyphosis`. The pointwise standard-error bands are computed using the *delta method* applied to the asymptotic standard errors for the fitted logits. The contour plot shows the 10%, 30% and 50% contours of the fitted prevalence surface. The 30% contour describes a reasonable decision boundary for classifying the binary observations, and indicates a good fit to the data.

The tests of significance use the χ^2 distribution as a reference. There are several levels of approximation here. Firstly, the χ^2

approximation is asymptotic in nature and may be poor for small samples. Secondly, our additive operator is not a projection, but rather a weighted sum of projections. The degrees of freedom $\Delta df^{\,\mathrm{err}}$ is asymptotically the mean difference in deviance when the smaller model is correct. Even this last statement is not strictly correct, since we need to assume that the biases, which exist in both models, approximately cancel under the null hypothesis. We give heuristic arguments in earlier chapters to justify this approach; here we perform a small simulation to check the accuracy of these approximations.

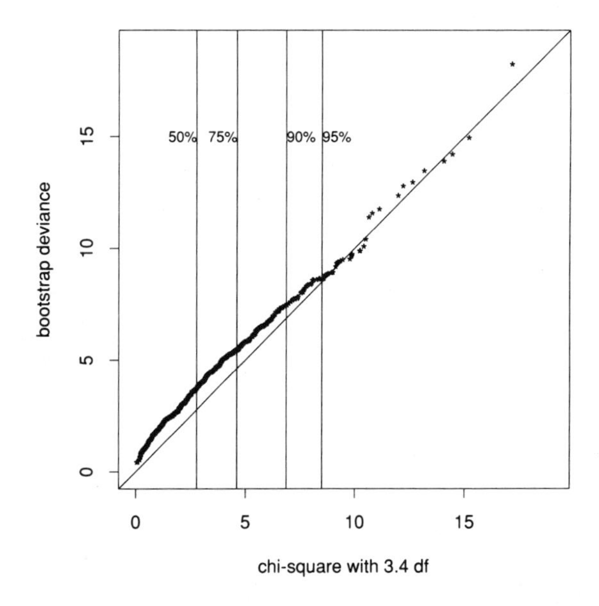

Fig. 10.8. *Quantile-quantile plot of the distribution of the bootstrapped deviance versus a* $\chi^2_{3.4}$ *distribution. The bootstrap estimates the distribution of the change in deviance when adding the term* `s(number,3)` *to the null model* `s(age,3) + s(start,3)`. *The* χ^2 *distribution is indexed by* $df^{\,\mathrm{err}} = 3.4$.

We can estimate the exact tail probabilities by a bootstrap, or simulation technique. We focus on the `number` term in the full model (iv) in Table 10.1. The null hypothesis is no `number` effect. Fixing the predictors, we generate data from the fitted model `s(age,3) + s(start,3)`. We do this simply by generating a sample of 81 binary observations with probabilities given by the fitted values from this null model. For each new data set we refit the null model, and also fit the model `s(age,3) + s(start,3) + s(number,3)`. In each case the decrease in deviance due to adding `number` to the model is computed. This entire process is repeated 500 times. The QQ-plot in Fig. 10.8 shows that the bootstrap distribution has a slightly different shape to the $\chi^2_{3.4}$ distribution; indeed, its mean is 4.3 versus an expected 3.4. However the 95% quantiles are in close agreement. The attained significance levels for the actual deviance drop of 5.2 due to `number`, is 0.80 under the $\chi^2_{3.4}$ distribution and 0.70 under the bootstrap distribution.

10.3 Atmospheric ozone concentration

The goal in our second case study is to predict the level of atmospheric ozone concentration from eight daily meteorological measurements made in the Los Angeles basin in 1976. Although measurements were made every day that year, some observations were missing; we use the 330 complete cases since not all the methods we compare have built in missing data capabilities. The data, displayed in Fig. 10.9, were given to us by L. Breiman; he was a consultant on a project from which these data are taken.

The response, referred to as `ozone`, is actually the log of the daily maximum of the hourly-average ozone concentrations in Upland, California; the log transformation, suggested by the ACE and AVAS procedures in Chapter 7 (See Fig. 7.2), tends to stabilize the variance. We refer to the predictors using abbreviated names, and for reference their full names are:

`vh`: 500 millibar pressure height, measured at the Vandenberg air force base.
`wind`: wind speed (mph) at Los Angeles airport (LAX).
`humidity`: humidity (%) at LAX.
`temp`: Sandburg Air Force Base temperature (°F)
`ibh`: inversion (temperature inversion) base height (feet)
`dpg`: Pressure gradient (mm Hg) from LAX to Daggert
`ibt`: inversion base temperature (°F) at LAX
`vis`: visibility (miles) at LAX
`doy`: day of the year

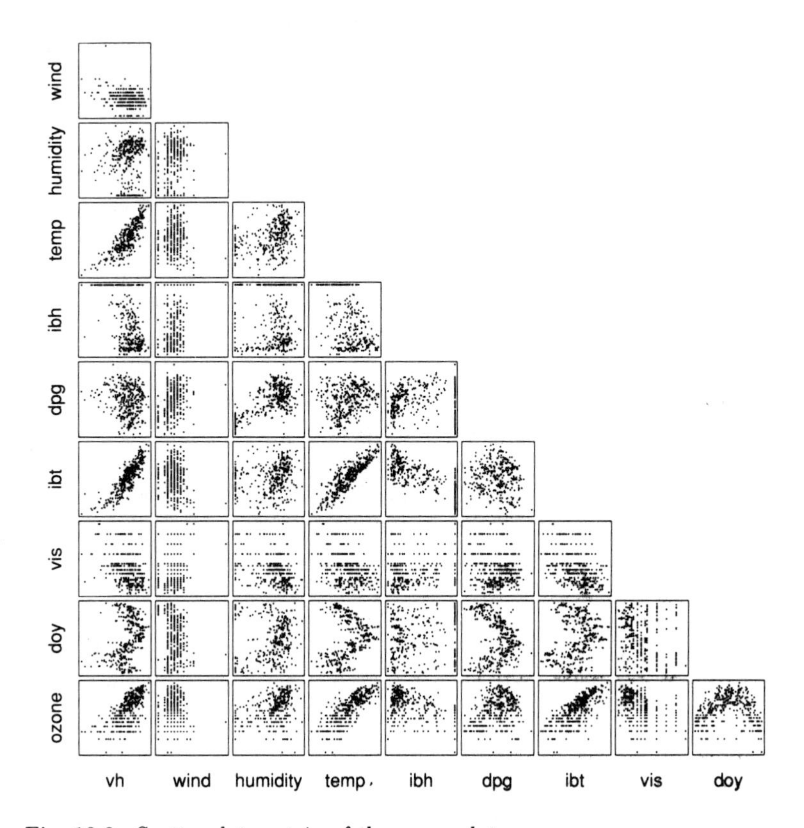

Fig. 10.9. *Scatterplot matrix of the ozone data.*

A number of authors have previously analysed these data, including Breiman and Friedman (1985) using ACE, Buja, Hastie and Tibshirani (1989) using an additive model, Friedman and Silverman (1989) using the TURBO adaptive regression splines method, and Hawkins (1989), using the more traditional Box and Tidwell (1962) method for finding transformations.

Our objective here is to make a comparative assessment of each of these methods in terms of their predictive ability. The methods that we compare here are:

(i) *Linear regression* (*LINEAR*). For a baseline of comparison, we apply linear least-squares regression to all nine predictors. Since many nonlinear effects have been found in previous analyses, we expect that linear regression will fare poorly.

(ii) *Linear regression after transformations* (*BOX-TIDWELL*). In the discussion of the *TURBO* paper by Friedman and Silverman (1989), Hawkins (1989) applied the Box–Tidwell method for finding transformations to these data. Power transformations were used for all of variables except for `Daggert pressure gradient` (quadratic transformation) and `day of year` (sine and cosine bases at three frequencies). Presumably these choices were at least partly based on the *TURBO* fit to the data. We thank D. Hawkins for clarifying the procedure that he used so that we could implement it.

(iii) *Additive model* (*ADDITIVE*). A simple additive model fit to all nine predictors with a nominal four *df* for each.

(iv) *Additive model with backward selection* (*STEP-ADDITIVE*). A backward stepwise strategy applied to a full-additive model fit, as described in Section 9.4.1. The regimen of degrees of freedom for each variable is 0, 1 or 4.

(v) *Additive regression splines* (*TURBO*), as described in sections 9.3.4 and 9.4.2. This is the same as the multivariate adaptive regression splines (MARS) method, but with no interactions allowed.

(vi) *Automatic backfitting* (*BRUTO*), as described in section 9.4.3.

Smoothing splines are used for smoothing in (iii), (iv) and (vi). The fitted functions from each of these methods is shown in Fig. 10.10. Denoting the response `ozone` by Y and the nine predictors by $\boldsymbol{X}$, the measure of performance in this study is the prediction error

$$PSE = E\{Y - \hat{Y}(\boldsymbol{X})\}^2$$

where $\hat{Y}(\boldsymbol{X})$ indicates the predicted value at $\boldsymbol{X}$, and the expectation is over new realizations $(Y, \boldsymbol{X})$ drawn from their true joint distribution. Let ASR be the average squared residual, or the *re-substitution prediction error*, computed by applying the prediction rule to data itself. Because it uses the same data both for fitting and assessment, ASR is probably too small, as in the smoothing parameter selection problem discussed in Chapter 3.

In order to obtain a better estimate of PSE, we follow the approach of Efron (1983) and estimate the *optimism*

$$OP = PSE - ASR$$

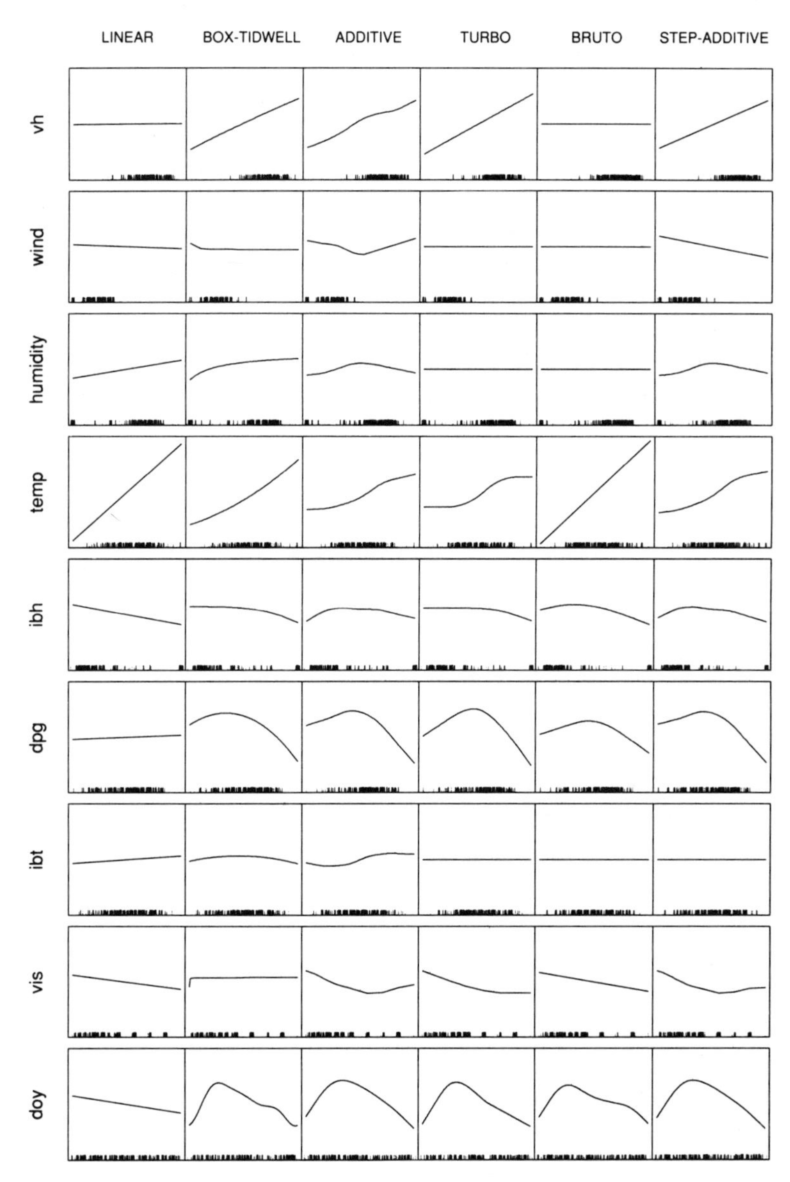

Fig. 10.10. *Estimated functions for the ozone data, for each of the six methods described in the text. All the plots are on the same scale, so that the vertical range covered by each function is an indication of its importance.*

The estimate of *OP* is then added to *ASR* to obtain an estimate of *PSE*.

To estimate *OP*, we use the so-called ".632" bootstrap method of Efron (1983). Of the many methods described in Efron's paper, including cross-validation, the .632 method performed the best. A bootstrap sample is generated from the original data; that is, a sample of size n is drawn with replacement. The prediction rule (like one of the additive model estimators above) is then fitted using this bootstrap sample, and is used to predict the responses in the original data at their corresponding predictor values. The .632 estimator differs from standard bootstrap approaches in that predictions are only made for those observations *not* in that particular bootstrap sample. The idea is that prediction at a given point in the original sample is easier if that point is in the bootstrap sample that was used to derive the rule. The name ".632" is taken from the fact that in a sample of size n taken with replacement from n items, the probability of any given item appearing is $1-(1-1/n)^n \rightarrow 1-e^{-1} \approx 0.632$ as $n \rightarrow \infty$.

In detail, let ASR_j^0 be the prediction error computed for the jth bootstrap sample by

(i) estimating a prediction rule using this jth bootstrap sample, and

(ii) averaging the squared error in predicting the responses for observations not in this bootstrap sample but in the original sample.

This is repeated B times, and then the average

$$ASR^0 = \sum_{j=1}^{B}(ASR_j^0)/B$$

is computed. Finally, the .632 estimate of optimism is

$$\widehat{OP} = 0.632(ASR^0 - ASR)$$

and the final error estimate is

$$\widehat{PSE} = ASR + \widehat{OP}$$

The derivation of $\widehat{OP}$ is complicated, and is described in Efron (1983).

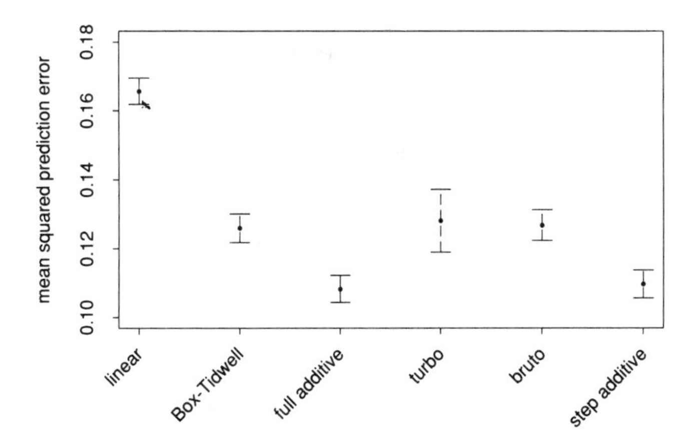

Fig. 10.11. *Bootstrap estimates of prediction error for each of the six methods described in the text. The bars around each estimate represent the estimated Monte Carlo standard error. Note that the vertical axis starts at 0.1.*

We carried out this prediction error estimation for the six procedures given earlier. The estimated prediction errors are shown in Fig. 10.11; the bars around the estimates represent the estimated Monte Carlo standard error.

Note that some of the predictions involve extrapolation of one or more of the estimated functions. Since such extrapolations are likely to be unstable, we decided a priori to leave these out of the prediction error estimates. As it turned out, extrapolations represented a small proportion of the total predictions, and the results with or without the extrapolations were virtually identical.

The results can be summarized as follows:

(i) All of the nonlinear methods improve upon the linear model

(ii) The additive and stepwise-additive methods perform the best, followed by the Box–Tidwell, *BRUTO* and *TURBO* techniques.

(iii) Despite the fact that the stepwise-additive method, *TURBO*, and *BRUTO* all dropped some terms, the full-additive fit does as well or better in terms of prediction error. Presumably additional variance is introduced due to the flexibility of the selection processes.

(iv) Perhaps surprisingly, the additive and stepwise-additive methods did slightly better than the Box–Tidwell technique, despite the fact that they are automatic and do not need to be fine-tuned to specific datasets.

These findings may seem alarming, but we must bear in mind that this is one dataset with a particular signal to noise ratio. Furthermore, one may argue that since these data have been visible for a while, our choice of $df_j = 4$ for all terms in the full additive fit may have been influenced by past experience. With these caveats in mind, we are still left with some indication that the full additive fit with moderate *df* for each term will not be misleading for a large data set. This simulation certainly should not discourage us and others from refining the existing model selection strategies, and inventing new ones.

APPENDIX A

Data

Here we give listings of two of the smaller datasets that are analysed in this book. These datasets and most of the others discussed in the book are included with the PC software described in Appendix C.

Kyphosis in Laminectomy patients

Data were collected on 83 patients undergoing corrective spinal surgery (Bell *et al.*, 1989). The objective was to determine important risk factors for kyphosis, or the forward flexion of the spine of at least 40 degrees from vertical, following surgery. The risk factors are age in years, the starting vertebrae level of the surgery and the number of vertebrae levels involved. These data are analysed in some detail in Chapter 10.

Kyphosis in laminectomy patients

obs #	age	start	number	kyphosis
1	71	5	3	absent
2	158	14	3	absent
3	128	5	4	present
4	2	1	5	absent
5	1	15	4	absent
6	1	16	2	absent
7	61	17	2	absent
8	37	16	3	absent
9	113	16	2	absent
10	59	12	6	present
11	82	14	5	present
12	148	16	3	absent
13	18	2	5	absent
14	1	12	4	absent
15	243	8	8	absent

Kyphosis in laminectomy patients (continued)

obs #	age	start	number	kyphosis
16	168	18	3	absent
17	1	16	3	absent
18	78	15	6	absent
19	175	13	5	absent
20	80	16	5	absent
21	27	9	4	absent
22	22	16	2	absent
23	105	5	6	present
24	96	12	3	present
25	131	3	2	absent
26	15	2	7	present
27	9	13	5	absent
28	12	2	14	present
29	8	6	3	absent
30	100	14	3	absent
31	4	16	3	absent
32	151	16	2	absent
33	31	16	3	absent
34	125	11	2	absent
35	130	13	5	absent
36	112	16	3	absent
37	140	11	5	absent
38	93	16	3	absent
39	1	9	3	absent
40	52	6	5	present
41	20	9	6	absent
42	91	12	5	present
43	73	1	5	present
44	35	13	3	absent
45	143	3	9	absent
46	61	1	4	absent
47	97	16	3	absent
48	139	10	3	present
49	136	15	4	absent
50	131	13	5	absent
51	121	3	3	present
52	177	14	2	absent
53	68	10	5	absent
54	9	17	2	absent
55	139	6	10	present

Kyphosis in laminectomy patients (continued)

obs #	age	start	number	kyphosis
56	2	17	2	absent
57	140	15	4	absent
58	72	15	5	absent
59	2	13	3	absent
60	120	8	5	present
61	51	9	7	absent
62	102	13	3	absent
63	130	1	4	present
64	114	8	7	present
65	81	1	4	absent
66	118	16	3	absent
67	118	16	4	absent
68	17	10	4	absent
69	195	17	2	absent
70	159	13	4	absent
71	18	11	4	absent
72	15	16	5	absent
73	158	14	5	absent
74	127	12	4	absent
75	87	16	4	absent
76	206	10	4	absent
77	11	15	3	absent
78	178	15	4	absent
79	157	13	3	present
80	26	13	7	absent
81	120	13	2	absent
82	42	6	7	present
83	36	13	4	absent

Diabetes data

These data come from a study (Sockett *et al.*, 1987) of the factors affecting patterns of insulin-dependent diabetes mellitus in children. The objective was to investigate the dependence of the level of serum C-peptide on various other factors in order to understand the patterns of residual insulin secretion. The response measurement is the logarithm of C-peptide concentration (pmol/ml) at diagnosis, and the predictor measurements age and base deficit, a measure of acidity. These two predictors are a subset of those studied in Sockett *et al.* (1987). These data are used to illustrate the various smoothers in Chapter 2, as well as additive models in Chapter 4.

Diabetes data

obs #	age	base deficit	C-peptide	obs #	age	base deficit	C-peptide
1	5.2	-8.1	4.8	23	11.3	-3.6	5.1
2	8.8	-16.1	4.1	24	1.0	-8.2	3.9
3	10.5	-0.9	5.2	25	14.5	-0.5	5.7
4	10.6	-7.8	5.5	26	11.9	-2.0	5.1
5	10.4	-29.0	5.0	27	8.1	-1.6	5.2
6	1.8	-19.2	3.4	28	13.8	-11.9	3.7
7	12.7	-18.9	3.4	29	15.5	-0.7	4.9
8	15.6	-10.6	4.9	30	9.8	-1.2	4.8
9	5.8	-2.8	5.6	31	11.0	-14.3	4.4
10	1.9	-25.0	3.7	32	12.4	-0.8	5.2
11	2.2	-3.1	3.9	33	11.1	-16.8	5.1
12	4.8	-7.8	4.5	34	5.1	-5.1	4.6
13	7.9	-13.9	4.8	35	4.8	-9.5	3.9
14	5.2	-4.5	4.9	36	4.2	-17.0	5.1
15	0.9	-11.6	3.0	37	6.9	-3.3	5.1
16	11.8	-2.1	4.6	38	13.2	-0.7	6.0
17	7.9	-2.0	4.8	39	9.9	-3.3	4.9
18	11.5	9.0	5.5	40	12.5	-13.6	4.1
19	10.6	-11.2	4.5	41	13.2	-1.9	4.6
20	8.5	-0.2	5.3	42	8.9	-10.0	4.9
21	11.1	-6.1	4.7	43	10.8	-13.5	5.1
22	12.8	-1.0	6.6				

APPENDIX B

Degrees of freedom

In this appendix we discuss an approximation for the degrees of freedom df^{err}, based on the trace of the smoother matrix.

Consider the smoother matrix for a cubic smoothing spline with penalty factor λ, denoted by $\mathbf{S}_\lambda$, or $\mathbf{S}$ for short. For model comparisons, the quantity $2\,\text{tr}(\mathbf{S}) - \text{tr}(\mathbf{S}\mathbf{S}^T)$ is needed; this is difficult to compute, while the quantity $\text{tr}(\mathbf{S})$ is easily obtained. Hence we seek an approximation for $2\,\text{tr}(\mathbf{S}) - \text{tr}(\mathbf{S}\mathbf{S}^T)$ based on $\text{tr}(\mathbf{S})$. Figure 3.5 of Chapter 3 suggests an approximation of the form $\{2\,\text{tr}(\mathbf{S}) - \text{tr}(\mathbf{S}\mathbf{S}^T)\} - 2 = k\{\text{tr}(\mathbf{S}) - 2\}$. Empirical inspection suggests the factor $k = 1.25$. The resulting approximation is

$$2\,\text{tr}(\mathbf{S}) - \text{tr}(\mathbf{S}\mathbf{S}^T) \approx 1.25\,\text{tr}\,\mathbf{S} - 0.5 \qquad (B.1)$$

Practical experimentation has verified that $(B.1)$ is an excellent approximation. For a more systematic study of this approximation, we make use of the formulae

$$\begin{aligned} \text{tr}(\mathbf{S}) &\approx 2 + \sum_3^n \frac{1}{1+\lambda\rho_i} \\ \text{tr}(\mathbf{S}\mathbf{S}^T) &\approx 2 + \sum_3^n \frac{1}{(1+\lambda\rho_i)^2} \end{aligned} \qquad (B.2)$$

where $\rho_i = c\pi^4(i-1.5)^4$, $i = 3, \ldots, n$ (Utreras, 1981; Silverman, 1984). Let Q be the corresponding formula for $2\,\text{tr}(\mathbf{S}) - \text{tr}(\mathbf{S}\mathbf{S}^T)$, based on $(B.2)$.

The approximations $(B.2)$ are functions of n and $r = c\lambda$ only. Using these in place of $\text{tr}(\mathbf{S})$ and $\text{tr}(\mathbf{S}\mathbf{S}^t)$, by direct numerical inspection, approximation $(B.1)$ has a maximum error of 0.25 over $20 \le n \le 10000$, and the range of r such that $2 \le Q \le 15$.

The preceding discussion pertained to the unweighted cubic smoothing spline. When observation weights are present, the error degrees of freedom df^{err} is $2\,\text{tr}(\mathbf{S}) - \text{tr}(\mathbf{S}^T\mathbf{W}\mathbf{S}\mathbf{W}^{-1})$, where $\mathbf{W}$ is a diagonal matrix of weights and $\mathbf{S}$ computes a weighted cubic smoothing spline (section 6.8.3). The weighted smoothing spline arises most often in non-Gaussian exponential family models, such as the binary logistic model, where the weights are derived from the fitted values. It is of interest to approximate this quantity as a function of the more easily computable quantity $\text{tr}(\mathbf{S})$. Proceeding as in the unweighted case, we consider the approximation

$$2\,\text{tr}(\mathbf{S}) - \text{tr}(\mathbf{S}^T\mathbf{W}\mathbf{S}\mathbf{W}^{-1}) \approx 1.25\,\text{tr}\,\mathbf{S} - 0.5 \qquad (B.3)$$

Due to the presence of the weights, it is not possible to use $(B.2)$ to assess this approximation. We have however examined the approximation numerically in the binary logistic model. The weight matrix $\mathbf{W}$ is diagonal in that case, with entries proportional to $\hat{p}_i(1-\hat{p}_i)$ where $\hat{p}_i$ is the estimated probability of success for the ith observation. The approximation $(B.3)$ performs quite well over a wide choice of weights, with a maximum error of about 0.37; the error tends to be largest when the ratio of the maximum weight to the minimum weight is large. These approximations are used in the GAIM and S software described in Appendix C.

APPENDIX C

Software

This appendix describes two software systems that are available for fitting generalized additive models.

GAIM

We have written a program called "GAIM", for *Generalized Additive Interactive Modelling*, for use on personal computers under DOS. It is also usable (with limited features) on non-DOS computers. The program has two main components. The first part is a Fortran module that reads a *model* file and data, and produces an analysis-of-deviance table as well as coordinates for plotting the function estimates and their standard errors. This program can be used on any computer having at least 640K of memory. On non-DOS computers a Fortran compiler is also needed. The program uses cubic smoothing splines in its additive-model fits. The choice of models includes exponential family models with flexible specification of the error and link components, as well as the proportional-hazards model. The individual predictors can be modelled by a linear term or a smooth term; categorical predictors are allowed, in which case a coefficient can be fitted for each level.

For those familiar with the old "GAIM", a program that we have distributed informally for a few years, this new version is completely rewritten, has many new features, and is much more efficient.

The second part of the program is a menu interface and graphics package for use under DOS. It was written by Tony Almudevar and features menu-driven input of the model specifications and plotting commands. It requires a micro-computer with DOS and at least 640K of memory. A numeric co-processor is not required but the program runs slowly without one.

Details about GAIM can be obtained from either author.

S software

Here we describe some functions written in the S computing and data analysis language (Becker, Chambers and Wilks, 1988). A forthcoming volume (Chambers, ed., 1990) will describe a collection of S functions for fitting statistical models in S; here we focus on the `glm` and `gam` functions for fitting generalized linear and additive models.

The S language provides a flexible environment for working with, computing on, and displaying data. It features a large variety of data structures that go far beyond the usual scalar, vector and matrix structures common in many packages. Functions are easy to write in S, and those supplied are adequate for most linear algebra operations and standard statistical computations. A large portion of the functions are concerned with graphics, which is one of the most attractive feature of S; it is possible to create static plots and graphs of almost arbitrary complexity.

The syntax of the `glm` and `gam` functions is the same, so we will describe the latter:

```
gam(formula, link, variance, ...)
```

The `formula` argument is probably the most interesting, and has the form `response~` ⟨`additive predictor`⟩. Irrespective of the error model, `response` describes what variable is to be used for the response, and ⟨`additive predictor`⟩ describes symbolically the composition of the additive model. The formula language has a syntax similar to that used in the GLIM language (Baker and Nelder, 1978; Wilkinson and Rogers, 1973), but with added flexibility. We describe it by example.

- `log(ozone) ~ dpg + ibt + ibh`
  The response is the log of `ozone`, and the regression function is linear in `dpg`, `ibt` and `ibh`. All four variables `ozone`, `dpg`, `ibt` and `ibh` are assumed to exist in one of the attached S directories. The predictors can either be a vector, a matrix, a category (a factor in GLIM parlance) or a logical variable. A single coefficient is associated with a vector, one for each of the columns of a matrix, and one for each column of the matrix of contrasts generated for a category or a logical variable.
- `death ~ age * weight`
  The terms on the right parse to the equivalent form `age +`

`weight + age:weight` and consists of a main effect term for `age` and `weight`, plus the interaction `age:weight`. In most cases the interaction amounts to forming the tensor product of the column(s) generated by each constituent term. The interactions can be of arbitrary order. All the operations provided in the GLIM package are available here, such as nesting — `age/weight` — and grouping terms with parentheses — `diagnosis*(weight + age)`.

- `death` $\sim$ `poly(age, 2)` $+$ `log(weight)`
  `poly(age,2)` generates a matrix of evaluated orthogonal polynomials in `age` of degree two. Any S function can be used in the formula, as long as it evaluates to one of the four data types listed above. For example `bs(age, df=5)` creates a basis matrix for a piecewise-cubic regression spline, with five degrees of freedom (excluding the intercept). The two interior knots are placed at the tertiles, and in general for k *df*, $k-3$ interior knots are placed at the $100/(k-2)$th, ..., $100(k-3)/(k-2)$th percentiles of the argument, in this case `age`. Alternatively, an arbitrary knot sequence can be supplied in the `knot` argument to `bs`. Example of other useful functions are `cut` for creating categories, all the standard transformation functions such as `log`, `sqrt`, `sin` etc, and `lag` for creating lagged versions of its argument. The identity function is useful for protecting its arguments from the formula syntax-analyser; for example `I(weight/age)` means a new variable which is the ratio of `age` to `weight`, rather than `age` nested within `weight`.
- `kyphosis` $\sim$ `s(age)` $+$ `s(start, 3)` $+$ `number`
  Some S functions, such as `s` above, are special. It indicates that smoothing splines are to be used in a backfitting algorithm to fit both `age` and `start`, while `number` is fit linearly. The smoothing parameter λ for `age` will be selected so that $df = \mathrm{tr}(\mathbf{S}_\lambda) - 1 = 4$, the default nominal degrees of freedom based on the *marginal* smoother $\mathbf{S}_\lambda$ for smoothing against `age`. In the term for `start`, we use the `df` argument of `s` to specify a value of 3 for *df*. Other smoothers can be used, such as locally-weighted lines (`lo`) and kernel smoothers `k`. It is simple to add new smoothers, and flag them as special in this way. For example, `offset(old.fit)` is also special; it flags the variable `old.fit` as an *offset* in the additive model, and so its coefficient is fixed at one.

The `link` and `variance` arguments allow for arbitrary link/variance combinations. For example, the call

```
fit1←gam(kyphosis ~ s(age)+I(start>12), link=logit)
```

assumes binomial data, and uses the *logit* link and by default the binomial variance function. It will fit a smooth term in `age` and separate intercept for `start` larger than 12. The result of the fit will be stored in the object `fit1`.

Both `link` and `variance` arguments are passed as S lists, whose elements are themselves S functions. For example, the `logit` object contains a function for computing the logit link, its inverse, its derivative, and an expression for obtaining initial fitted values.

Other arguments (...) to `gam` include `weight` for giving prior weights, `subset` for specifying a subset of the data for fitting the model, and `na.action`, a function for dealing with missing data.

The object returned by `gam` is an S list object, with elements such as `coefficients`, `deviance`, `fitted.values` and others that describe the fitted model. Other functions exist for summarizing and displaying the fit. Suppose as above the fitted `gam` object is named `fit1`, then `print(fit1)` will give a brief summary of the fitted model, `summary(fit1)` gives a more detailed summary, and `plot(fit1)` produces plots of the terms comprising the fit. Many of the plots in this monograph, for example those in Figs 10.3–6 were produced using this software.

This appendix is intended to give a taste of the software; a detailed description is given in Chambers (ed. 1990).

References

Akaike, H. (1973) Information theory and an extension of the maximum likelihood principle, in *Second International Symposium on Information Theory.* (eds. B.N. Petrov and F. Csàki) Akademia Kiadó, Budapest, pp. 267-81.

Allen, D.M. (1974) The relationship between variable selection and data augmentation and a method of prediction. *Technometrics* **16**, 125–7.

Anderson, T.W. (1958) *An Introduction to Multivariate Statistical Analysis.* J. Wiley & Sons, New York.

Andrews, D.F. (1971) A note on the selection of data transformations. *Biometrika* **58**, 249–54.

Andrews, D.F. and Herzberg A.M. (1985) *Data: a Collection of Problems from Many Fields for the Student and Research Worker.* Springer-Verlag, New York.

Anscombe, F. and Tukey, J.W. (1963) The examination and analysis of residuals. *Technometrics* **5**, 141–60.

Ansley, C.F. and Kohn, R.E. (1987) Efficient generalized cross-validation for state space models. *Biometrika* **74**, 139–48.

Atilgan, T. (1988) Basis selection for density estimation and regression. AT&T Bell Laboratories technical memorandum.

Atkinson, A.C. (1985) *Plots, Transformations and Regression.* Oxford, Clarendon Press.

Azzalini, A., Bowman, A.W. and Härdle, W. (1989) On the use of nonparametric regression for model checking. *Biometrika* **67**, 1–12.

Bacchetti, P. (1989) Additive isotonic models. *J. Am. Statist. Assoc.* **84**, 289–94.

Baker, R.J. and Nelder, J.A. (1978) *The GLIM System.* Release 3, *Generalized Linear Interactive Modelling.* Numerical Algorithms Group, Oxford.

Barlow, R., Bartholomew, D., Bremner, J. and Brunk, H. (1972) *Statistical Inference Under Order Restrictions.* J. Wiley & Sons, New York.

Barry, D. (1986) Nonparametric Bayesian regression. *Ann. Statist.* **14**, 934–53.

Bates, D.M., Lindstrom, M.J., Wahba, G. and Yandell, B.S. (1987) GCVPACK-routines for generalized cross-validation. *Commun. Statist.-Simula.* **16**, 263–97.

Becker, R., Chambers, J.M. and Wilks. A. (1988) *The S language.* Wadsworth, Belmont CA.

Becker, R., Clark, L. and Pregibon, D. (1989) Tree-based models. Proceedings of the ASA Stat. Comp. Sec., Washington, D.C.

Bell, D., Walker, J., O'Connor, G., Orrel, J. and Tibshirani, R. (1989) Spinal deformation following multi-level thoracic and lumbar laminectomy in children. Submitted for publication.

Bellman, R.E. (1961) *Adaptive Control Processes.* Princeton University Press.

Bickel, P.J., Klaassen, C., Ritov, Y. and Wellner, J. (1990) *Efficient and Adaptive Estimation for Semi-Parametric Models.* Manuscript in preparation.

Bickel, P.J. and Doksum, K.A. (1981) An analysis of transformations revisited. *J. Am. Statist. Assoc.* **76**, 296–311.

Bishop, Y.M.M., Fienberg, S.E. and Holland, P.W. (1975) *Discrete Multivariate Analysis.* MIT Press, Cambridge, Mass.

Bloomfield, P. (1976) *Fourier Analysis of Time Series: an Introduction.* J. Wiley & Sons, New York.

de Boor, C. (1978) *A Practical Guide to Splines.* Springer-Verlag, New York.

Box, G.E.P. and Cox, D.R. (1964) An analysis of transformations. *J. R. Statist. Soc. B* **26**, 211–52.

Box, G.E.P. and Cox, D.R. (1982) An analysis of transformations revisited, rebutted. *J. Am. Statist. Assoc.* **77**, 209–10.

Box, G.E.P. and Hill, W.J. (1974) Correcting inhomogeneity of variance with power transformation weighting. *Technometrics* **16**, 385–9.

Box, G.E.P. and Tidwell, P.W. (1962) Transformations of the independent variables. *Technometrics* **4**, 531–50.

Brant, R. and Tibshirani, R. (1990) Missing covariates and likelihood models. Manuscript in preparation.

Brasher, P. (1989) Residuals for Cox's proportional hazards model. Ph.D. dissertation, University of Toronto.

Breiman, L. (1989) Discussion of "Linear smoothers and additive models" by Buja *et al.Ann. Statist.* **17**, 510–5.

Breiman, L. (1989a) Fitting additive models to regression data. Tech. rep., Dept. of Statistics, Univ. of California, Berkeley.

Breiman, L. (1989b) The Π method for estimating multivariate functions from noisy data. Tech. rep., Dept. of Statistics, Univ. of California, Berkeley.

Breiman, L., Friedman, J.H., Olshen, R. and Stone, C.J. (1984) *Classification and Regression Trees.* Wadsworth, Belmont, CA.

Breiman, L. and Friedman, J.H. (1985) Estimating optimal transformations for multiple regression and correlation (with discussion). *J. Am. Statist. Assoc.* **80**, 580–619.

Breiman, L. and Peters, S. (1988) Comparing automatic bivariate smoothness (a public service enterprise). Technical report No. 161, Dept. of Statistics, Univ. of California, Berkeley.

Breslow, N.S. (1987) Statistical design and analysis of epidemiological studies. Tech. rep. No. 81, Dept. of Biostatistics, University of Washington.

Breslow, N.S. and Day, N.E. (1980) *Statistical Methods in Cancer Research.* **1:** *The Analysis of Case-Control studies.* I.A.R.C, Lyon.

Brillinger, D.R. (1977) Comment on "Consistent nonparametric regression" by C.J. Stone. *Ann. Statist.* **5**, 549–645.

Brillinger, D.R. and Preisler, H.K. (1984) An exploratory analysis of the Joyner-Boore attenuation data. *Bull. Seismological Soc. Amer.* **74**, 1441–50.

Bruntz, S.M., Cleveland, W.S., Kleiner, B. and Warner, J.L. (1974) The dependence of ambient ozone on solar radiation, temperature, and mixing height. In *Symposium on atmospheric diffusion and air pollution.* Boston; American Meteorological Society 125–8.

Buja, A. (1989) Remarks on functional canonical variates, alternating least squares methods, and ACE. Submitted for publication.

Buja, A., Donnell, D. and Stuetzle, W. (1986) Additive principal components. Technical Report, Department of Statistics, University of Washington, Seattle.

Buja, A., Hastie, T. and Tibshirani, R. (1989) Linear smoothers and additive models (with discussion). *Ann. Statist.* **17**, 453–555.

Buja, A. and Kass, R. (1985) Discussion of "Estimating optimal transformations for multiple regression and correlation" by L. Breiman and J. Friedman, *J. Am. Statist. Assoc.* **80**, 580–619.

Burman, P. (1985) Estimation of generalized additive models. Unpublished Ph.D. thesis, Rutgers University, New Brunswick, N.J.

Carroll, R.J. and Ruppert, D. (1981) On prediction and the power family. *Biometrika* **68**, 609–15.

Carroll, R.J. and Ruppert, D. (1984) Power transformations when fitting theoretical models to data. *J. Am. Statist. Assoc.* **79**, 321–8.

Carroll, R.J. and Ruppert, D. (1988) *Transformations and Weighting in Regression.* J. Wiley & Sons, New York.

Chambers, J.M. (1990) (Ed.) *Statistical Models in S.* In preparation.

Chen, Z., Gu, C. and Wahba, G. (1989) Discussion of " Linear smoothers and additive models." by Buja *et al.Ann. Statist.* **17**, 515–21.

Clark, R.M. (1977) Non-parametric estimation of a smooth regression function. *J. R. Statist. Soc. B* **39**, 107–13.

Clayton, D. and Cuzick, J. (1985) The EM algorithm for Cox's regression model using GLIM. *Appl. Statist.* **34**, 148–56.

Cleveland, R.B., Cleveland, W.S., McRae, J.E. and Terpenning, I. (1990) STL: A seasonal trend decomposition procedure based on Loess. To appear, *Journal of Official Statistics.*

Cleveland, W.S. (1979) Robust locally-weighted regression and smoothing scatterplots. *J. Am. Statist. Assoc.* **74**, 829–36.

Cleveland, W.S. and Devlin, S.J. (1988) Locally-weighted regression: an approach to regression analysis by local fitting. *J. Am. Statist. Assoc.* **83**, 597–610.

Cleveland, W.S., Devlin, S.J. and Grosse, E.H. (1988) Regression by local fitting: methods, properties and computational algorithms. *Journal of Econometrics* **37**, 87–114.

Cleveland, W.S., Devlin, S.J. and Terpenning, I. (1982) The SABL seasonal adjustment and calendar adjustment procedures. *Time Series Analysis: Theory and Practice* **1**, 539–64, edited by O.D. Anderson. North-Holland, New York.

Cleveland, W.S., Freeny, A. and Graedal, T.E. (1983) The seasonal decomposition of atmospheric CO_2: information from new approaches to the decomposition of time series. *J. of Geophysical Res.* **88**, No. C15, 10935–46.

Cook, R.D. (1977) Detection of influential observations in linear regression. *Technometrics* **19**, 15–8.

Cook, R.D. and Weisberg, S. (1982) *Residuals and Influence in Regression.* Chapman and Hall, London.

Cox, D.D (1983) Asymptotics for M-type smoothing splines. *Ann. Statist.* **11**, 530–51.

Cox, D.D. (1989) Discussion of "Linear smoothers and additive models." by Buja *et al.Ann. Statist.* **17**, 522–5.

Cox, D.D. and O'Sullivan, F. (1985) Analysis of penalized likelihood type estimators with applications to generalized smoothing in Sobolev Spaces, Tech. Rep. No. 51, Dept. of Statistics, University of California, Berkeley.

Cox, D.D and O'Sullivan, F. (1989) Generalized non-parametric regression via penalized likelihood. Unpublished.

Cox, D.R. (1972) Regression models and life tables (with discussion) *J. R. Statist. Soc. B* **74**, 187–220.

Cox, D.R. and Oakes, D. (1984) *Analysis of Survival Data.* Chapman and Hall, London.

Cox, D.R. and Reid, N. (1987) Orthogonal parameters and approximate conditional inference (with discussion). *J. R. Statist. Soc. B* **49**, 1–39.

Craven, P. and Wahba, G. (1979) Smoothing noisy data with spline functions. *Numerische Mathematik* **31**, 377–403.

Crowley, J. and Storer, B. (1983) Discussion of "A Re-analysis of the Stanford heart transplant data", by M. Aitkin, N. Laird and B. Francis. *J. Am. Statist. Assoc.* **78**, 264–91.

Dagum, E. (1978) Modelling, forecasting and seasonally adjusting economic time series with the X-11 ARIMA method. *The Statistician* **27**, 203–16.

Demmler, A. and Reinsch, C. (1975) Oscillation matrices with spline smoothing. *Numerische Mathematik* **24**, 375–82.

Denby, L. (1986) Smooth regression functions. AT&T Bell Laboratories Statistical Report 26.

Deutsch, F. (1983) von Neumann's alternating method: the rate of convergence. In *Approximation Theory IV.* (C.K. Chui, L.L. Schumaker, and J.D. Ward, eds.), 427–34 Academic Press, New York.

Devlin, S.J. (1986) Locally-weighted multiple regression: statistical properties and a test of linearity. Bell Communications Research technical memorandum.

Devroye, L. P. (1981) On the almost everywhere convergence of nonparametric regression function estimates. *Ann. Statist.* **9**, 1310–9.

Devroye, L. P. and Wagner, T.J. (1980) Distribution-free consistency results in nonparametric discrimination and regression function estimation. *Ann. Statist.* **8**, 231–9.

DiCiccio, T. and Romano, J. (1988) A review of bootstrap confidence intervals. *J. R. Statist. Soc. A* **151**, 338–54.

Duan, N. (1983) Smearing method: a nonparametric retransformation method. *J. Am. Statist. Assoc.* **78**, 605–10.

Duan, N. and Li,. K.C. (1988) Slicing regression: a link-free regression method. UCLA Statistical series No. 25.

Efron, B. (1982) *The Jackknife, Bootstrap, and Other Resampling Plans.* Siam monograph No. 38, CBMS-NSF. Philadelphia.

Efron, B. (1983) Estimating the error rate of a prediction rule: improvements on cross-validation. *J. Am. Statist. Assoc.* **78**, 316–31.

Efron, B. (1986) How biased is the apparent error rate of a prediction rule? *J. Am. Statist. Assoc.* **81**, 461–70.

Efron, B. and Tibshirani, R. (1986) Bootstrap methods for standard errors, confidence intervals, and other measures of statistical accuracy. *Statist. Sci.* **1**, 54–77.

Engle, R.F., Granger, C.W.J., Rice, J.A. and Weiss, A. (1986) Semiparametric estimates of the relation between weather and electricity sales. *J. Am. Statist. Assoc.* **81**, 310–20.

Epanechnikov, V.A. (1969) Nonparametric estimation of a multivariate probability density. *Theor. Prob. Appl.* **14**, 153–8

Eubank, R.L. (1984) The hat matrix for smoothing splines. *Statist. and Prob. Letters* **2**, 9–14.

Eubank, R.L. (1985) Diagnostics for smoothing splines. *J. R. Statist. Soc. B* **47**, 332–41.

Eubank, R.L. (1988) *Smoothing Splines and Nonparametric Regression.* Marcel Dekker, New York and Basel.

Eubank, R.L. and Gunst, R.F. (1986) Diagnostics for penalized least-squares estimators, *Statist. and Prob. Letters* **4**, 265–72.

Ezekiel, M. (1941) *Methods of Correlation Analysis.* 2nd edition. J. Wiley & Sons, New York.

Finney, D.J. (1971) *Probit Analysis.* 3rd edition. Cambridge University Press.

Fowlkes, E.B. (1987) Some diagnostics for binary logistic regression via smoothing. *Biometrika* **74**, 503–15.

Fowlkes, E.B. and Kettenring, J. (1985) Discussion of "Estimating optimal transformations for multiple regression and correlation" by L. Breiman and J. Friedman. *J. Am. Statist. Assoc.* **80**, 580–619.

Fraser, D.A.S. (1967) Data transformations and the linear model. *Ann. Math. Statist.* **38**, 1456–65.

Friedman, J.H. (1984) A variable span smoother. Tech. rep. LCS5, Dept. of Statistics, Stanford University.

Friedman, J.H. (1990) Multivariate additive regression splines. To appear, *Ann. Statist.*

Friedman, J.H. and Silverman, B.W. (1989) Flexible parsimonious smoothing and additive modelling (with discussion). *Technometrics* **31**, 3–39.

Friedman, J.H. and Stuetzle, W. (1981) Projection pursuit regression. *J. Am. Statist. Assoc.* **76**, 817–23.

Friedman, J.H. and Stuetzle, W. (1982) Smoothing of scatterplots. Department of Statistics Technical Report, Orion 3, Stanford University.

Gasser, Th. and Muller, H.G. (1979) Kernel estimation of regression functions. In *Smoothing Techniques for Curve Estimation*, Gasser and Rosenblatt, eds. Lecture notes in Math. 757, 23-68, Springer-Verlag, New York.

Gasser, Th. and Muller, H.G. (1984) Estimating regression functions and their derivatives by the kernel method. *Scand. J. Statist.* **11**, 171–85.

Gasser, Th., Muller, H.G. and Mammitzsch, V. (1985) Kernels for nonparametric curve estimation. *J. R. Statist. Soc. B* **47**, 238–52.

Gennings, C. and Hastie, T. (1990) Generalized additive models with cross-product terms: a nonparametric approach to modelling drug interactions. Unpublished manuscript.

Gentleman, R. (1988) Non-linear covariates in the proportional hazards model. Tech. rep. STAT-88-14, Univ. of Waterloo.

Golub, G.H., Heath, M. and Wahba, G. (1979) Generalized cross-validation as a method for choosing a good ridge parameter. *Technometrics* **21**, 213–23.

Golub, G.H. and Van Loan, C.F. (1983) *Matrix Computations.* Johns Hopkins University Press, Baltimore, MD.

Good, I.J. and Gaskins, R.A. (1971) Non-parametric roughness penalties for probability densities. *Biometrika* **58**, 255–77.

Goodall, C. (1990) A survey of smoothing techniques. In *Modern Methods in Data Analysis.* Fox and Long, Eds. Sage.

Green, P.J. (1984) Iteratively reweighted least-squares for maximum likelihood estimation, and some robust and resistant alternatives (with discussion). *J. R. Statist. Soc. B* **46**, 149–92.

Green, P.J. (1985) Linear models for field trials, smoothing and cross-validation. *Biometrika* **72**, 527–37.

Green, P.J. (1987) Penalized likelihood for general semi-parametric regression models. *Int. Statist. Rev.* **55**, 245–60.

Green, P.J., Jennison, C. and Seheult, A. (1985) Analysis of field experiments by least squares smoothing. *J. R. Statist. Soc. B* **47**, 299–315.

Green, P.J. and Yandell, B. (1985) Semi-parametric generalized linear models. *Proceedings 2nd International GLIM Conference.* Lancaster, Lecture notes in Statistics No. 32 44–55 Springer-Verlag, New York.

Greenacre, M.J. (1984) *Theory and Application of Correspondence Analysis.* Academic Press, New York.

Grenander, U. and Szego, G. (1958) *Toeplitz Forms and Their Applications.* University of California Press, Los Angeles.

Gu, C. (1989) Penalized likelihood regression: a Bayesian analysis. Tech. rep. Dept. of Statistics, University of Wisconsin.

Gu, C., Bates, D.M., Chen, Z. and Wahba, G. (1988) The computation of GCV functions through Householder tridiagonalization with application to fitting of interaction spline models. Technical report No. 823, Dept. of Statistics, Univ. of Wisconsin, Madison.

Gu, C. and Wahba, G. (1988) Minimizing GCV/GML scores with multiple smoothing parameters via the Newton method. Tech. rep. 847, Univ. of Wisconsin, Madison.

Hampel, F.R., Ronchetti, E.M., Rousseeuw, P.J. and Stahel, W.A. (1986) *Robust Statistics: the Approach Based on Influence Functions.* J. Wiley & Sons, New York.

Härdle, W. (1986) Resistant smoothing using the fast Fourier transform. *Appl. Statist.* **36**, 104–11.

Härdle, W. (1989) Investigating smooth multiple regression by the method of average derivatives. *J. Am. Statist. Assoc.* **84**, 986–95.

Härdle, W. (1990) *Applied Non-parametric Regression.* Oxford University Press.

Härdle, W., Hall. P. and Marron, S. (1988) How far are the optimally chosen smoothing parameters from their optimum? *J. Am. Statist. Assoc.* **83**, 86–95.

Härdle, W., and Marron, S. (1985) Optimal bandwidth selection in nonparametric regression function estimation. *Ann. Statist.* **13**, 1465–81.

Hastie, T. (1988) Pseudo smoothers and additive model approximations. Submitted for publication.

Hastie, T. (1989) Discussion of "Flexible parsimonious smoothing and additive modelling" by J. Friedman and B. Silverman. *Technometrics* **31**, 3–39.

Hastie, T., Botha, J.L. and Schnitzler, C.M. (1989) Regression with an ordered categorical response. *Statistics in Medicine* **8**, 785–94.

Hastie, T. and Pregibon, D. (1988) A new algorithm for matched case-control studies with applications to additive models. Proceedings COMPSTAT 88, Copenhagen.

Hastie, T. and Tibshirani, R. (1984) Generalized additive models. Tech, rep. 98, Dept. of Statistics, Stanford University.

Hastie, T. and Tibshirani, R. (1985) Generalized additive models: some applications. *Proceedings 2nd International GLIM Conference.* Lancaster. Springer lecture notes in Statistics No. 32, Berlin, Heidelberg.

Hastie, T. and Tibshirani, R. (1986) Generalized additive models (with discussion) *Statist. Sci.* **1**, 297–318.

Hastie, T. and Tibshirani, R. (1987a) Generalized additive models: some applications, *J. Am. Statist. Assoc.* **82**, 371–86.

Hastie, T. and Tibshirani, R. (1987b) Non-parametric logistic and proportional-odds regression. *Appl. Statist.* **36**, 260–76.

Hastie, T. and Tibshirani, R. (1988) Comment on "Monotone splines in action" by J. Ramsay. *Statist. Sci.* **3**, 450–6.

Hastie, T. and Tibshirani, R. (1990) Exploring the nature of covariate effects in the proportional hazards model. To appear, *Biometrics* .

Hawkins, D. (1989) Discussion of "Flexible parsimonious smoothing and additive modelling" by J. Friedman and B. Silverman. *Technometrics* **31**, 3–39.

Heckman, N.E. (1986) Spline smoothing in a partly linear model. *J. R. Statist. Soc. B* **48**, 244–8.

Heckman, N.E. (1988) Minimax estimation in a semi-parametric model. *J. Am. Statist. Assoc.* **83**, 1090–6.

Herman, A.A. and Hastie, T. (1990) An analysis of gestational age, neonatal size and neonatal death using non-parametric logistic regression. To appear, *J. Clin. Epidem.*

Hinkley, D.V. and Runger, G. (1984) Analysis of transformed data (with discussion). *J. Am. Statist. Assoc.* **79**, 302–8.

Hoaglin, D.C., Mosteller, F. and Tukey, J.W. (1983) *Understanding Robust and Exploratory Data Analysis.* J. Wiley & Sons, New York.

Huber, P.J. (1964) Robust estimation of a location parameter. *Ann. Math. Statist.* **35**, 73–101.

Huber, P.J. (1979) Robust smoothing. In *Robustness in statistics*, 33–47. R.C. Launer and G.N. Wilkinson Eds. Academic Press, New York.

Huber, P.J. (1981) *Robust Statistics.* Wiley, New York.

Hurn, M.W., Barker, N.W. and Magath, T.D. (1945) The determination of prothrombin time following the administration of dicumarol with specific reference to thromboplastin. *J. Lab. Clin. Med.* **30**, 432–47.

Jørgensen, B. (1984) The delta algorithm and GLIM. *Int. Statist. Rev.* **52**, 283–300.

Kalbfleisch, J.D. and Prentice, R.L. (1980) *The Statistical Analysis of Failure Time Data.* J. Wiley & Sons, New York.

Keeling, C.D., Bacastow, R.B., Whorf, T.P. (1982) Measurements of the concentration of carbon dioxide at Mauna Loa observatory, Hawaii. *Carbon Dioxide Review*: 1982, edited by W.C. Clark, 317–85. Oxford University Press.

Kelly, C. and Rice, J.A. (1988) Monotone smoothing with application to dose response curves and the assessment of synergism. Technical report, Dept. of Math., Univ. of California, San Diego.

Kimeldorf, G.S. and Wahba, G. (1970) A correspondence between Bayesian estimation on stochastic processes and smoothing by splines. *Ann. Math. Statist.* **2**, 495–502.

Kimeldorf, G.S. and Wahba, G. (1971) Some results on Tchebysheffian spline functions.*J. Math. Anal. Appl.* **33**, 82–95.

Knuth, D.E. (1986) *The TEXbook.* Reading, Mass. Addison-Wesley, Reading, MA.

Kohn, R.E. and Ansley, C.F. (1989) Discussion of " Linear smoothers and additive models." by Buja *et al.Ann. Statist.* **17**, 535–40.

Krasker, W.S. and Welsch, R. E. (1982) Efficient bounded-influence regression estimation. *J. Am. Statist. Assoc.* **77**, 595–604.

Kruskal, J. (1965) Analysis of factorial experiments by estimating monotone transformations of the data. *J. R. Statist. Soc. B* **27**, 251–63.

Lambert, D. (1989) Nondetects, detection limits, and the probability of detection. AT&T Bell Laboratories technical memorandum.

Landwehr, J.M., Pregibon, D. and Shoemaker, A. (1984) Graphical methods for assessing logistic regression models. *J. Am. Statist. Assoc.* **79**, 61–3.

Launer, R.L. and Wilkinson, G.N. (eds) (1979) *Robustness in Statistics.* Academic Press, New York.

Lawless, J.F. (1982) *Statistical Models and Methods for Lifetime Data.* J. Wiley & Sons, New York.

de Leeuw, J., Young, F.W. and Takane, Y. (1976) Additive structure in qualitative data: an alternating least-squares method with optimal scoring features. *Psychometrika* 1976, 471–503.

Li, K.C. (1984) Consistency for cross-validated nearest-neighbour estimates in nonparametric regression. *Ann. Statist.* **12**, 230–40.

Li, K.C. (1986) Asymptotic optimality for C_L, and generalized cross-validation in ridge regression with application to spline smoothing. *Ann. Statist.* **14**, 1101–12.

Li, K.C. and Duan, N. (1989) Regression analysis under link violation. *Ann. Statist.* **17**, 1009–52.

Linsey, J.K. (1972) Fitting response surfaces with power transformations. *Appl. Statist.* **21**, 234–7.

Little, R. J. A. and Rubin, D. B. (1986) *Statistical Analysis with Missing Data.* J. Wiley & Sons, New York.

Mack, Y. (1981) Local properties of K-NN regression estimates. *SIAM J. Alg. Disc. Meth.* **2**, 311–23.

Mallows, C.L. (1973) Some Comments on C_p. *Technometrics* **15**, 661–7.

Mallows, C.L. (1980) Some theory of nonlinear smoothers. *Ann. Statist.* **8**, 695–715.

Mallows, C.L. (1986) Augmented partial residuals. *Technometrics* **28**, 313–20.

Mardia, K.V., Kent, J.T. and Bibby, J.M. (1979) *Multivariate Analysis.* Academic Press, New York.

Matheron, G. (1973) The intrinsic random functions and the applications. *Adv. Appl. Prob.* **5**, 439–68.

McCullagh, P. (1980) Regression models for ordinal data (with discussion) *J. R. Statist. Soc. B* **42**, 109–42.

McCullagh, P. and Nelder, J.A. (1989) *Generalized Linear Models (2nd ed.).* Chapman and Hall, London.

Miller, R.G. (1981) *Survival Analysis.* J. Wiley & Sons, New York.

Mo, M. (1989) Robust additive regression. Ph.D. dissertation, Dept. of Statistics, Univ, of California, Berkeley.

Morgan, J.N. and Sonquist, J.A. (1963) Problems in the analysis of survey data, and a proposal. *J. Am. Statist. Assoc.* **58**, 415–34.

Mosteller, F. and Tukey, J.W. (1977) *Data Analysis and Regression.* Addison-Wesley, Reading, MA.

Müller, H.G. (1985) Empirical bandwidth choice for nonparametric regression by means of pilot estimators. *Statistics and decisions.* **2**, 193–206.

Müller, H.G. (1987) Weighted local regression and kernel methods for nonparametric curve fitting. *J. Am. Statist. Assoc.* **82**, 231–8.

Nadaraya, E.A. (1964) On estimating regression. *Theor. Prob. Appl.* **9**, 141–2.

Nelder, J.A. (1989) Contribution to GLIM Newsletter No. 18. Numerical Algorithms Group, Royal Statistical Society.

Nelder, J.A. and Wedderburn, R.W.M. (1972) Generalized linear models. *J. R. Statist. Soc. A* **135**, 370–84.

Nychka, D. (1988) Bayesian "confidence" intervals for smoothing splines. *J. Am. Statist. Assoc.* **83**, 1134–43.

O'Sullivan, F. (1983) The analysis of some penalized likelihood estimation schemes. Statistics Department Technical Report No. 726, University of Wisconsin, Madison.

O'Sullivan, F. (1985) Discussion of "Some aspects of the spline smoothing approach to nonparametric regression curve fitting" by B.W. Silverman. *J. R. Statist. Soc. B* **36**, 111–47.

O'Sullivan, F. (1986) Estimation of densities and hazards by the method of penalized likelihood. *Department of Statistics Technical Report No. 58.* University of California, Berkeley.

O'Sullivan, F. (1988) Nonparametric estimation of relative risk using splines and cross-validation. *SIAM J. Sci. and Statist. Comput.* **9**, 531–42.

O'Sullivan, F., Yandell, B. and Raynor, W. (1986) Automatic smoothing of regression functions in generalized linear models. *J. Am. Statist. Assoc.* **81**, 96–103.

Ortega, J.M. and Rheinboldt, W.C. (1970) *Iterative Solution of Nonlinear Equations in Several Variables.* Academic Press, New York.

Papadakis, J. (1937) Methode statistique pour des experiences sur champ. *Bulliten Institute Plantes a Salonique*, 23.

Peto, R. (1972) Contribution to the discussion "Regression models and life tables" by D.R. Cox. *J. R. Statist. Soc. B* **74**, 205–7.

Prakasa Rao, B.L.S. (1983) *Nonparametric Functional Estimation.* Academic Press, New York.

Pregibon, D. (1979) Data analytic methods for generalized linear models. Ph.D. dissertation, University of Toronto.

Pregibon, D. (1980) Goodness of link tests for generalized linear models. *Appl. Statist.* **29**, 15–24.

Pregibon, D. (1982) Resistant fits for some commonly used logistic models with medical applications. *Biometrics* , **38**, 485–98.

Pregibon, D. and Vardi, Y. (1985) Discussion of "Estimating optimal transformations for multiple regression and correlation" by L. Breiman and J. Friedman. *J. Am. Statist. Assoc.* **80**, 580–619.

Priestley, M.B. and Chao, M.T. (1972) Non-parametric function fitting. *J. R. Statist. Soc. B* **4**, 385–92.

Ramsay, J. (1988) Monotone splines in action (with discussion). *Statist. Sci.* **3**, 450–6.

Rao, C.R. (1973) *Linear Statistical Inference and its Applications (second edition).* J. Wiley & Sons, New York.

Reinsch, C. (1967) Smoothing by spline functions. *Numer. Math.* **10**, 177–83.

Rice, J.A. (1984) Bandwidth choice for nonparametric regression. *Ann. Statist.* **12**, 1215–30.

Rice, J.A. (1986) Comment on "A statistical perspective of ill-posed inverse problems" by O'Sullivan, F. *Statist. Sci.* **1**, 502–27.

Rice, J.A. and Rosenblatt, M. (1983) Smoothing splines, regression, derivatives and convolution. *Ann. Statist.* **11**, 141–56.

Rosenblatt, M. (1971) Curve Estimates. *Ann. Math. Statist.* **42**, 1815–42.

Rousseauw, J., du Plessis, J., Benade, A., Jordaan, P., Kotze, J., Jooste, P. and Ferreira, J. (1983) Coronary risk factory screening in three rural communities. *South African medical journal* **64**, 430–6.

Rousseeuw, P. (1984) Least median of squares regression. *J. Am. Statist. Assoc.* **79**, 871–80.

Schoenberg, I.J. (1964) Spline functions and the problem of graduation. *Proc. Nat. Academy. Sci.* USA. **52**, 947–50.

Shiau, J. and Wahba, G. (1988) Rates of convergence for some estimates of a semi-parametric model. *Commun. Statist.-Simula.* **17**, 111–3.

Shiskin, J., Young, A.H. and Musgrave, J.C. (1967) An X-11 variant of the census method 11 seasonal adjustment program. Bureau of Census Technical paper, U.S. Dept. of Commerce, Washington.

Silverman, B.W. (1982) Kernel density estimation using the fast Fourier transform. *Appl. Statist.* **31**, 93–9.

Silverman, B.W. (1984) A fast and efficient cross-validation method for smoothing parameter choice in spline regression. *J. Am. Statist. Assoc.* **79**, 584–9.

Silverman, B.W. (1984a) Spline smoothing: the equivalent variable kernel method. *Ann. Statist.* **12**, 898–916.

Silverman, B.W. (1985) Some aspects of the spline smoothing approach to nonparametric regression curve fitting (with discussion). *J. R. Statist. Soc. B* **47**, 1–52.

Sleeper, L. and Harrington, D. (1989) B-spline models for non-linear covariate effects in Cox's regression model. Unpublished.

Smith, P.L. (1982) Curve fitting and modelling with splines using statistical variable selection techniques. NASA report 166034, Langely Research Centre. Hampton, VA.

Sockett, E. B., Daneman, D., Clarson, C. and Ehrich, R. M. (1987) Factors affecting and patterns of residual insulin secretion during the first year of type I (insulin dependent) diabetes mellitus in children. *Diabet.* **30**, 453–9.

Speckman, P.E. (1988) Regression analysis for partially linear models. *J. R. Statist. Soc. B* **50**, 413–36.

Srivastava, M.S. and Khatri, C.G. (1979) *An Introduction to Multivariate Statistics.* North Holland, New York.

Staniswalis, J. (1989) The kernel estimate of a regression function in likelihood-based models. *J. Am. Statist. Assoc.* **84**, 276–83.

Stein. M (1988) Asymptotically efficient spatial interpolation with a misspecified covariance function. *Ann. Statist.* **16**, 55–63.

Stone, C. J. (1977) Consistent nonparametric regression (with discussion). *Ann. Statist.* **5**, 549–645.

Stone, C. J. (1980) Optimal rates of convergence for nonparametric estimators. *Ann. Statist.* **8**, 1348–60.

Stone, C. J. (1982) Optimal global rates of convergence for nonparametric regression. *Ann. Statist.* **10**, 1040–53.

Stone, C.J. (1985) Additive regression and other nonparametric models. *Ann. Statist.* **13**, 689–705.

Stone, C.J. (1986) The dimensionality reduction principle for generalized additive models. *Ann. Statist.* **14**, 590–606.

Stone, C.J. and Koo, C.Y. (1985) Additive splines in statistics. Proceedings of the Stat. Comp. Sec., ASA 45-8.

Stone, M. (1974) Cross-validatory choice and assessment of statistical predictions (with discussion). *J. R. Statist. Soc. B* **36**, 111–47.

Stone, M. (1977) An asymptotic equivalence of choice of model by cross-validation and Akaike's criterion. *J. R. Statist. Soc. B* **39**, 44–7.

Thisted, R.A. (1988) *Elements of Statistical Computing.* Chapman and Hall, London.

Tibshirani, R. (1984) Local Likelihood Estimation. Ph.D. dissertation, Dept. of Statistics, Stanford University.

Tibshirani, R. (1988) Estimating optimal transformations for regression via additivity and variance stabilization. *J. Am. Statist. Assoc.* **83**, 394–405.

Tibshirani, R. and Hastie, T. (1987) Local likelihood estimation. *J. Am. Statist. Assoc.* **82**, 559–68.

Titterington, D.M. (1985) Common structure of smoothing techniques in statistics. *Int. Statist. Rev.* **53**, 141–70

Tukey, J.W. (1977) *Exploratory Data Analysis.* Addison-Wesley, Reading, MA.

Tukey, J.W. (1982) The use of smelting to guide re-expression. In *Modern Data Analysis.* Launer and Siegel (eds.), Academic Press, New York.

Utreras, F.D. (1979) Cross-validation techniques for smoothing spline functions in one or two dimensions, In *Smoothing techniques for curve estimation.* (Gasser and Rosenblatt, eds.) Lecture notes in Math. 757,196–232, Springer-Verlag, New York.

Utreras, F.D. (1980) Sur le choix du parameter d'adjustement dans le lissage par fonctions spline. *Numer. Math* **34**, 15–28.

Wahba, G. (1975) Smoothing noisy data with spline functions. *Numer. Math.* **24**, 383–93.

Wahba, G. (1978) Improper priors, spline smoothing and the problem of guarding against model errors in regression. *J. R. Statist. Soc. B* **40**, 364–72.

Wahba, G. (1980) Spline bases, regularization, and generalized cross-validation for solving approximation problems with large quantities of noisy data, *Proceedings of the International Conference on Approximation theory in honour of George Lorenz.* Jan 8-10, Austin TX, (ed. Ward Chaney) Academic Press, New York.

Wahba, G. (1983) Bayesian "confidence intervals" for the cross-validated smoothing spline. *J. R. Statist. Soc. B* **45**, 133–50.

Wahba, G. (1986) Partial and interaction splines for semi-parametric estimation of functions of several variables. *Department of Statistics Technical report No. 784.* University of Wisconsin, Madison.

Wahba, G. (1990) *Spline Functions for Observational Data.* CBMS-NSF Regional Conference series, SIAM. Philadelphia.

Wahba, G. and Wold, S. (1975) A completely automatic French curve: fitting splines functions by cross-validation. *Commun. Statist* **4**, 1–17.

Walker, E. and Birch, J. (1988) Influence measures in ridge regression. *Technometrics* **30**, 221–7.

Watson, G.S. (1964) Smooth regression analysis. *Sankhya* A **26**, 359–72.

Watson, G.S. (1984) Smoothing and interpolation by Kriging and with splines. *J. Int. Assoc. Math. Geol.* **16**, 601–15.

Wecker, W.E. and Ansley, C.F. (1983) The signal extraction approach to nonlinear regression and spline smoothing. *J. Am. Statist. Assoc.* **78**, 81–9.

Wedderburn, R.W.M. (1974) Quasi-likelihood functions, generalized linear models and the Gauss-Newton method. *Biometrika* **61**, 439–47.

Wegman, E.J. and Wright, I.W. (1983) Splines in statistics. *J. Am. Statist. Assoc.* **78**, 351–65.

Weinert, H.L., Byrd, R.H. and Sidhu, G.S. (1980) A stochastic framework for recursive computation of spline functions: part II, smoothing splines. *J. Optim. Theor, Appl.* **30**, 255–68.

Whitehead, J. (1980) Fitting Cox's regression model to survival data using GLIM. *Appl. Statist.* **29**, 268–75.

Whittaker, E. (1923) On a new method of graduation. Proceedings of the Edinburgh Mathematics Society **41**, 63–75.

Wilkinson, G.N. and Rogers, C.E. (1973) Symbolic description of factorial models for analysis of variance. *Appl. Statist.* **22**, 392–9.

Williams, W.G., Rebeyka, I.M., Tibshirani, R., Coles, J., Lightfoot, N.E., Freedom, R.M. and Trusler, G. (1990) Warm induction cardioplegia in the infant: a technique to avoid rapid cooling myocardial contracture. To appear *J. Thorac. and Cardio. Surg.*

Wold, S. (1974) Spline functions in data analysis. *Technometrics* **16**, 1–11.

Wong, W.H. (1983) On the consistency of cross-validation in kernel nonparametric regression. *Ann. Statist.* **11**, 1136–41.

Yandell, B.S. and Green, P.J. (1986) Semi-parametric generalized linear model diagnostics. *Proc. Statist. Comp. Sec.* 48–53, ASA annual meeting, Chicago.

Young, F.W. de Leeuw, J. and Takane, Y. (1976) Regression with qualitative and quantitative variables: an alternating least-squares method with optimal scaling features. *Psychometrika* 505–29.

Author index

Akaike, H., 170, 311
Allen, D.M., 79, 311
Almudevar, T., 307
Anderson, T.W., 196, 311
Andrews, D.F., 196, 234, 311
Anscombe, F., 196, 311
Ansley, C.F., 78, 130, 131, 311, 318, 323
Atilgan, T., 247, 277, 311
Atkinson, A.C., 196, 311
Azzalini, A., 170, 311

Bacastow, R.B., 224, 226, 318
Bacchetti, P., 170, 311
Baker, R.J., 102, 308, 311
Barker, N.W., 141, 318
Barlow, R., 196, 311
Barry, D., 278, 311
Bartholomew, D., 196, 311
Bates, D.M., 79, 170, 311, 317
Becker, R., xv, 77, 102, 278, 308, 312
Bell, D., 282, 301, 312
Bellman, R.E., 83, 312
Benade, A., 6, 321
Bibby, J.M., 196, 319
Bickel, P.J., 131, 196, 312
Birch, J., 278, 323
Bishop, Y.M.M., 117, 312
Bloomfield, P., 79, 312
Botha, J.L., 232, 317
Bowman, A.W., 170, 311
Box, G.E.P., 196, 295, 312
Brant, R., 170, 312
Brasher, P., 231, 312
Breiman, L., 5, 7, 35, 102, 130, 179, 196, 197, 274, 277, 278, 294, 295, 312
Bremner, J., 196, 311
Breslow, N.S., 95, 202, 231, 313
Brillinger, D.R., 170, 196, 313
Brunk, H., 196, 311
Bruntz, S.M., 257, 313
Buja, A., 29, 78, 80, 129, 130, 131, 132, 133, 169, 196, 278, 295, 313
Burman, P., 170, 313
Byrd, R.H., 78, 323

Carroll, R.J., 196, 313
Chambers, J.M., xv, 77, 102, 278, 285, 308, 310, 312, 313
Chao, M.T., 77, 320
Chen, Z., 112, 131, 170, 278, 313, 317
Clark, L., 278, 312
Clark, R.M., 77, 78, 313
Clarson, C., 6, 304, 321
Clayton, D., 232, 313
Cleveland, R.B., 224, 232, 313
Cleveland, W.S., 17, 29, 35, 75, 77, 224, 226, 232, 257, 277, 278, 279, 313, 314
Coles, J., 7, 143, 148, 323
Cook, R.D., 196, 256, 314
Cox, D.D., 56, 131, 150, 170, 314
Cox, D.R., 196, 231, 312, 314
Craven, P., 76, 80, 314
Crowley, J., 231, 314
Cuzick, J., 232, 313

Dagum, E., 232, 314
Daneman, D., 6, 304, 321
Day, N.E., 95, 202, 231, 313
de Boor, C., 25, 76, 312
de Leeuw, J., 196, 319, 324
Demmler, A., 57, 80, 314
Denby, L., 102, 131, 154, 169, 314
Deutsch, F., 132, 314

Devlin, S.J., 35, 77, 232, 257, 278, 313, 314
Devroye, L.P., 77, 315
DiCiccio, T., 78, 315
Doksum, K.A., 196, 312
Donnell, D., 131, 313
Duan, N., 170, 195, 197, 315, 319
du Plessis, J., 6, 321

Efron, B., 78, 296, 298, 315
Ehrich, R.M., 6, 304, 321
Engle, R.F., 102, 131, 315
Epanechnikov, V.A., 18, 315
Eubank, R.L., 35, 57, 76, 77, 131, 278, 315
Ezekiel, M., 77, 315

Ferreira, J., 6, 321
Fienberg, S.E., 117, 312
Finney, D.J., 242, 315
Fowlkes, E.B., 170, 196, 315
Fraser, D.A.S., 196, 315
Freedom, R.M., 7, 143, 148, 323
Freeny, A., 226, 314
Friedman, J.H., 5, 7, 78, 102, 130, 179, 196, 197, 249, 274, 275, 277, 278, 295, 296, 312, 315, 316

Gaskins, R.A., 170, 316
Gasser, Th., 77, 316
Gennings, C., 278, 316
Gentleman, R., 170, 231, 316
Golub, G.H., 80, 132, 316
Good, I.J., 170, 316
Goodall, C., 79, 316
Graedal, T.E., 226, 314
Granger, C.W.J., 102, 131, 315
Green, P.J., 37, 102, 131, 132, 154, 169, 170, 171, 278, 316, 324
Greenacre, M.J., 198, 316
Grenander, U., 234, 317
Grosse, E.H., 35, 77, 257, 314
Gu, C., 112, 131, 157, 160, 170, 262, 278, 313, 317
Gunst, R.F., 278, 315

Hall, P., 52, 78, 317
Hampel, F.R., 75, 317
Härdle, W., 35, 37, 52, 77, 78, 102, 170, 311, 317
Harrington, D., 232, 278, 321
Hastie, T., 7, 29, 77, 78, 80, 102, 129, 130, 131, 132, 133, 167, 169, 170, 172, 214, 231, 232, 253, 255, 278, 295, 313, 316, 317, 318, 322
Hawkins, D., 295, 296, 318
Heath, M., 80, 316
Heckman, N.E., 131, 169, 318
Herman, A.A., 170, 318
Herzberg, A.M., 234, 311
Hill, W.J., 196, 312
Hinkley, D.V., 196, 318
Hoaglin, D.C., 196, 318
Holland, P.W., 117, 312
Huber, P.J., 75, 79, 236, 277, 279, 318
Hurn, M.W., 141, 318

Jennison, C., 37, 102, 316
Jooste, P., 6, 321
Jordaan, P., 6, 321
Jørgensen, B., 171, 207, 231, 318

Kalbfleisch, J.D., 214, 215, 231, 318
Kass, R., 196, 313
Keeling, C.D., 224, 226, 318
Kelly, C., 189, 318
Kent, J.T., 196, 319
Kettenring, J., 196, 315
Khatri, C.G., 196, 321
Kimeldorf, G.S., 76, 77, 131, 318
Klaassen, C., 131, 312
Kleiner, B., 257, 313
Knuth, D.E., xv, 318
Kohn, R.E., 78, 131, 311, 318
Koo, C.Y., 247, 277, 322
Kotze, J., 6, 321
Krasker, W.S., 277, 318
Kruskal, J., 196, 319

Lambert, D., 170, 319
Landwehr, J.M., 102, 319
Launer, R.L., 277, 319
Lawless, J.F., 231, 319
Li, K.C., 76, 78, 170, 315, 319
Lightfoot, N.E., 7, 143, 148, 323
Lindstrom, M.J., 79, 311
Linsey, J.K., 196, 319
Little, R.J.A., 170, 319

Mack, Y., 78, 319
Magath, T.D., 141, 318
Mallows, C.L., 79, 102, 319
Mammitzsch, V., 77, 316
Mardia, K.V., 196, 319
Marron, S., 52, 78, 317
Matheron, G., 78, 319
McCullagh, P., 102, 141, 164, 169, 172, 219, 220, 232, 319
McRae, J.E., 224, 232, 313
Miller, R.G., 231, 319
Mo, M., 277, 319
Morgan, J.N., 278, 319
Mosteller, F., 196, 318, 319
Müller, H.G., 35, 77, 78, 316, 320
Musgrave, J.C., 131, 232, 321

Nadaraya, E.A., 37, 77, 320
Nelder, J.A., 102, 141, 164, 169, 172, 219, 220, 232, 308, 311, 319, 320
Nychka, D., 60, 320

O'Connor, G., 282, 301, 312
O'Sullivan, F., 49, 102, 131, 150, 169, 170, 205, 231, 233, 314, 320
Oakes, D., 231, 314
Olshen, R., 102, 274, 278, 312
Orrel, J., 282, 301, 312
Ortega, J.M., 151, 320
Owen, A., 197

Papadakis, J., 130, 320
Peters, S., 35, 312
Peto, R., 212, 320
Prakasa Rao, B.L.S., 78, 320
Pregibon, D., 102, 170, 196, 197, 198, 231, 240, 242, 277, 278, 312, 317, 319, 320
Preisler, H.K., 196, 313
Prentice, R.L., 214, 215, 231, 318
Priestley, M.B., 77, 320

Ramsay, J., 170, 189, 321
Rao, C.R., 56, 321
Raynor, W., 102, 169, 170, 320
Rebeyka, I.M., 7, 143, 148, 323
Reid, N., 196, 314
Reinsch, C., 36, 57, 76, 80, 314, 321
Rheinboldt, W.C., 151, 320
Rice, J.A., 64, 76, 77, 102, 131, 169, 189, 315, 318, 321
Ritov, Y., 131, 312
Rogers, C.E., 308, 323
Romano, J., 78, 315
Ronchetti, E.M., 75, 317
Rosenblatt, M., 76, 77, 321
Rousseauw, J., 6, 321
Rousseeuw, P.J., 75, 317, 321
Rubin, D.B., 170, 319
Runger, G., 196, 318
Ruppert, D., 196, 313

Schnitzler, C.M., 232, 317
Schoenberg, I.J., 76, 321
Seheult, A., 37, 102, 316
Shiau, J., 131, 321
Shiskin, J., 131, 232, 321
Shoemaker, A., 102, 319
Sidhu, G.S., 78, 323
Silverman, B.W., 29, 35, 37, 55, 76, 77, 78, 81, 249, 277, 295, 296, 305, 316, 321
Sleeper, L., 232, 278, 321
Smith, P.L., 277, 321
Sockett, E.B., 6, 304, 321
Sonquist, J.A., 278, 319
Speckman, P.E., 131, 154, 169, 170, 321
Srivastava, M.S., 196, 322
Stahel, W.A., 75, 317
Staniswalis, J., 170, 322
Stein, M., 78, 322
Stone, C.J., 78, 102, 131, 148, 170, 247, 274, 277, 278, 312, 322
Stone, M., 79, 322
Storer, B., 231, 314
Stuetzle, W., 78, 102, 130, 131, 313, 316
Szego, G., 234, 317

Takane, Y., 196, 319, 324
Terpenning, I., 224, 232, 313, 314
Thisted, R.A., 132, 322
Tibshirani, R., 7, 29, 77, 78, 80, 102, 129, 130, 131, 132, 133, 143, 148, 167, 169, 170, 172, 193, 197, 214, 231, 232, 278, 282, 295, 301, 312, 313, 315, 317, 318, 322, 323
Tidwell, P.W., 295, 312
Titterington, D.M., 278, 322
Trusler, G., 7, 143, 148, 323
Tukey, J.W., 79, 196, 311, 318, 319, 322

Utreras, F.D., 57, 76, 305, 322

Van Loan, C.F., 132, 316
Vardi, Y., 196, 198, 320

Wagner, T.J., 77, 315
Wahba, G., 60, 76, 77, 78, 79, 80, 112, 131, 160, 170, 262, 278, 311, 313, 314, 316, 317, 318, 321, 322, 323
Walker, E., 278, 323
Walker, J., 282, 301, 312
Warner, J.L., 257, 313
Watson, G.S., 37, 77, 78, 323
Wecker, W.E., 78, 130, 323
Wedderburn, R.W.M., 102, 169, 172, 320, 323
Wegman, E.J., 77, 323
Weinert, H.L., 78, 323
Weisberg, S., 196, 314
Weiss, A., 102, 131, 315
Wellner, J., 131, 312
Welsch, R.E., 277, 318
Whitehead, J., 233, 323
Whittaker, E., 76, 323
Whorf, T.P., 224, 226, 318
Wilkinson, G.N., 277, 308, 319, 323
Wilks, A., xv, 77, 102, 278, 308, 312
Williams, W.G., 7, 143, 148, 323
Wold, S., 76, 77, 323
Wong, W.H., 78, 102, 323
Wright, I.W., 77, 323

Yandell, B.S., 79, 102, 131, 154, 169, 170, 278, 311, 316, 320, 324
Young, A.H., 131, 232, 321
Young, F.W., 196, 319, 324

Subject index

ACE, see Alternating Conditional Expectation
ADE (Average Derivative Estimation), 167
AIC (Akaike's Information Criterion), 158, 160
ASR (Average Squared Residual), 44
AVAS (Additivity and Variance Stabilization), 190–193
Adaptive backfitting (BRUTO), 262, 296
Adaptive regression splines (TURBO), 249–251, 261, 296
Additive model, 3–6, 86–95, 105–106
  balanced, 104, 129, 135
  Bayesian interpretation, 129
  diagnostics for, 256
  estimating equations for, 106
  examples, 86–89
  hierachical, 266
  interactions in, 94, 264
  overinterpretation of, 161
  penalized least squares, 110
  pseudo, 254
  theory of, 105–135
  see also generalized additive model
Additive predictor, 5, 8,97, 140, 174, 203, 246
Additivity and Variance Stabilization (AVAS), 190–193
Adjusted dependent variable, 138–139, 154, 156, 171, 206
Akaike's Information Criterion (AIC), 158, 160
Algorithm
  ACE, 176, 178
  AVAS, 192
  BRUTO, 262
  CART, 271, 278
  Gauss-Newton, 167
  Gauss-Seidel, 108, 132
  Gram-Schmidt, 117
  MARS, 275
  Newton-Raphson, 138, 171, 206, 213
  TURBO, 249–251, 261
  backfitting, 4, 90–91, 103–104, 106
  conjugate gradient, 231
  delta, 171, 206
  for cubic smoothing splines, 28
  for kernel smoothing, 19, 37
  for running lines smoothing, 36
  local scoring, 98, 100–102, 140, 206, 241
  modified weights, 237, 279
  resistant fitting, 236–241
Alternating Conditional Expectation (ACE), 175–186
  anomalies, 184–186
  and canonical correlation, 198
  and correspondence analysis, 197
  and Box-Cox, 189
  theory of, 179
Analysis,
  correspondence, 197
  of deviance, 155, 142, 216, 289

Applications
  CO_2 time series, 224–231
  Los Angeles ozone, 7, 176, 294–300
  New York ozone, 257, 268, 271
  clotting time of blood, 141
  diabetes data, 1, 4,6, 41, 304
  heart attacks, 6, 95–101
  kyphosis, 7, 282–294, 301–303
  mouse leukemia, 214–218
  vasoconstriction, 242
  warm cardioplegia, 7, 143–148
Asymptotics
  of additive models, 131, 157
  of smoothers, 40, 68
Average Derivative Estimation (ADE), 167

B-splines, 25–26, 28–29, 246–249, 291
BRUTO (adaptive backfitting), 262, 296
Backfitting, 4, 90–91, 103–104, 106
  convergence, 122
  modifications of, 124, 208
  two smoother case, 118
Backward stepwise selection, 260, 215, 296
Balanced additive model, 104, 129, 135
Bandwidth, 18, 77–78, 228
  see also smoothing parameter
Basis function, 23–25, 28, 90, 165, 246–247, 249–250, 254
Bayesian model
  for additive models, 129, 157, 278
  for smoothing, 55, 60, 71, 76, 78, 81, 103
Bernoulli distribution, 95, 136
  see also Binomial model
Bias,
  of a smoother, 46, 60, 251–253
  in logistic regression, 164
Bias-variance trade-off, 11, 40
  for linear smoothers, 44
Bin smoothers, 14
Binary data, 95, 136
Binary tree, 271
Binomial model
  generalized additive, 95–101
  generalized linear, 136, 139
Bootstrap
  for confidence bands, 65
  hypothesis testing, 293
  prediction error, 298
  standard errors, 146
Bounded influence regression, 75
Box-Cox procedure, 187, 189
  generalizations, 187
Boxplot, 283–284

CART (Classification and Regression Trees), 271, 278
CV (Cross-Validation), 42, 46, 159, 273
C_p statistic, 48–49, 53, 160
Calendar effect, 228
Canonical correlation
  relation to ACE, 183, 186, 196, 198–199
Canonical link function, 137–138
Cardinal splines, 24, 247
Case control data, 202–210
Case studies, 281–300
Categorical variable, 13, 90, 94
Centered smoother matrix, 115
Circulant matrix, 230
Classification and Regression trees (CART), 271, 278
Collinearity, 115
Complimentary log-log transformation, 166
Concurvity, 5, 105, 115, 120, 123–124, 245, 282
Conditional expectation, 39, 91, 106–108, 149, 174–176, 191, 266
  alternating (ACE), 175–186
Conditional likelihood, 204–205, 232–233
Conditioned plot, 99, 270
Confidence bands
  global, 61–64, 128
  pointwise, 60, 127, 156
Conjugate gradient algorithm, 231
Consistency
  of additive model estimators, 131, 157
  of smoothers, 40, 68
Contour plot, 92, 160, 163

Convergence
asymptotic, of additive model, 131, 157
asymptotic, of smoothers, 40, 68
numerical, of ACE, 180, 182
numerical, of AVAS, 194
numerical, of backfitting, 118, 122
numerical, of local scoring, 151, 213
rate of smoothers, 69
Cook's distance, 256
Correlation
canonical, 183, 186, 196, 198–199
maximal, 189
Correspondence analysis
relation to ACE, 197
Cox's proportional hazards model, 211–218
Cross-validation (CV), 42, 46, 159, 273
for linear smoothers, 46
generalized, 49–52, 159
Cubic smoothing splines, 27, 47, 49, 54, 57, 76, 120, 124, 150, 206
derivation, 36
computation, 28–29
Cubic splines, 38
Curse of dimensionality, 83–84, 92

Degrees of freedom (*df*),
approximations for, 305
for additive models, 128, 133
for generalized additive models, 157
of a smooth, 52–55
Deletion formula, 47
Delta algorithm, 171, 206–207
Deviance (D, Dev), 138
analysis of, 155, 142, 216, 289
cross-validated, 159
Diagnostics for additive models, 256–258
Dimensionality, curse of, 83–84, 92
Dispersion parameter, 138, 155
Distance
Cook's, 256
Mahalanobis, 33
Dummy variable, 90, 146, 218, 265

Effective number of parameters, see degrees of freedom,

Eigenanalysis,
cubic smoothing spline matrix, 57–58
linear filters, 59, 229–231
Endpoint effect, 15, 19, 22
Equivalent kernel, 20, 22, 29, 38, 78, 92, 252–253
Error variance
estimation of,48, 66
Estimating equations for additive models, 106
properties of solutions, 121
penalized likelihood and, 110
solution of, 118, 126
Estimation
of additive model, 89–94
of link function, 166
of generalized additive model, 140
Exponential family, 137, 139

F-tests, 65, 155
FFT (Fast Fourier Transform), 19, 37
Fast Fourier Transform (FFT), 19, 37
Filter
linear, 57–59, 229–231
Fisher information, 171
Fisher scoring, 137, 171
Formula language in S, 283, 289–290, 308–310
Forward stepwise selection, 260
Fourier analysis
of time series, 59

GAIM (Generalized Additive Interactive Modelling), 102, 307
GAM function in S language, 308–310
GCV (Generalized Cross Validation), 49–52, 159
GLIM (Generalized Linear Interactive Modelling),102
Gamma model
generalized additive, 141–143
generalized linear, 139
Gauss-Newton algorithm, 167
Gauss-Seidel algorithm, 108, 132
Gaussian (normal) model
generalized additive, 139–140
Generalized Additive Interactive Modelling (GAIM), 102, 307
Generalized Linear Interactive Modelling (GLIM), 102

Generalized additive model, 95–101, 136, 140
Generalized cross-validation, 49–52, 159
Generalized linear model, 137–139
Generalized residual, 213
Generalized ridge regression (pseudo additive model), 254–258
Gram-Schmidt algorithm, 117

Hanning, 20
Hat matrix, 126
Hazard function, 211, 233
Hierachical additive model, 266
Hilbert space (L_2 function space), 107, 112, 148, 179

I-splines, 189
IRLS (Iteratively Reweighted Least Squares), 97, 138, 171
Impulse response function, 59
Incidence, 203
Indicator function, 100, 152
Inference
  in additive models, 128
  in generalized additive models, 155–158
  in response transformation model, 195
  in smoothing, 65–67
Influence, 74–75, 242, 244
Information
  Fisher, 171
Interaction spline, 34, 113, 275, 278
Interactions
  examination of residuals, 146, 268
  hierarchical models, 266
  modelling, 264
  simple, 264
Interpolating spline, 38
Inverse Gaussian model, 139
Isotonic regression, 196
Iteratively Reweighted Least Squares (IRLS), 97, 138, 171

Jackknifed fit, 46, 79

Kernel smoothers, 18–19
  fast Fourier transform, 19, 37
Knots
  selection and placement, 22–24, 247, 249
Kriging, 71
Kullback-Leibler distance, 148, 158, 172

LOESS (Locally weighted regression), 29–31, 224
L_2 function spaces, see Hilbert space
Lagrange multiplier, 36, 181
Laplacian, 33
Least median of squares regression, 75
Leverage points, 75
Life
  meaning of, 336
Likelihood
  conditional, 204–205, 232–233
  expected log, 148, 172, 189–190
  local, 167
  partial, 212
  profile log, 188
  penalized, 149–151, 205, 212
  quasi, 101, 233
Linear filters, 57–59, 229–231
Linear model, 1, 3,83
Linear predictor, 5, 8,96, 137, 246
Linear regression, 1, 3,25, 83, 295
Linear smoothers, 44, 57, 105, 108
Link function, 136–137, 139
  estimation of, 166
Local scoring algorithm, 98, 100–102, 140, 206, 241
  convergence, 151
  derivation, 148–150
Local-likelihood estimation, 167
Locally weighted regression (LOESS), 29–31, 224
Logistic model
  generalized additive, 95–101, 143–148, 161–165, 283–294
  generalized linear, 95–96, 139
Logistic regression, 95–96, 139
Logit, 96, 137

M-estimates, 75, 236
MARS (Multivariate Adaptive Regression Splines), 275–277
MORALS, 196
MSE (mean squared error), 42
Mahalanobis distance, 33
Main effect, 266–267, 275
Mallows's C_p, 48–49, 53, 160
Matched case-control data, 201–210, 232
Matched set, 202

Matrix, 38, 44, 115
  circulant,230
  norm, 119
Maximal correlation, 180
Mean-squared error (MSE), 42
Missing values, 166, 228
Model selection techniques, 159–161, 259–263
Model
  Gaussian, 139–140
  Poisson, 139, 232–233
  binomial, 95–101, 136, 139
  gamma, 139, 141–143
  inverse Gaussian, 139
  log-linear, 101, 117, 136
  logistic, 95–101, 139, 144, 161–165, 283–294
  multinomial, 220, 233
  proportional hazards, 211–218
  proportional odds, 219–223
  transformation, 174–200
Modified-backfitting algorithm, 124, 208
Modified-weights algorithm, 237, 279
Monotone transformation, 184–185, 194
Moving-average (running-mean), 9, 15, 40
Multinomial, 220, 233
Multiple predictor smoothing, 32–34, 83, 266, 275
Multiple regression, 3, 82–86
Multivariate Adaptive Regression Splines (MARS), 275

Natural splines, 24, 27–28, 36, 200
Nearest-neighbour, 15, 31, 32–33, 68
Neighbourhood, 15–17
  nearest-neighbour, 15, 31
  symmetric nearest-neighbour, 15, 31
Newton-Raphson algorithm, 138, 171, 206, 213
Nonlinear smoothers, 20, 70
Normal model, see Gaussian model
Notation, 7

Odds-ratio, 95, 202, 219
Optimal rate of convergence, 68, 131
Optimal transformation, 179–180, 197
Optimism
  in prediction error, 296

Ordinal data, 219
Orthogonal parameters, 196
Orthogonal projection, 58, 107, 109, 115, 125, 132

PE (prediction error), 58
PSE (prediction squared error), 42, 46, 296
Parametric
  additive models, 245
  regression, 14, 22
  transformation, 183, 187
Partial likelihood, 212
Partial residual, 91, 102, 165, 244, 285–286
Partially-splined model, 152
Penalized least-squares, 27, 47, 72, 80, 110
Penalized likelihood, 149–151, 205, 212
Piecewise-cubic polynomial, 22–23, 27
Pivotal quantity
  approximate, 62, 64, 128
Plot
  box, 283–284
  conditioned, 99, 270
  contour, 92, 160, 163
  quantile-quantile (Q-Q), 63, 293
  rug, 96
  scatter, 2
  scatterplot matrix, 282–283
Poisson model, 139, 232–233
Posterior distribution,
  in Bayesian additive model, 129, 157
  in Bayesian model for smoothing, 55, 60, 81
Power transfer function, 59, 231
Powering up an operator, 180, 183
Prediction error (PE), 158
Prediction squared error (PSE), 42, 46, 296
Prediction from a transformation model, 194
Predictor,
  additive, 5, 8,97, 140, 174, 203, 246
  linear, 5, 8,96, 137, 246
Prevalence, 7, 95, 99, 203, 292
Prior distribution, 55, 71, 129
Profile log-likelihood, 188

Projection operators, 58, 107, 109, 115, 125, 132
Projection pursuit regression, 85
Proportional-hazards model (Cox model), 211–218
Proportional-odds model, 219–223
Pseudo additive models, 254–258

Q-Q plot (Quantile-quantile plot), 63, 293
Quantile-quantile plot (Q-Q plot), 63, 293
Quasi-likelihood, 101, 233

RSS (Residual Sum of Squares), 54
Randomization
  role of, 146
Rates of convergence
  of smoothers, 68
  of additive-model estimators, 131
Recursive partitioning regression, see regression trees
Regression
  bounded-influence, 75
  isotonic, 196
  least median-of-squares, 75
  linear, 1, 3,25, 83, 295
  logistic, 95–96, 139
  multiple, 3, 82–86
  parametric, 14, 22
  ridge, 90, 254
  slicing, 167
  transform-both-sides, 196
  transformations for, 187
Regression splines, 22, 246
  adaptive knot selection, 249
  simple knot selection, 247
Regression trees, 84, 271, 278
Relative risk, 211, 233
Representers of evaluation,
Reproducing kernel Hilbert spaces, 112
Residual sum of squares (RSS), 54
Residuals
  examination of, 146, 268
  generalized, 213
  partial, 91, 102, 105, 244, 285–286
Resistant fitting
  of additive models, 236
  of generalized additive models, 240
  in smoothing, 74
Resistant smoothing, 74
Response transformation, 74–200
Retrospective study, 95, 202
Ridge regression, 90, 254
  generalized, 254
Robust fitting
  of additive models, 236
  of generalized additive models, 240
  in smoothing, 274
Rug plot, 96
Running-lines smoothers, 15
  locally weighted, 29–31, 224
  computation, 17, 35
Running-mean smoothers (moving average), 9, 15, 40
Running-median smoothers, 20, 70

S language, 102, 278, 308
STL procedure, 224–231
Scale, see dispersion parameter
Scatterplot matrix, 282–283
Scatterplot smoothing, 9, 13, 39–81
Score equations, 106, 110, 118, 121, 126
Seasonal decomposition of time series (STL), 224–231
Sequence smoothing, 20
Semi-parametric models, 37, 118, 152
Shrinking smoother, 124
σ^2- estimation of, 48, 66
Slicing regression, 167
Smearing estimate, 194
Smoother matrix, 38, 44
  centered, 115
Smoothing,
  LOESS, 29–31, 224
  asymptotic behaviour of, 40, 68
  bias in, 46, 60, 251–253
  bin, 14
  cubic smoothing spline, 27, 47, 49, 54, 57, 76, 120, 124, 150, 206
  cubic spline, 38
  kernel, 18–19, 37
  linear, 44, 57, 105, 108
  locally-weighted, 29–31, 224
  multi-predictor, 32–34, 83, 266, 275
  nonlinear, 20, 70
  resistant, 74
  running-line, 15
  running-mean (moving-average), 9,

15, 40
running-median, 20, 70
scatterplot, 9, 13, 39–81
shrinking, 124
state-space approach, 78
surface, 32, 83, 206, 275
variance, 60
weighted, 1, 17, 18, 20, 29–31
Smoothing parameter, 11, 15, 18, 27, 30, 40
asymptotics, 40, 68
bias-variance tradeoff, 11, 40, 44
selection of, 40, 52, 159, 262
Software, 307–310
GAIM, 102, 307
GLIM, 102
S, 102, 278, 309
XploRe, 102
Span, 17
see also smoothing parameter
Slicing regression, 167
Spline
B, 25–26, 28–29, 246–249, 291
I, 189
cardinal, 24, 247
cubic smoothing, 27, 47, 49, 54, 57, 76, 120, 124, 150, 206
interaction, 34, 113, 275, 278
interpolating, 38
natural, 24, 27–28, 36, 200
partial, 152
regression, 22, 246
tensor-product, 34, 275
thin plate, 33
Splitting, 20
Standard error bands, 60, 127, 156
State-state smoothing, 78
Step size optimization, 151, 207
Stepwise selection, 260, 215, 296
Supersmoother, 70
Surface smoothing, 32, 83, 266, 275
Surrogate variable, 274
Survival data, 211–218
Symmetry
of a neighbourhood, 15, 31
of smoother matrix, 121

TURBO (adaptive regression splines), 261
Target value, 15, 16, 40
Tensor-product splines,
Thin-plate splines, 34, 275
Tied predictor values, 74
Time series,
Fourier analysis of, 59
digital filter for, 59
moving average for, 20
running median for, 20, 70
seasonal decomposition of, 224–231
Trade-off
Bias-variance, 11, 40, 44
Transfer function, 57–59
Transform-both-sides regression, 196
Transformation
monotone, 184–185, 194
of response, 174–200
optimal, 179–180, 197
parametric, 183, 187
variance-stabilizing, 191–194
Trend component, 71, 224–226
Tri-cube function, 30
Twicing, 20, 80

Uniqueness
of estimating-equation solutions, 121
Updating formula, 35

Variance
heterogeneous, 72
Variance of a smooth, 60
Variance-stabilizing transformation, 191–194

Weights
in additive models, 124
in generalized additive models, 141
in generalized linear models, 138
in least-squares fitting, 30, 73
in smoothing, 72
Window, 18
see also smoothing parameter

X-11 method, 232
XploRe, 102